Let's Get IoT-*fied*!

Internet of Things (IoT) stands acclaimed as a widespread area of research and has definitely enticed the interests of almost the entire globe. IoT appears to be the present as well as the future technology. This book attempts to inspire readers to explore and become accustomed to IoT. Presented in a lucid and eloquent way, this book adopts a clear and crisp approach to impart the basics as expeditiously as possible. It kicks off with the very fundamentals and then seamlessly advances in such a way that the step-by-step unique approach, connection layout, and the verified codes provided for every project can enhance the intuitive learning process and will get you onboard to the world of product building. We can assure that you will be definitely raring to start developing your own IoT solutions and to get yourself completely lost in the charm of IoT.

Let's start connecting the unconnected! It's time to get IoT-*fied*.

Let's Get IoT-*fied*!
30 IoT Projects for All Levels

Anudeep Juluru
Shriram K. Vasudevan
T. S. Murugesh

CRC Press
Taylor & Francis Group
Boca Raton London New York

CRC Press is an imprint of the
Taylor & Francis Group, an **informa** business

First edition published 2023
by CRC Press
6000 Broken Sound Parkway NW, Suite 300, Boca Raton, FL 33487-2742

and by CRC Press
4 Park Square, Milton Park, Abingdon, Oxon, OX14 4RN

CRC Press is an imprint of Taylor & Francis Group, LLC

© 2023 Anudeep Juluru, Dr. Shriram K. Vasudevan and Dr. T. S. Murugesh

Library of Congress Cataloging-in-Publication Data
Names: Vasudevan, Shriram K., author. | Juluru, Anudeep, author. |
Murugesh, T. S., author.
Title: Let's get IoT-fied! : 30 IoT projects for all levels / Shriram K.
Vasudevan, Anudeep Juluru, TS Murugesh.
Description: First edition. | Boca Raton, FL : CRC Press, 2023. | Includes
bibliographical references and index.
Identifiers: LCCN 2022012520 (print) | LCCN 2022012521 (ebook) |
ISBN 9780367706067 (hbk) | ISBN 9780367706074 (pbk) | ISBN 9781003147169 (ebk)
Subjects: LCSH: Internet of things—Amateurs' manuals. | Electronic
apparatus and appliances—Amateurs' manuals.
Classification: LCC TK5105.8857 .V37 2023 (print) | LCC TK5105.8857
(ebook) | DDC 004.67/8—dc23/eng/20220524
LC record available at https://lccn.loc.gov/2022012520
LC ebook record available at https://lccn.loc.gov/2022012521

ISBN: 9780367706067 (hbk)
ISBN: 9780367706074 (pbk)
ISBN: 9781003147169 (ebk)

DOI: 10.1201/9781003147169

Typeset in Times
by codeMantra

Dedicated to

Anudeep Juluru - G V Subba Rao

Dr. Shriram K. Vasudevan - Muralidharan

Dr. T. S. Murugesh - My Mentors Dr. M. Ramaswamy & Dr. A. Murugappan

Contents

Preface

This book teaches readers how to build IoT products through practical projects, ranging from beginners to advanced level, so that they can build knowledge through hands-on tasks rather than theoretical knowledge.

Internet of Things (IoT) stands acclaimed as a widespread area of research and it appears to be the present as well as the future technology. This book is an out of the box attempt suitable for all readers to explore further, indulge and lose themselves into the charm of IoT. The contents are expressed in a lucid and eloquent way starting from the very basics, seamlessly advances in such a way that the step-by-step unique approach, connection layout and the verified codes provided for each and every project can enhance the intuitive learning process. This book will definitely make you onboard to the world of product building with the development of your own IoT solutions guaranteed.

Importantly, this book is authored by right blend of experience and expertise from the industry and academia. Anudeep is working in the core IoT sector, Dr. Shriram is an Intel IoT Innovator and Intel IoT Innovator Champion, and Dr. Murugesh is a seasoned professor with over 23 years of teaching and research experience. This book will be a certain platform for anyone to learn, practice and become an expert in IoT.

Let's start connecting the unconnected! It's time to get IoT-*fied*.

We welcome any constructive criticism to: shriramkv@gmail.com.

About the Authors

Anudeep Juluru is an Embedded Software Engineer and IoT expert. He is the Founder and CEO of Loopus WearTech Pvt. Ltd., an IoT-based wearable startup. He is a gold medal awardee for the Best Innovative Startup 2019 received from the then Governor of New Delhi, Anil Baijal, and the Deputy Chief Minister of New Delhi, Manish Sisodia. He has authored a digital learning courseware on IoT comprising three modules under Wiley Publishers and more than 12 research papers in coveted international journals and conferences like IEEE, Springer and Inderscience. He is an innovator and has developed various IoT products for solving real-world problems. He was awarded by the Indian Navy, DPSRU and several companies like Titan, Cisco, Sabre, GE, Rakuten, Tata Sons, NEC, L&T, Syndicate Bank, ELGi, Infosys, Bounce, Hacker Earth, Axis Bank and Intel. He is the winner of HackHarvard Global Hackathon, Harvard University, USA; Technology Infusion Grand Challenge, La Trobe University, Australia; LTTS Techgium 2018–2019, L&T; Ageing better with ICTs, WSIS Forum; Rakathon 2018, Rakuten; Tata Crucible hackathon, Tata Sons; and more than 15 hackathons.

Dr. Shriram K. Vasudevan obtained his B.E degree in Electronics and Instrumentation from Annamalai University in 2004, M.Tech in Embedded Systems from SASTRA University in 2007, and Ph.D (Embedded Systems) from PRIST University in 2015. He also got his M.B.A degree awarded in HRM and Marketing from Periyar Maniammai University in the year 2008. He is a has a blend of industrial and teaching experience over the course of 15 years. He is strongly passionate to take up challenging tasks. He has authored/co-authored 42 books for reputed publishers across the globe. He has authored 132 research papers in revered international journals and 30 papers in international/national conferences. He is currently associated with a multinational software company as a Project Manager. Earlier, he was associated with Wipro Technologies, VIT University, and Aricent Technologies. He holds the Corporate Fellow Membership in IETE and a – IETE, ACM Distinguished Speaker, CSI Distinguished Speaker, NASSCOM Prime Ambassador, NVIDIA DLI, and Intel Software Innovator. He has a YouTube channel, Shriram Vasudevan, through which he teaches thousands of people all around the world.

He is recognized/honoured for his technical expertise by Factana, Accumulate, Telecommunications, and the Digital Government Regulatory Authority (United Arab Emirates), ZyBooks (Wiley Brand), AOTMP, Adani Digital, NASSCOM Foundation, World Summit on the Information Society (WSIS), De-Nora, IIT Kharagpur E Cell, Huawei, NVIDIA, Cubestop, IETE, Datastax, Honda, Wiley, AGBI, ACM, Uletkz, The Hindu (Tamil), Exact Sciences Corp, Proctor and Gamble Innovation Centre (India), Dinamalar, AWS (Amazon Web Services), Sabre Technologies, IEEE Compute, Syndicate Bank, MHRD, Elsevier, Bounce, IncubateInd, Smart India Hackathon, Stop the Bleed, HackHarvard (Harvard University), Accenture Digital (India), NEC (Nippon Electric Company, Japan), Thought Factory (Axis Bank Innovation Lab), Rakuten (Japan), Titan, Future Group, Institution of Engineers of India (IEI), Ministry of Food Processing Industries (MoFPI – Govt. of India), Intel, Microsoft, Wipro, Infosys, IBM India, SoS Ventures (USA), VIT University, Amrita University, Computer Society of India, TBI – TIDE, ICTACT, Times of India, Nehru Group of Institutions, Texas Instruments, IBC Cambridge, Cisco, CII (Confederation of Indian Industries), Indian Air Force, DPSRU Innovation & Incubation Foundation, ELGi Equipments (Coimbatore), etc. He is also listed in many famous biographical databases.

He has delivered talks at various international conferences and forums of high repute. Shriram has also been granted many patents. He is a hackathon enthusiast and won about 50 hackathons including HackHarvard 2019.

Dr. T. S. Murugesh obtained his B.E degree in Electronics and Instrumentation in 1999, M.E degree in Process Control and Instrumentation in 2005 and Ph.D in Instrumentation Engineering in 2015 from Annamalai University, Annamalai Nagar, Tamil Nadu. He possesses a vast experience of above 22 years in academia in the field of Analog and Digital Electronics, Automation and Control, IoT, System design, Instrumentation and Computational Bio-engineering. After a tenure of almost 19 years with the Department of Electronics and Instrumentation Engineering, belonging to the Faculty of Engineering and Technology, Annamalai University, he is currently posted as an Associate Professor in the Department of Electronics and Communication Engineering, Government College of Engineering Srirangam, Tiruchirappalli, Tamil Nadu.

He has delivered invited lectures at National level in various institutions like Sastra University, Annamalai University, Manakula Vinayagar Institute of Technology, Puducherry, Government College of Engineering Srirangam, Madurai Institute of Engineering and Technology, etc. He has delivered invited lectures in Faculty Development Programmes organized by the Faculty Training Centre, Government College of Technology, Coimbatore in association with Government College of Engineering Thanjavur, and another one with Government College of Engineering Salem, and also in a national-level webinar conducted on behalf of "Unnat Bharat Abhiyan", a flagship programme of Ministry of Education, Government of India. He has also delivered an invited lecture at International level in the 5th Virtual Congress on Materials Science and Engineering, "Materials Info 2022". To his credit, he has around three-dozen peer-reviewed indexed publications including Springer Nature, Elsevier and Inderscience, has organized a one-week AICTE Training And Learning (ATAL) Academy sponsored FDP, conducted few workshops at national level, is a reviewer in IEEE and few other peer-reviewed journals, etc. He has acted as a Primary Evaluator for Government of India's Smart India Hackathon 2022 (Software & Hardware Edition) as well as Toycathon 2021 and also as a Judge in the Grand Finale for the Government of India's "Toycathon 2021", an inter-ministerial initiative organized by Ministry of Education's Innovation Cell with support from AICTE (All India Council for Technical Education). He is a hackathon enthusiast and his team has won the second Prize in the IFG x TA Hub Hackathon 2022. He is a Conference Committee Member as well as a Publishing Committee Member in the International Association of Applied Science and Technology. He also holds the professional body membership of Institution of Engineers (India).

LED Control with Different Evaluation Boards (Arduino Uno and NodeMCU)

1

We welcome readers to the world of product building. This book is crafted in simple technical language enabling the readers to learn with ease. The first project in the list is LED (light-emitting diode) control using the most popular evaluation boards in IoT (Internet of Things). If asked, the expert's choice would be either Arduino or ESP in IoT-based evaluation boards. So, in this project, you'll learn about these evaluation boards by controlling an LED. This project is divided into two parts:

- Controlling a LED using Arduino Uno
- Controlling a LED using NodeMCU

To build anything, fundamentals should be very strong. In that aspect, this is a very important project in the whole book as it teaches you about the basics of the evaluation boards used. Please do follow this project carefully so that you don't miss any basic points.

1.1 INSTALLATION OF ARDUINO IDE

In this book, for uploading code to Arduino boards or NodeMCU, we will use Arduino IDE. For installation of Arduino IDE, download the free software from https://www.arduino.cc/en/software. Download the version as per the configuration of your machine. If you are not aware of the configuration of your machine, make sure you visit the properties of your machine to get the details. After opening the Arduino software webpage, you'll see the options as shown in Figure 1.1.

Before downloading, the Arduino website provides the option for donating to its open-source project. You can donate and download or just download as shown in Figure 1.2. It is not mandatory to donate for downloading the Arduino IDE.

Arduino's Software page also provides other options like web editor, Hourly Builds, previous releases, experimental software and many more as shown in Figure 1.3. If you want to try them, click the appropriate links and follow the guidelines mentioned on the website.

After downloading the installation file, run it. The installation steps are pretty much straightforward. Figure 1.4 shows the License Agreement which needs to be agreed upon for proceeding further. Figure 1.5 shows the list of components that can be included as part of the installation. Figure 1.6 shows the path at which the Arduino IDE has to be installed on your PC (personal computer). Click the *Browse* button for selecting your preferred location. After the installation is completed, you'll get a screen similar to the one presented in Figure 1.7.

DOI: 10.1201/9781003147169-1

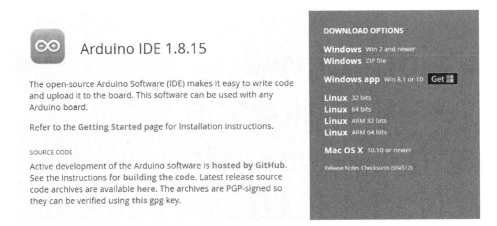

Arduino IDE 1.8.15

The open-source Arduino Software (IDE) makes it easy to write code and upload it to the board. This software can be used with any Arduino board.

Refer to the **Getting Started** page for Installation instructions.

SOURCE CODE

Active development of the Arduino software is **hosted by GitHub**. See the instructions for **building the code**. Latest release source code archives are available **here**. The archives are PGP-signed so they can be verified using **this** gpg key.

DOWNLOAD OPTIONS

Windows Win 7 and newer
Windows ZIP file

Windows app Win 8.1 or 10   Get

Linux 32 bits
Linux 64 bits
Linux ARM 32 bits
Linux ARM 64 bits

Mac OS X 10.10 or newer

Release Notes Checksums (sha512)

FIGURE 1.1   Installation page

Support the Arduino IDE

Since the release 1.x release in March 2015, the Arduino IDE has been downloaded **52,295,201** times — impressive! Help its development with a donation.

$3 $5 $10 $25 $50   Other

JUST DOWNLOAD   CONTRIBUTE & DOWNLOAD

Learn more about **donating to Arduino**.

FIGURE 1.2   Donation and download options

After completing the installation, on opening the Arduino IDE, you will see the screen as shown in Figure 1.8. Each of the buttons at the top-left corner is used for a different purpose. The purpose of each button is explained in Figure 1.9. The button at the top-right corner is used for opening the serial monitor which is very useful in visualizing the data from the sensors.

Many examples will be downloaded along with the installation of Arduino IDE. You can find them in *File > Examples* as represented in Figure 1.10. We recommend everyone to explore these examples as they would serve as a fantastic guide.

Arduino Web Editor

Start coding online and save your sketches in the cloud. The most up-to-date version of the IDE includes all libraries and also supports new Arduino boards.

CODE ONLINE   **GETTING STARTED**

Hourly Builds

Download a **preview of the incoming release** with the most updated features and bugfixes.

DOWNLOAD OPTIONS

Windows
Mac OS X 10.10 or newer
Linux: 32 bits, 64 bits, ARM, ARM64

LAST UPDATE: 08 Apr 2021, 18:07:23 GMT

Previous Releases

Download the previous version of the current release, the classic 1.0.x, or old beta releases.

DOWNLOAD OPTIONS

Previous Release (1.8.14)

Arduino 1.0.x
Arduino 1.5.x beta
Arduino 1.9.x beta

Arduino with Chromebook

To program Arduino from a Chromebook, you can use the **Arduino Web Editor** on Arduino Cloud. The desktop version of the IDE is not available on Chrome OS.

Experimental Software

 Arduino IDE 2.0 beta (2.0.0-beta.7)

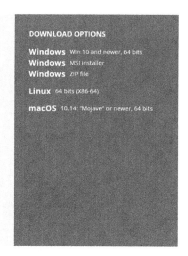

The new major release of the Arduino IDE is faster and even more powerful! In addition to a more modern editor and a more responsive interface it features autocompletion, code navigation, and even a live debugger.

Note: this software is still in **beta** status, which means that it's almost complete but there might be minor issues. Help us test it and report your feedback in the **forum**!

You can also find more information about the release **here**.

SOURCE CODE

The Arduino IDE 2.0 is open source and its source code is hosted on **GitHub**.

DOWNLOAD OPTIONS

Windows Win 10 and newer, 64 bits
Windows MSI installer
Windows ZIP file

Linux 64 bits (X86-64)

macOS 10.14: "Mojave" or newer, 64 bits

Nightly Builds

Download a **preview of the incoming release** with the most updated features and bugfixes.

Windows
macOS Version 10.14: "Mojave" or newer, 64 bits
Linux 64 bits (X86-64)

FIGURE 1.3   Different download options

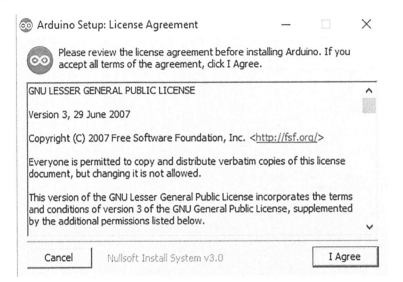

FIGURE 1.4   License agreement

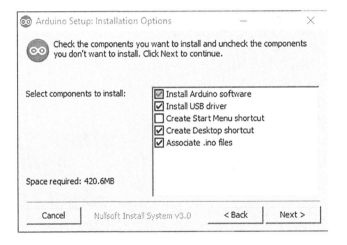

FIGURE 1.5   Selection of components

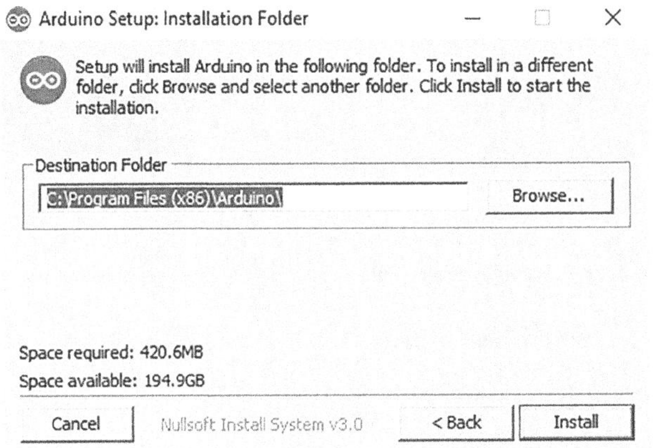

FIGURE 1.6   Folder path for installation

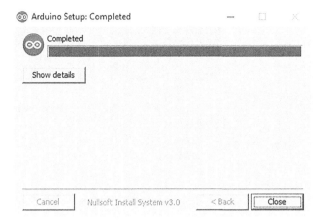

FIGURE 1.7   Installation complete

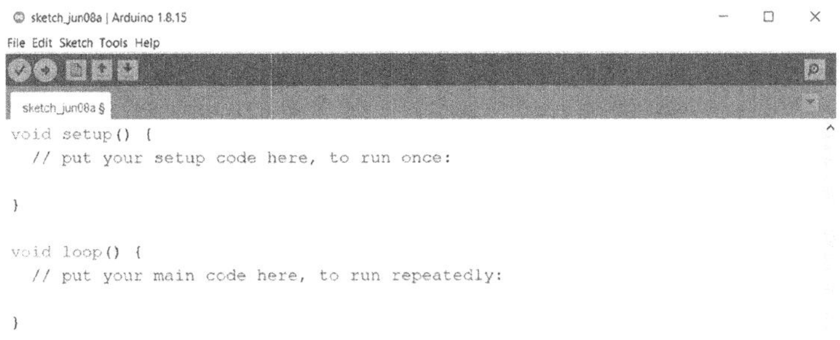

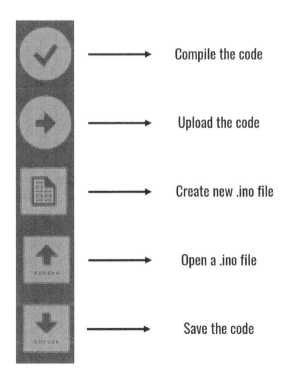

FIGURE 1.8   First launch of Arduino IDE

Compile the code

Upload the code

Create new .ino file

Open a .ino file

Save the code

FIGURE 1.9   Useful buttons in Arduino IDE

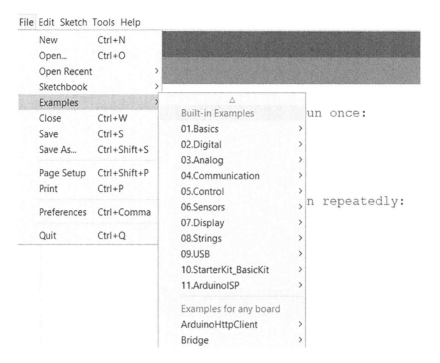

FIGURE 1.10   Examples in Arduino IDE

The programming language written in Arduino IDE is similar to C++. So, if you know C++, then it would be very easy to understand the code written in Arduino IDE. Don't worry! If you don't know C++, knowing at least any one of the computer programming languages will help in easily understand the code used in this book. Even if you don't know any programming language, you can understand most of the code used in this book as a clear explanation is provided for each line of the code used. But it is advisable to have basic knowledge of at least one programming language.

1.2 BREADBOARD

A breadboard (shown in Figure 1.11) is a solderless board used as a construction base for most prototyping electronic circuits. As the breadboard does not require soldering, it can be reused until it is broken or spoiled. This makes the breadboard convenient for creating temporary prototypes and experimenting with circuit design.

If you have designed and built an electronic circuitry on a breadboard and find out that everything works as expected, then you can build the same circuitry on a printed circuit board (PCB). They are semi-permanent soldered boards that cannot be reused. PCBs will help in bringing down the size of the same circuitry and will ensure every soldered connection is perfectly connected.

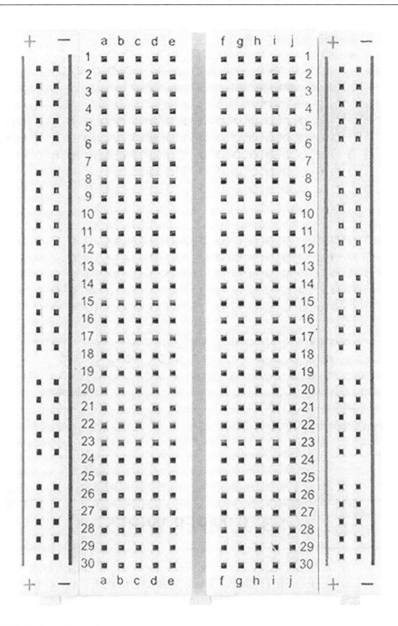

FIGURE 1.11   400 pin breadboard

The holes on the top of the breadboard are connected by strips of metal underneath the board. The holes of two rows at the top and bottom are connected horizontally while the remaining holes (in the middle) are connected vertically as shown in Figure 1.12. To connect sensors or modules using the breadboard, jumper wires or single-strand wires can be used.

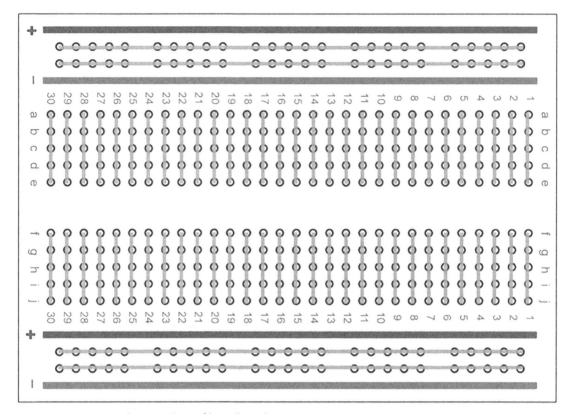

FIGURE 1.12 Internal connections of breadboard

1.3 JUMPER WIRES

Jumper wires are specialized electrical wires with connectors or pins at each end. These are used for interconnection between breadboards, evaluation boards, sensors, modules or any other electronic devices and are available in various combinations as shown in Figure 1.13:

- Male to Male Jumper Wires – shortly called as M2M Jumper wires. These jumper wires have pins at both ends.
- Male to Female Jumper Wires (or Female to Male Jumper Wires) – shortly called as M2F Jumper wires. These jumper wires have pins at one end and a connector at the other end.
- Female to Female Jumper Wires – shortly called as F2F Jumper wires. These jumper wires have connectors at both ends.

Jumper wires are generally used for prototyping purposes. These are reusable and need not be thrown after using once. These wires occupy a lot of space and sometimes become clumsy. With an increase in the number of sensors or modules in the project, the number of jumper wires used also increases.

To reduce the size of your prototype, you can also use single-strand wires (shown in Figure 1.14). These can only be used in the place of Male to Male Jumper wires. But these are not reusable and need to be cut using a wire stripper according to your requirement.

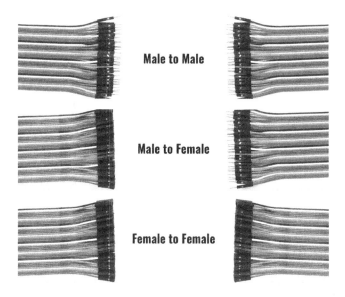

FIGURE 1.13   Jumper wires

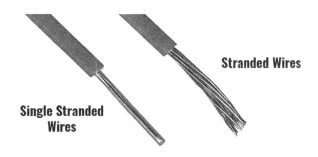

FIGURE 1.14   Single stranded wires

1.4 RESISTORS

A resistor is a passive two-terminal electronic component which reduces the flow of electrons (electric current) through a circuit. An active component is an electronic component that can generate energy for supplying to an electrical circuit, whereas a passive component can only receive energy but cannot amplify or generate energy. Resistors are mainly used for limiting the current flowing through a circuit, reducing the voltage across a component or as a pull-up/pull-down resistor. The electrical resistance of a resistor is measured in Ohms denoted by the symbol Ω. Generally, resistors are available in two types: through-hole and surface-mount. A 1 kΩ resistor in both through-hole and surface-mount types is shown in Figure 1.15.

The resistance value of a through-hole resistor can be easily calculated just by seeing them without using a multimeter. Though it doesn't have a display to show its resistance, it has markings of different colours around it known as bands. Most of the through-hole resistors will have 4 bands, but we'll also see 5-band and 6-band resistors.

In a 4-band resistor, the first two bands indicate the two most significant digits of the resistance. The third bit is a multiplier which multiplies the first two digits with a power of ten. The final band indicates the tolerance of the resistor. Tolerance is the deviation of the actual resistance from the nominal value. The tolerance band will be usually either gold or silver, and it will be clearly separated from the remaining bands.

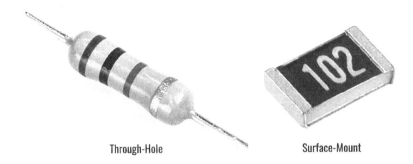

Through-Hole   Surface-Mount

FIGURE 1.15   1 kΩ resistor

In a 5-band resistor, there will be an extra significant digit band between the first two bands and the multiplier. In a 6-band resistor, there will be an extra band at the end to indicate the temperature coefficient of the resistor. The temperature coefficient is the change in resistance value with the change in temperature (°C).

For calculating the resistance of a through-hole resistor, refer to the colour code table provided in Figure 1.16. If you have a 4-band resistor with colours GREEN-BROWN-ORANGE-SILVER, then its resistance is calculated as follows.

- 4-band resistor resistance = GREEN (5) BROWN (1) * ORANGE (10^3) ± 10% = 5.1 kΩ with 10% tolerance

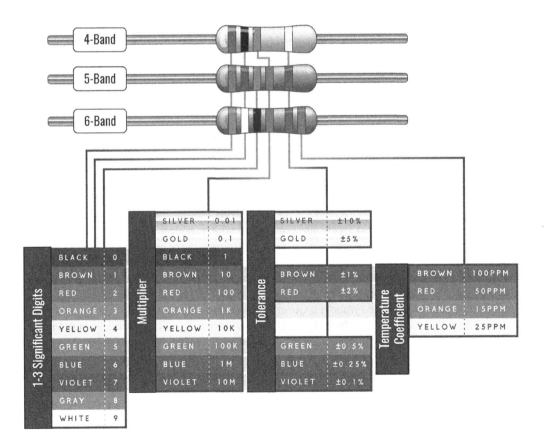

FIGURE 1.16   Resistance colour code table

Similarly, you can also calculate the resistance of any 5-band or 6-band through-hole resistor using the colour code table. Mnemonics are helpful in easily remembering the order of the colour codes. There are different mnemonics used in different places. For example, *Bye Bye Rosie Off You Go to Bristol Via Great Western* is a commonly used mnemonic in the UK, and *B B ROY of Great Britain had a Very Good Wife who wore Gold and Silver Necklace* is a commonly used mnemonic in India. To know all the commonly used mnemonics for remembering the colour codes, refer to this link (https://en.wikipedia.org/wiki/List_of_electronic_color_code_mnemonics; last edited May 18, 2022).

The resistance value of a surface-mount resistor can be calculated using a different marking system. Explanation about the marking system used for surface-mount resistors is beyond the scope of this book.

1.5 CAPACITORS

A capacitor is also a two-terminal passive electronic component which can store energy. Capacitors and resistors form the fundamental passive components of any IC. Capacitors are mainly used for signal filtering, voltage regulation, local energy storage and as bypass capacitors. The amount of energy that can be stored in a capacitor is measured in capacitance. The unit of capacitance is Farad denoted by the symbol F.

A capacitor is made using two metal plates with a dielectric layer in between them. The two metal plates are placed very close to each other, and the dielectric in between them is to make sure that they don't touch each other. Each of the two metal plates is connected to a terminal wire for connecting with other electronic components.

There are many types of capacitors available in the market depending on their size, maximum voltage, tolerance, leakage current, etc. Among them, the most commonly used capacitors are ceramic or disc capacitors and electrolytic capacitors which are shown in Figure 1.17. Usually, we'll use the capacitors of capacitance in the range of micro (10^{-6}), nano (10^{-9}) and pico (10^{-12}) farad. Even milli farad is considered as huge capacitance. For using capacitors in the range of Farad and kiloFarad, there will be another type of special capacitors known as supercapacitors or ultracapacitors.

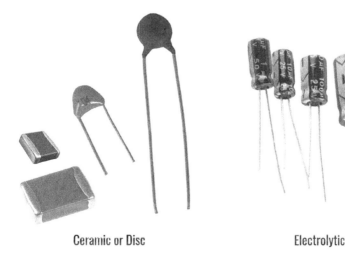

Ceramic or Disc   Electrolytic

FIGURE 1.17 Different types of capacitors

1.6 UART

UART (Universal Asynchronous Receiver-Transmitter) is one of the simplest and oldest forms of device to device communication protocols used in electronics. It is a hardware communication protocol that uses asynchronous serial communication with a configurable speed. In asynchronous communication, the transmitter and receiver don't share a common clock, whereas in synchronous communication, the transmitter and receiver will share a common clock for transferring data. There are mainly two ways in which data is transferred between electronic devices: serial transmission and parallel transmission. In serial transmission, the data is transferred bit by bit using a single line, whereas in parallel communication, multiple bits are transferred using multiple lines as shown in Figure 1.18.

UART is a hardware circuit or a standalone IC whose main purpose is to serially transmit or receive data. It uses only two lines, Receiver (Rx) and Transmitter (Tx), to communicate with another UART as shown in Figure 1.19. The transmitting UART converts the parallel data obtained from a controller data bus into serial data and transmits it to the receiving UART which will again convert it into parallel data for sending to its controller data bus as shown in Figure 1.20.

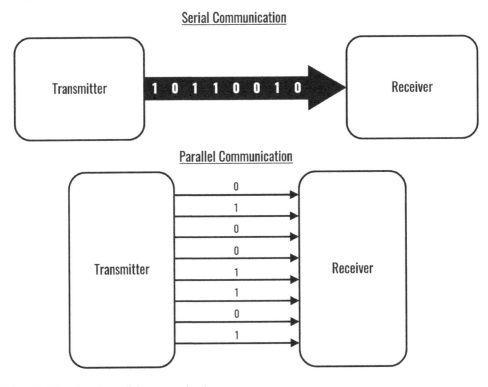

FIGURE 1.18   Serial and parallel communication

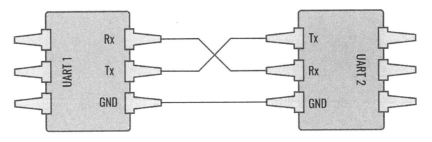

FIGURE 1.19   Connection between two UARTs

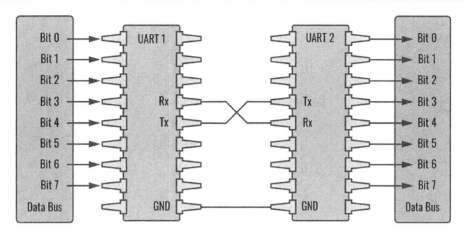

FIGURE 1.20 UART communication with data bus

Before starting a UART communication, both the transmitter and receiver must be set with the same baud rate (bits/second). It is the rate at which data is transferred on a communication line. The transmitter and receiver must be set with a standard baud rate but not with any random baud rate. The standard baud rates that can be used for UART communication are 110, 300, 600, 1200, 2400, 4800, 9600, 14400, 19200, 38400, 57600, 115200, 128000 and 256000. If the transmitter and receiver are not set with the same baud rate, then the data won't be sampled properly at the receiving end.

As you can see from Figure 1.20, devices in a UART communication don't share a clock signal and transfer data asynchronously. So, the data transfer in UART will be in the form of packets. Each data packet consists of a start bit, data frame, a parity bit and stop bits as shown in Figure 1.21. Usually, the UART data transmission line will be HIGH when no data is transmitted. For transferring data, the transmitter UART will pull the transmission line from HIGH to LOW. On detecting the voltage change from HIGH to LOW, the receiver UART begins to read the data at the baud rate frequency. The actual data to be transferred will be stored in the data frame which can be of 5–9 bits long. The parity bit is used by the UART receiver for determining whether any data has changed during transmission. After the UART receiver reads the data frame, it counts the number of 1s (HIGH) and checks whether it is an odd number or an even number. If the number of 1s is odd, then the parity bit should be 1 or else the parity bit should be 0. If the parity bit matches the condition, then the transmission is free of errors or else the bits in the data frame have changed during the transmission. Finally, the stop bit is used to end the transmission by pulling the UART transmission from LOW to HIGH for a 1 or 2 bits transfer duration.

Start Bit (1 Bit)	Data Frame (5 to 9 Data Bits)	Parity Bit (0 or 1 Bit)	Stop Bit (1 or 2 Bits)

FIGURE 1.21 UART data packet

1.7 LED (LIGHT-EMITTING DIODE)

While many of us know what an LED is all about, it is good to recollect some quick information on the same. LED is nothing but a diode that emits light when an electric potential is applied across it. Similar to any other diode, LED also has a positive (long) and a negative (short) pin. While using an LED with any microcontroller or an evaluation board, remember to connect a resistor in series with the LED. This will prevent the LED from damage and is known as current limiting resistors. Most of the LEDs operate at 1.2–3.6 V. Figure 1.22 shows the pin configuration of an LED.

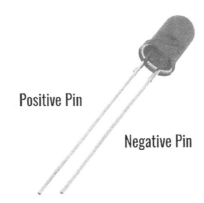

Positive Pin

Negative Pin

FIGURE 1.22   LED pins

1.8 ARDUINO UNO REV3

Arduino Uno is one of the most famous and most used evaluation boards in IoT. It is very easy to use and can be interfaced with most of the sensors and modules in IoT. It has 14 digital GPIO (General Purpose Input/Output) pins and 6 analog input pins. A GPIO pin is an uncommitted digital signal pin on a microcontroller or evaluation board or IC (integrated circuit) which can be used as an input or output but not both at the same time. The direction of the GPIO pin (Input or Output) is controllable by the user. GPIO pins have no predefined purpose and are unused by default.

Arduino Uno can provide an output voltage of 5 V and 3.3 V for powering external sensors and modules. Some of the important specifications of Arduino Uno are tabulated in Table 1.1. The original (not cloned) Arduino Uno is shown in Figure 1.23. To know more about Arduino Uno, you can visit their official website (https://store.arduino.cc/usa/arduino-uno-rev3).

TABLE 1.1   Specifications of Arduino Uno

MICROCONTROLLER	*ATMEGA328P*
Operating voltage	5 V
Input voltage (recommended)	7–12 V
Input voltage (limit)	6–20 V
Digital I/O pins	14
PWM digital I/O pins	6
Analog input pins	6
DC current per I/O pin	20 mA
DC current for 3.3 V pin	50 mA
Flash memory	32 KB (ATmega328P) of which 0.5 KB used by bootloader
SRAM	2 KB (ATmega328P)
EEPROM	1 KB (ATmega328P)
Clock speed	16 MHz
LED_BUILTIN	13
Length	68.6 mm
Width	53.4 mm
Weight	25 g

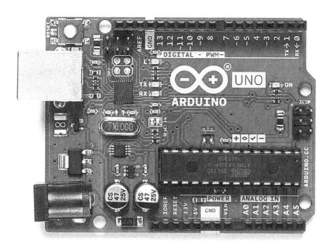

FIGURE 1.23   Arduino Uno Rev 3

1.9 LED CONTROL USING ARDUINO UNO

1.9.1 Motive of the project

To control a LED using Arduino Uno.

1.9.2 Hardware required

TABLE 1.2   Hardware required for the project

HARDWARE	QUANTITY
Arduino Uno	1
Resistor (220 Ω)	1
LED	1
Breadboard	1
Jumper wires	As required

Note: To prevent an LED from damage due to high voltage, a resistor is connected in series. Even though we have used a 220 Ω resistor between the LED −ve pin and GND, you can use any resistor (like 100 Ω, 1 kΩ, 10 kΩ, etc.).

1.9.3 Connections

You can refer to the list of components required from Table 1.2, followed by Table 1.3, which presents the pin-wise connections. Also, the connection diagram is presented in Figure 1.24, which shall help to get a visualization of the connections.

TABLE 1.3   Connection details

ARDUINO UNO	RESISTOR (220 Ω)	LED
GND	Pin 1	
9		+ve pin (long pin)
	Pin 2	−ve pin (short pin)

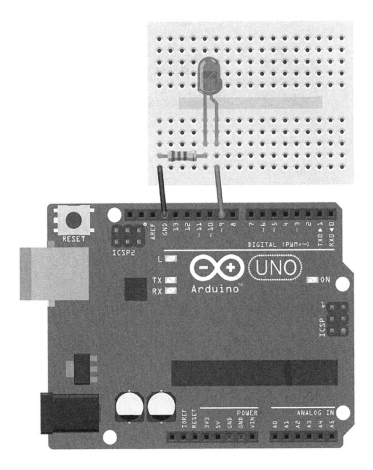

FIGURE 1.24   Connection layout

Note: It is always preferred to use a red wire for VCC connections and black wire for GND connections. This will help in easy debugging and will prevent accidental shorting of the circuitry. We will also follow the same in all the circuits used in this book and recommend you to follow the same. The connections between 5 V and 3.3 V of evaluation boards or microcontrollers to the VCC or +ve pins of the sensors or modules are referred to as VCC connections.

1.9.4 Arduino IDE code for blinking LED using Arduino Uno

Similar to C or C++, in Arduino IDE a single line can be commented using // and multiple lines can be commented using /**/. All the information written in the same line after // is treated as the comment. Similarly, all the information written between /* and */ is treated as a comment.

For example,

```
// This is an in-line comment

/*
This is a multiple line comment section.
This is another line of Comment.
*/
```

LEDPin is used for defining the connection between the LED and Arduino. As we are controlling the LED from the digital pin 9 of Arduino, we will assign the LEDPin as 9.

```
int LEDPin = 9; //Arduino Digital Pin 9
```

void setup() and void loop() are the two important functions for any Arduino Code without which the code doesn't even compile. void setup() executes only once. As the name says, it is used to perform the initial setup. void loop() runs infinitely as long as the board is powered. It is mainly used for controlling sensors or modules.

Here, we are using void setup() to define the pinMode of LEDPin and to initially keep the LED OFF. pinMode is used to define the direction of a GPIO pin. If you want to take input from a component, then the GPIO pin needs to be defined as INPUT. If you want to write on to a component or control a component, then the GPIO pin needs to be defined as OUTPUT. As we are controlling the LED, we define LEDPin as INPUT. digitalWrite() is used to digitally write onto a component. You can only digitally write LOW or HIGH onto a digital pin.

```
void setup() {
pinMode(LEDPin,OUTPUT);     //Set the direction of LEDpin
digitalWrite(LEDPin,LOW); //Initially, keep the LED OFF
}
```

Here, we are using void loop() to blink the LED. delay(x) is used to pause the program for x milliseconds. Initially, the LED is set to ON state for 1,000 milliseconds (1 second), and then it is set to OFF state for 1,000 milliseconds (1 second). By running this code, you can blink an LED using Arduino Uno.

```
void loop() {
   digitalWrite(LEDPin,HIGH);     //Write HIGH to ON the LED
   delay(1000);                   //Wait for a second
   digitalWrite(LEDPin,LOW);      //Write LOW to OFF the LED
   delay(1000);                   //Wait for a second
}
```

Finally, you need to upload code to the Arduino Uno board for the hardware to work. Before uploading the code, select the appropriate board and COM port in the Arduino IDE.

COM port (communication port or serial port) is a serial communication interface through which information can be transferred sequentially one bit after another. It is different from the parallel ports which can transfer multiple bits at the same time. COM ports enable serial communication between a serial device and a PC. Whenever you connect a serial device to a PC, it assigns a COM port to the serial device for communication. To know the COM port assigned by your PC to the connected Arduino Uno board follow the below steps.

If you are using a Windows machine, you can find the COM port of the connected board by going to the *Device Manager* as shown in Figure 1.25.

In OS X, you can find the COM port of the connected device by typing ls /dev/* command in the terminal. Note the port number listed for /dev/tty.usbmodem<x> or /dev/tty.usbserial<x>. The port number of the connected device is <x>.

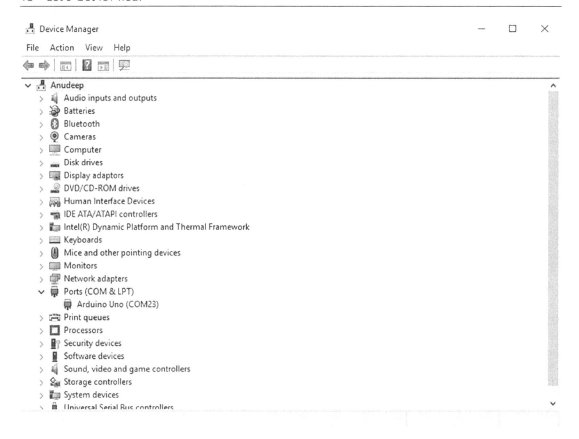

FIGURE 1.25   COM ports

For Linux OS (Operating System), you can find the COM port of the connected device by typing `ls /dev/tty*` in the terminal. Note the port number listed for `/dev/ttyUSB<x>` or `/dev/ttyACM<x>`. The port number of the connected device is `<x>`.

After finding the COM port, you can proceed with the board and port selection as shown in Figures 1.26 and 1.27, respectively.

Now, you are all set to upload the code to the connected board. Immediately after uploading the code to the Arduino Uno, you can see the LED blinking (1 second ON and then 1 second OFF).

1.9.5 Arduino IDE code for brightness control of a LED using Arduino Uno

`LEDPin` is used for defining the connection between the LED and Arduino. As we are controlling the LED from the digital pin 9 of Arduino, we will assign the `LEDPin` as 9. The amount of brightness of the LED at any point of time is stored in the variable `brightness`. The amount of brightness of LED to be increased or decreased per step is defined by the variable `fadeAmount`.

```
int LEDPin = 9; //The PWM pin to which the LED is connected.
int brightness = 0;   //Initially LED brightness.
int fadeAmount = 5;   //Quantity of LED fade per step!
```

Here, we are using `void setup()` to define the `pinMode` of `LEDPin`. As we are controlling the LED, the direction of `LEDPin` is defined as `OUTPUT`.

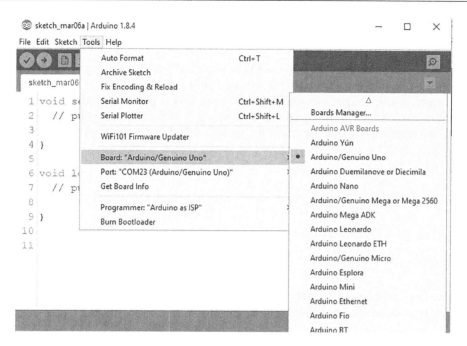

FIGURE 1.26   Board selection

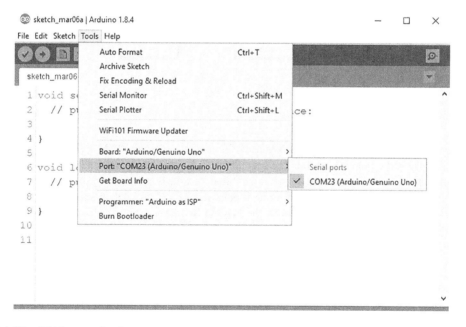

FIGURE 1.27   COM port selection

```
void setup() {
pinMode(LEDPin, OUTPUT); //Set the LED port as OUTPUT port.
}
```

Here, we are using void loop() to control the brightness of an LED. By seeing the name analog-Write(), you might be coming to an early conclusion that analogWrite() might be used to write analog values on to analog pin similar to the digitalWrite(). But the analogWrite() function has nothing to do with the analog pins or the `analogRead()` function which you will be seeing in the next project. `analogWrite()` is used to write values on PWM (pulse width modulation) pins. In most of the boards, PWM pins are indicated by a tilde (~) besides the pin number. In case of Arduino Uno, the digitals pins 3, 5, 6, 9, 10 and 11 are hardware-driven PWM pins. You can also configure other digital pins in Arduino Uno as PWM pins using software but it is not at all recommended.

PWM is a technique to generate analog output using digital pins. It generates a square wave by switching the signal ON and OFF. The ON time in one cycle of square wave is known as pulse width. Duty cycle is the proportion of time for which the signal is ON. For the signal shown in Figure 1.28, the duty cycle is 50% since $T_{ON} = T_{OFF}$.

$$Duty\ cycle = \frac{T_{ON}}{T_{ON} + T_{OFF}} \tag{1.1}$$

$$Duty\ cycle\ (in\ \%) = \frac{T_{ON}}{T_{ON} + T_{OFF}} * 100 \tag{1.2}$$

where
T_{ON} is the time for which the signal is ON in one cycle (pulse width)
T_{OFF} is the time for which the signal is OFF in one cycle.

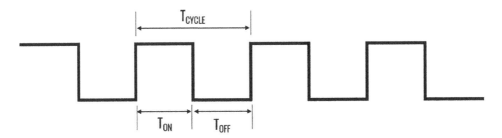

FIGURE 1.28   Square wave of 50% duty cycle

`analogWrite()` is an inbuilt function in Arduino IDE used to generate PWM signals. It takes two arguments: the first one is the pin number, and the second one is a value between 0 and 255. A square wave of 0% duty cycle (always OFF) is generated for 0 input value. Similarly, a square wave of 50% duty cycle and 100% duty cycle (always ON) is generated for an input value of 127 and 255, respectively.

Every time the `void loop()` executes, the brightness increases by the `fadeAmount`. After the brightness becomes greater than 255, then the direction of `fadeAmount` becomes negative. But if the brightness becomes less than 0, then `fadeAmount` again becomes positive. To see the brightness change properly, we have given a delay of 50 ms (milliseconds) for every step.

```
void loop() {
  // Set the brightness of LEDPin
  analogWrite(LEDPin, brightness);

  // Change the brightness for next time through the loop
  brightness = brightness + fadeAmount;

  // Reverse the direction of the fading at the ends of the fade
  if (brightness <= 0 || brightness >= 255) {
    fadeAmount = -fadeAmount;
  }

  delay(50);  // wait for 50ms to see the dimming effect per step
}
```

Now, upload the code to Arduino Uno by selecting the appropriate board and COM port. Immediately after uploading the code, you can see the brightness of the LED increasing and then decreasing continuously.

Now, it's time to move on. Let's perform the same tasks using NodeMCU.

1.10 NodeMCU

NodeMCU is another evaluation board similar to Arduino Uno. It is made using the ESP8266-based ESP-12F System on Chip (SoC) module. This module helps in computing the sensor data as well as connecting to WiFi. NodeMCU also has an onboard USB-mini connector, USB to UART converter and 5–3.3 V voltage regulator IC. The major advantage of NodeMCU is that it can be connected to WiFi and accessed remotely. It can also connect to the cloud for storing and retrieving data. The major disadvantages of NodeMCU are it has only 1 analog pin and it can only output a maximum voltage of 3.3 V. By default, more than 1 analog sensor cannot be connected to NodeMCU. But we can use multiplexing techniques to connect more than 1 analog sensor. Explaining the multiplexing techniques is out of the scope of this book. If you are curious about multiplexing or want to connect more than 1 analog sensor to NodeMCU, go to this link (https://diyi0t.com/increase-the-number-of-analog-inputs-with-an-analog-multiplexer/), which has an interesting mini-project that uses multiplexing techniques. Overcoming the other disadvantage of NodeMCU will be discussed in a later project. Figure 1.29 shows the clipart image of NodeMCU Amica and its pin configurations. For knowing more about the hardware configurations of NodeMCU, go to this link (https://components101.com/development-boards/nodemcu-esp8266-pinout-features-and-datasheet).

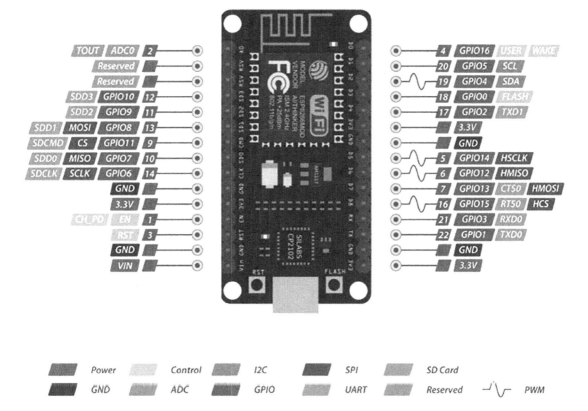

FIGURE 1.29 NodeMCU pin configuration

1.11　INSTALLATION OF SUPPORT FILES FOR NodeMCU

For uploading code to NodeMCU or any ESP device, you need to install support files in Arduino IDE. Go to the esp8266/Arduino GitHub repo (https://github.com/esp8266/Arduino) and scroll down to the *Installing with Boards Manager* section. Copy the link of the JSON file (https://arduino.esp8266.com/stable/package_esp8266com_index.json) as shown in Figure 1.30.

After copying the link, go to the *File > Preferences* tab or press *Ctrl+Comma* to open the preferences window in Arduino IDE as shown in Figure 1.31.

Paste the link of the JSON file (URL ending with *.json*) in the field which says *Additional Boards manager URLs*, and then click OK to save the preferences as shown in Figure 1.32. These .json files are nothing but board files that help in coding ESP-based boards with the same Arduino IDE. You can add as many board files as you want in this field. After saving the preferences, Go to *Tools > Board > Boards Manager* as shown in Figure 1.33.

After the *Boards Manager* window opens, search for *ESP8266* in the search bar. A board file named *esp8266 by ESP8266 Community* will be displayed as shown in Figure 1.34. By hovering the mouse on it, you can see the *Install* button along with the version. You can also select the version to be downloaded

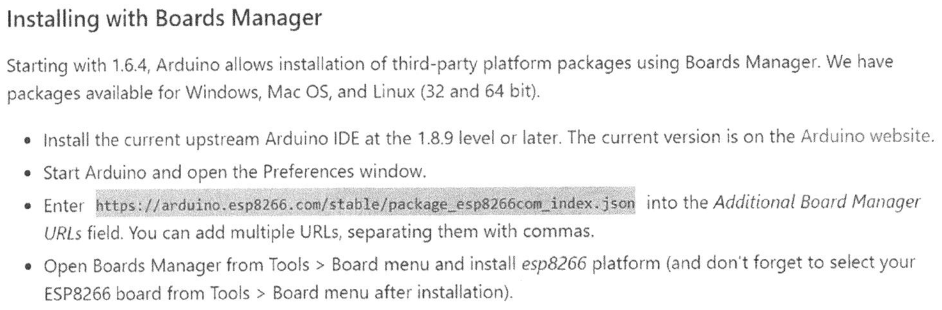

Installing with Boards Manager

Starting with 1.6.4, Arduino allows installation of third-party platform packages using Boards Manager. We have packages available for Windows, Mac OS, and Linux (32 and 64 bit).

- Install the current upstream Arduino IDE at the 1.8.9 level or later. The current version is on the Arduino website.
- Start Arduino and open the Preferences window.
- Enter `https://arduino.esp8266.com/stable/package_esp8266com_index.json` into the *Additional Board Manager URLs* field. You can add multiple URLs, separating them with commas.
- Open Boards Manager from Tools > Board menu and install *esp8266* platform (and don't forget to select your ESP8266 board from Tools > Board menu after installation).

FIGURE 1.30　Installation link for ESP8266 support files

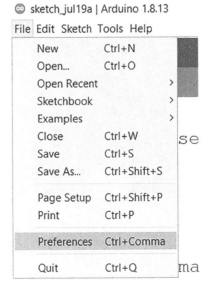

FIGURE 1.31　Select preferences

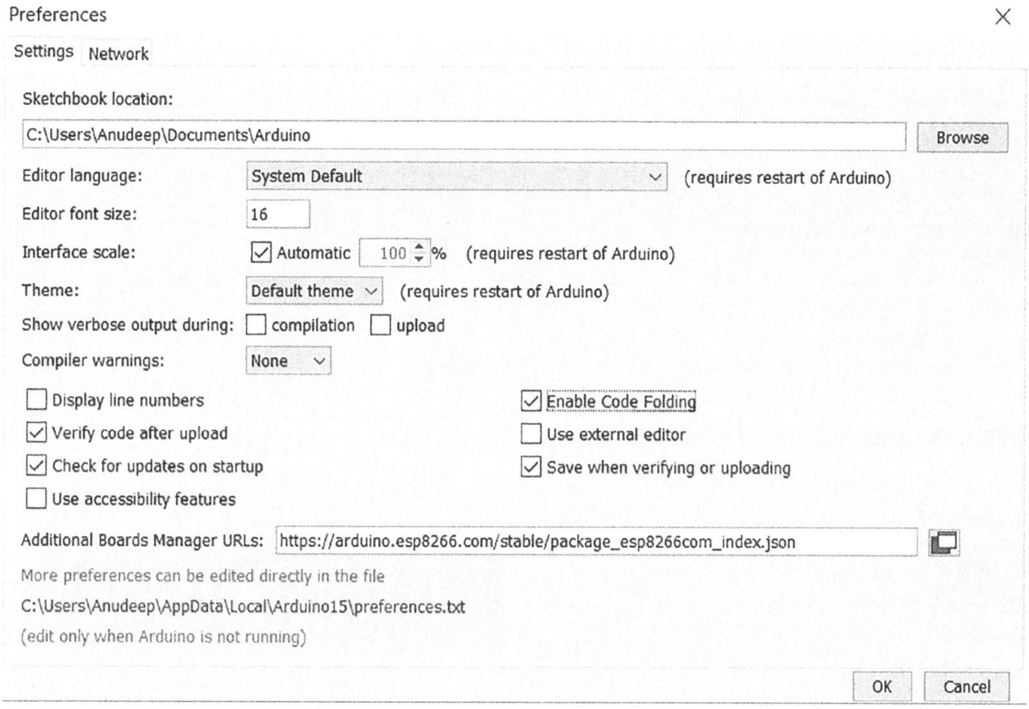

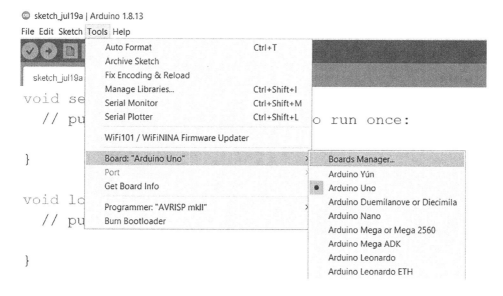

FIGURE 1.32   Preferences tab

FIGURE 1.33   Select boards manager

from the drop-down which appears beside the Install button. Choose the latest stable version and click *Install*. Wait for the library files to download. Once the download is completed, the board files are installed to Arduino IDE. Now you can upload code to NodeMCU.

After successful installation, open *Tools > Board: > ESP8266 Boards (2.7.2)* as shown in Figure 1.35. Here, you can see all the modules that are supported by this library, and click on *NodeMCU 1.0 (ESP-12E Module)* to upload code into the NodeMCU board used in this book. (This is the same NodeMCU board commonly used in IoT.)

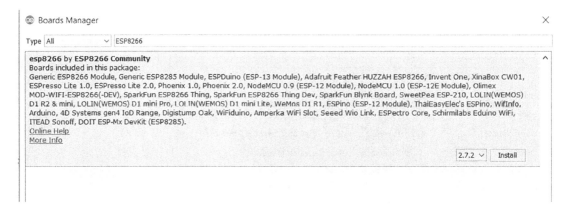

FIGURE 1.34   Installation of ESP8266 library files

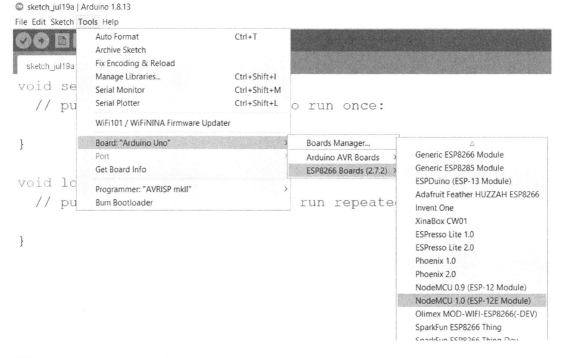

FIGURE 1.35   Successful installation and validation

1.12 LED CONTROL USING NodeMCU

1.12.1 Motive of the project

To control a LED using NodeMCU.

1.12.2 Connections

You can refer to the list of components required from Table 1.4, followed by Table 1.5, which presents the pin-wise connections. Also, the connection diagram is presented in Figure 1.36, which shall help to get a visualization of the connections.

TABLE 1.4   Hardware required for the project

HARDWARE	QUANTITY
NodeMCU	1
Resistor (220 Ω)	1
LED	1
Breadboard	1
Jumper wires	As required

TABLE 1.5   Connection details

NODEMCU	RESISTOR (220 Ω)	LED
GND	Pin 1	
D2		+ve pin (long pin)
	Pin 2	−ve pin (short pin)

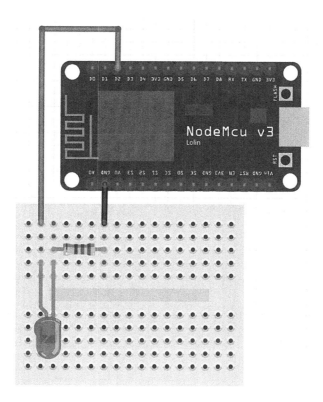

FIGURE 1.36   Connection layout

1.12.3 Arduino IDE code for blinking an LED using NodeMCU

Remember: Unlike Arduino Uno, the digital pin number on the NodeMCU board and its corresponding GPIO pin number need not be the same. So, you can use either one of them in your code to define the pin. For example, the D2 pin on NodeMCU has the GPIO number 4. So, you can use either one of the following ways **but not both**.

```
const int LEDpin = D2
or
const int LEDpin = 4
```

LEDpin is used for defining the connection between the LED and NodeMCU. As we are controlling the LED from the Digital Pin 2 of NodeMCU, we will assign the LEDpin as D2. The remaining code is similar to the one used in Section 1.9.4.

```
int LEDpin = D2; //NodeMCU D2 Pin

void setup() {
pinMode(LEDpin,OUTPUT);    //Determine the direction of IO pin
digitalWrite(LEDpin,LOW); //Initially, keep the LED OFF
}
void loop() {
    digitalWrite(LEDpin,HIGH);    // Write HIGH to ON the LED
    delay(1000);                      //Wait for a second
    digitalWrite(LEDpin,LOW);    // Write LOW to OFF the LED
    delay(1000);                      //Wait for a second
}
```

After writing the code in Arduino IDE, locate the COM port allocated to the NodeMCU. Select the appropriate board and COM port as shown in Figures 1.37 and 1.38.

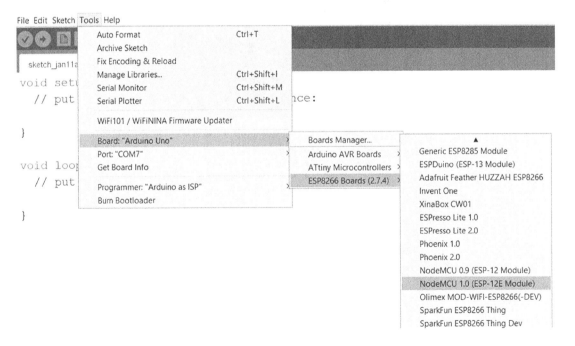

FIGURE 1.37 Board selection

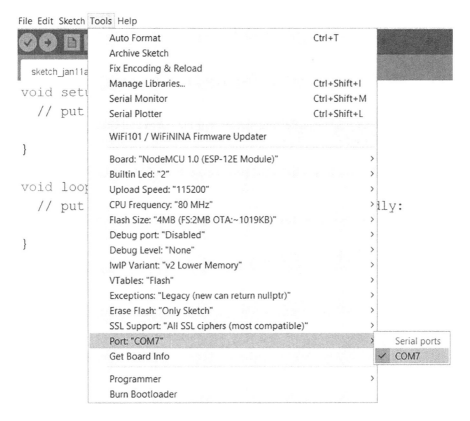

File Edit Sketch Tools Help

Auto Format	Ctrl+T
Archive Sketch	
Fix Encoding & Reload	
Manage Libraries...	Ctrl+Shift+I
Serial Monitor	Ctrl+Shift+M
Serial Plotter	Ctrl+Shift+L
WiFi101 / WiFiNINA Firmware Updater	
Board: "NodeMCU 1.0 (ESP-12E Module)"	>
Builtin Led: "2"	>
Upload Speed: "115200"	>
CPU Frequency: "80 MHz"	>
Flash Size: "4MB (FS:2MB OTA:~1019KB)"	>
Debug port: "Disabled"	>
Debug Level: "None"	>
lwIP Variant: "v2 Lower Memory"	>
VTables: "Flash"	>
Exceptions: "Legacy (new can return nullptr)"	>
Erase Flash: "Only Sketch"	>
SSL Support: "All SSL ciphers (most compatible)"	>
Port: "COM7"	>
Get Board Info	
Programmer	>
Burn Bootloader	

Serial ports
✓ COM7

FIGURE 1.38   COM port selection

Now, upload the code to NodeMCU by pressing the upload button at the top-left corner or by pressing *Ctrl + U*. Immediately after uploading the code, you can see the LED blinking (1 second ON and then 1 second OFF).

1.12.4 Arduino IDE code for brightness control of an LED using NodeMCU

The PWM pins of NodeMCU are represented by a tilde (~) besides the pin as shown in Figure 1.29. As D2 is a PWM pin, we have connected the LED to the D2 pin. LEDPin is used for defining the connection between the LED and NodeMCU. As we are controlling the LED from the D2 pin of NodeMCU, we will assign the LEDPin as D2. The remaining code is similar to the one used in Section 1.9.5.

```
int LEDPin = D2;      //The PWM pin to which LED is connected.
int brightness = 0;   //Initially LED brightness.
int fadeAmount = 5;   //Quantity of LED fade per step!

void setup() {
pinMode(LEDPin, OUTPUT); //Set the LED port as OUTPUT port.
}

void loop() {
  // Set the brightness of LEDPin
  analogWrite(LEDPin, brightness);

  // Change the brightness for next time through the loop
  brightness = brightness + fadeAmount;
```

```
    // Reverse the direction of the fading at the ends of the fade
    if (brightness <= 0 || brightness >= 1023) {
        fadeAmount = -fadeAmount;
    }
delay(50);  // wait for 50ms to see the dimming effect per step
}
```

Now, upload the code to NodeMCU by selecting the appropriate board and COM port. Immediately after uploading the code, you can see the brightness of the LED increasing and then decreasing continuously.

You can find all the resources of this project at this Github Link (https://github.com/ anudeep-20/30IoTProjects/tree/main/Project%201).

In this project, you have learnt to set up and use the two most commonly used evaluation boards in IoT. Controlling an LED using these evaluation boards is discussed in detail in this project. In the next project, you'll use this knowledge to control an LED depending on the environmental lighting conditions.

We are all set to go with the next one!

Smart Street Light Using IoT

2

Streetlights are something very essential for every city, town or village. It is not just a matter of convenience but also ensures safety and security. Having said that, many modern cities around the globe have automated the control of the street light with ON/OFF happening autonomously without any human intervention. This not only offers a lot of convenience to the maintenance team but also saves an intense amount of electricity. Often we do see or have seen street lights being ON even after the sun has started to shine brightly which can be curtailed and technology is the only go. The proposed technological approach is efficient, frugal and easy to implement.

2.1 MOTIVE OF THE PROJECT

In this project, we will control a street light depending on the ambient light. We'll use an LDR for detecting environmental light and an LED instead of a street light. The basic motive of the project is to switch ON the LED when there is no sufficient environmental light and switch OFF the LED when there is sufficient lighting.

2.2 LIGHT-DEPENDENT RESISTOR (LDR)

An LDR (light-dependent resistor) or a photoresistor is an electronic component that is sensitive to light. It is a passive component whose resistance decreases with the increase in the incident light intensity and vice versa. Similar to a normal resistor, it has two pins with no polarity (no positive and negative pins). Figure 2.1 shows a 20 mm LDR where 20 mm refers to the diameter of the LDR. Figure 2.2 shows the cadmium sulphide track on top of the LDR which is the main light-sensitive material along with the circuit symbol.

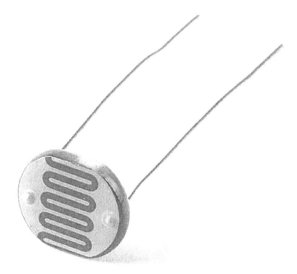

FIGURE 2.1   Light-dependent resistor

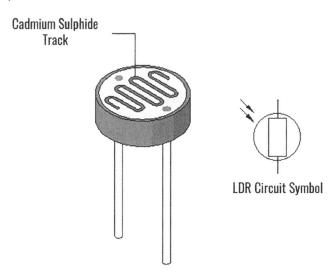

FIGURE 2.2   Components of LDR

2.3 HARDWARE REQUIRED

TABLE 2.1   Hardware required for the project

HARDWARE	QUANTITY
Arduino Uno	1
Light-dependent resistor (LDR)	1
Resistor (1 kΩ)	2
LED	1
Breadboard	1
Jumper wires	As required

2.4 CONNECTIONS

You can refer to the list of components required from Table 2.1, followed by Table 2.2, which presents the pin-wise connections. Also, the connection diagram is presented in Figure 2.3, which shall help you to get a visualization of the connections.

TABLE 2.2 Connection details

ARDUINO UNO	LDR	RESISTOR (1 KΩ)	RESISTOR (1 KΩ)	LED
5 V	Pin 1			
A0	Pin 2	Pin 1	Pin 1	−ve pin (long pin)
7				+ve pin (short pin)
Gnd		Pin 2	Pin 2	

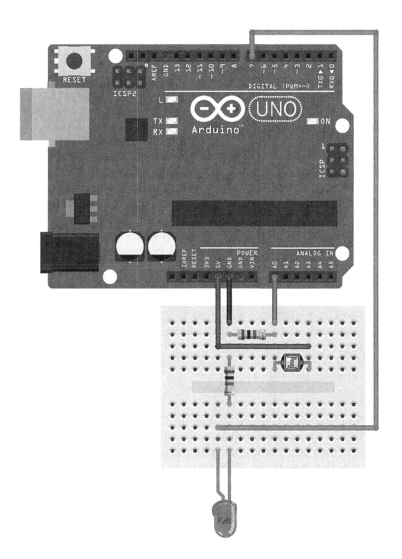

FIGURE 2.3 Connection layout

2.5 ARDUINO IDE CODE

LDR outputs different analog values depending on the intensity of light falling on it. As it outputs analog values, we have to connect the LDR to an analog pin as digital pins cannot acquire analog values. LEDs will glow only when there is sufficient voltage across their +ve and −ve pins. So, we can connect the LED to a digital pin.

LDRPin is used for defining the connection between LDR and Arduino Uno. LDR output value is stored in LDRValue and is initially set to 0. streetLight is used for defining the connection between the LED and Arduino.

```
int LDRPin = A0; //Select the input pin for LDR
int LDRValue = 0; //Variable to store the input value
int streetLight = 7; //Pin of light that supplies artificial sunlight
```

Serial.begin() is used for defining the communication speed between the Arduino Uno and PC. Here, the communication will begin at 9600 baud rate (bits/second). As we are taking values from the LDR, we define LDRPin as INPUT, and as we are controlling the LED, streetLight is defined as OUTPUT.

```
void setup() {
  Serial.begin(9600); //Begins the Serial Communication
  pinMode(LDRPin, INPUT); //Set A0 as Input Port for receiving LDR values
  pinMode(streetLight, OUTPUT); //Set Digital Pin 7 as Output Port
}
```

Note: In the previous project, you have come across the function analogWrite(). In this project, we will use the function analogRead(). Even though the names analogRead() and analogWrite() seem to be similar, they have nothing to do with each other. analogWrite() is used to output a PWM signal through a PWM pin in an evaluation board. But analogRead() converts the analog input from an analog pin into a digital output using the ADC (analog to digital converter) on an evaluation board. As Arduino Uno has a 10-bit ADC, you can read the values from 0 to 1023. You can know more about the analogRead() at this link (https://www.arduino.cc/reference/en/language/functions/analog-io/analogread/).

analogRead(LDRPin) reads the value from the LDR and stores it in LDRValue variable. Serial.print() is used to print information on the serial monitor. It is used for visualization and debugging purpose. Serial.println() is similar to Serial.print(), but the only difference is that the information printed after Serial.println() will be printed on a new line. Serial.println("<Information>"); is equivalent to

```
Serial.print("<Information>");
Serial.print("\n");
```

We have used an IF-ELSE statement to check whether the LDR output value is less than 700 or not. If it is less than 700, then there is no enough lighting and the LED gets switched ON. But if the LDR output value is not less than 700, then there is enough lighting and the LED gets switched OFF. A delay of 500 milliseconds is given after executing the IF-ELSE block.

```
void loop() {
  LDRValue = analogRead(LDRPin); // Reads the value from the sensor
  Serial.print("LDR Value: ");
  Serial.println(LDRValue); // Prints the value on serial monitor

  if (LDRValue < 700) { //Check the light intensity
    Serial.println("There is no enough environmental light. So, LED is
getting ON");
    digitalWrite(streetLight, HIGH);
  } else {
    Serial.println("There is enough environmental light. So, LED is getting
OFF");
    digitalWrite(streetLight, LOW);
  }
  delay(500);
}
```

Now, upload the code to the Arduino Uno after selecting the appropriate board and COM port. After uploading the code, open the serial monitor to visualize the data for which we need to select the appropriate baud rate. At the right bottom of the serial monitor, as shown in Figure 2.4, you will be having an option to adjust the baud rate using a dropdown. Adjust it to 9600 (which you have given in Serial. begin()) for visualizing the data properly. If the baud rate is not set properly, you will either see junk data or see no data on the serial monitor.

FIGURE 2.4   Baud rate selection

You also have settings like *AutoScroll*, *Show timestamp*, *Clear output*, etc. in the serial monitor. If the printed data contains only numbers or values, then it can be visualized graphically using serial plotter. It can be found in *Tools* as shown in Figure 2.5. A snapshot of the serial plotter is presented in Figure 2.6.

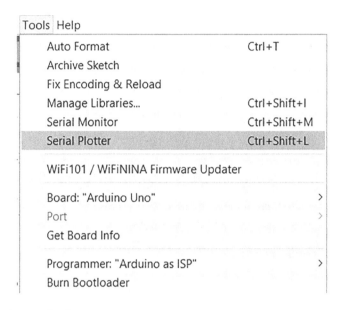

FIGURE 2.5   Serial plotter selection

FIGURE 2.6   Serial plotter GUI Hardware required for the project

Note: In the first six projects, you'll be building basic projects using different sensors and modules which are most commonly used in IoT. These projects will help you in getting familiarized with evaluation boards and sensors. Building these projects doesn't require any Internet usage or cloud storage. But from the seventh project, you'll be building projects which include Internet and cloud storage.

You can find all the resources of this project at this Github Link (https://github.com/anudeep-20/30IoTProjects/tree/main/Project%202).

In this project, you have learnt to control an LED depending on the ambient light. In the next project, you'll control various appliances/electronic components using potentiometers.

Let's build the next one!

Controlling Appliances Using Potentiometers

3

Having seen LEDs and street lights being controlled efficiently with the microcontrollers in the previous two projects, it is now time for the readers to know how to control electrical appliances with the potentiometers. You, me and everyone have been using the potentiometer every day in our life without knowing it as a potentiometer. We know it as a fan regulator or as the music system's volume adjustment knobs. Essentially, they are all potentiometers. This project will help you in building applications using potentiometers. It is divided into three parts:

- Controlling the brightness of an LED using potentiometer
- Controlling the brightness of an LED using microcontroller and potentiometer
- Controlling a remote antenna using potentiometer

3.1 POTENTIOMETER

Potentiometer is a three-terminal resistor with a slider or rotating knob that forms an adjustable voltage divider. If only two terminals are used, one end and the wiper, it acts as a variable resistor or rheostat. Potentiometer is also referred to as POT by some books and websites. In this project, we'll use the potentiometer as a variable resistor to control the electronic appliances.

Note: Even though Figure 3.1 shows the 1st pin as VCC and the 3rd pin as GND, these can be interchanged. The point to remember is that the outer pins of the potentiometer must be connected to VCC and GND and the middle pin to an adjustable wiper which outputs variable resistance. One can see the various types of potentiometers by referring to Figure 3.2.

For this project, you can use either a *rotary or reset* type of potentiometer.

DOI: 10.1201/9781003147169-3

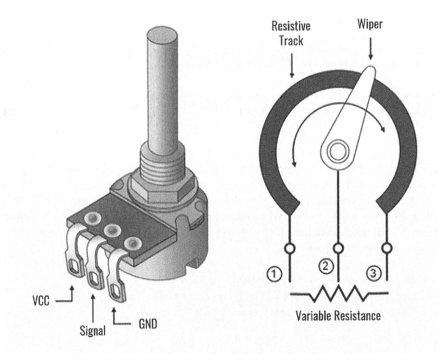

FIGURE 3.1   Potentiometer layout diagram

FIGURE 3.2   Types of potentiometers

3.2 CONTROLLING THE BRIGHTNESS OF AN LED USING POTENTIOMETER

3.2.1 Motive of the project

In this part of the project, you'll understand about varying the potential applied across an electronic component using the variable resistance of a potentiometer. We'll use an LED, potentiometer and a power source (microcontroller is not required) to control the brightness of the LED.

3.2.2 Hardware required

TABLE 3.1   Hardware required for controlling the brightness of an LED using potentiometer

HARDWARE	QUANTITY
Battery (9 V)	1
Resistor (220 Ω)	1
Potentiometer (10 kΩ)	1
LED	1
Breadboard	1
Jumper wires	As required

Note: Potentiometers can be of any value similar to a resistor. If a potentiometer is labelled as 10 kΩ, then the potentiometer can output resistance between 0 Ω and 10 kΩ. Even though we have used a 10 kΩ potentiometer for controlling the brightness of the LED, you can use any potentiometer (like 1 kΩ, 100 kΩ, 1 MΩ, etc). This applies to all other projects also unless otherwise mentioned.

3.2.3 Connections

You can refer to the list of components required from Table 3.1, followed by Table 3.2, which presents the pin-wise connections. Also, the connection diagram is presented in Figure 3.3, which shall help you to get a visualization of the connections.

After making the connections as mentioned in Table 3.2, you can rotate the potentiometer for increasing or decreasing the brightness of the LED. In our day-to-day life, we use a potentiometer for controlling the speed of the fan. Similarly, in this part of the project, you have learnt to control the brightness of an LED using a potentiometer.

TABLE 3.2   Connection details for controlling the brightness of an LED using potentiometer

BATTERY (9 V)	RESISTOR (220 Ω)	POTENTIOMETER (10 KΩ)	LED
−ve pin		1st pin	
	1st pin	2nd pin	
	2nd pin		−ve pin (short pin)
+ve pin		3rd pin	+ve pin (long pin)

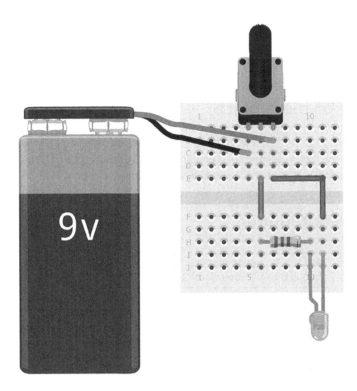

FIGURE 3.3 Connection layout for controlling the brightness of an LED using potentiometer

3.3 CONTROLLING THE BRIGHTNESS OF AN LED USING MICROCONTROLLER AND POTENTIOMETER

In the first part of the project, the brightness of the LED increases or decreases linearly with the rotation of the potentiometer. But if you want to control the brightness of the LED quite differently, then you need to employ a microcontroller.

A microcontroller can help in controlling the brightness of the LED in the following ways:

- Inversely control the brightness of the LED.
- Increase or decrease the brightness of the LED in steps.

3.3.1 Motive of the project

In this part of the project, you'll control the brightness of an LED in steps using a microcontroller.

3.3.2 Hardware required

TABLE 3.3 Hardware required for controlling the brightness of an LED using microcontroller and potentiometer

HARDWARE	QUANTITY
Arduino Uno	1
Resistor (220 Ω)	1
Potentiometer (10 kΩ)	1
LED	1
Breadboard	1
Jumper wires	As required

3.3.3 Connections

You can refer to the list of components required from Table 3.3, followed by Table 3.4, which presents the pin-wise connections. Also, the connection diagram is presented in Figure 3.4, which shall help get a visualization of the connections.

TABLE 3.4 Connection details for controlling the brightness of an LED using microcontroller and potentiometer

ARDUINO UNO	RESISTOR (220 Ω)	POTENTIOMETER (10 KΩ)	LED
5 V		3rd pin	
A5		2nd pin	
6			+ve pin (long pin)
	2nd pin		−ve pin (short pin)
GND	1st pin	1st pin	

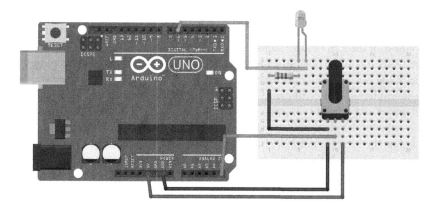

FIGURE 3.4 Connection layout for controlling the brightness of an LED using microcontroller and potentiometer

3.3.4 Arduino IDE code

Potentiometer outputs different analog values depending on the rotation of its knob. So, we need to connect the potentiometer to an analog pin of the microcontroller. For controlling the brightness of an LED, we need to connect it to a PWM pin.

 `pot` is used for defining the connection between potentiometer and Arduino Uno. The potentiometer output value is stored in `potVal` and is initially set to 0. LED is used for defining the connection between the LED and Arduino. For controlling the brightness of the LED, it must be connected to a PWM pin. So, the LED is connected to the digital pin 6 of Arduino Uno which is a PWM pin.

```
int pot = A5; //Analog pin to which Potentiometer is connected.
int potVal = 0;
int LED = 6; //The PWM pin to which LED is connected.
```

Whenever anything needs to be printed on the serial monitor, you need to use `Serial.begin()` for initializing the serial. As we need to read the values from the potentiometer, `pinMode` of the potentiometer is set as `INPUT`. As the brightness of the LED is controlled depending on the values of the potentiometer, the `pinMode` of LED is set as `OUTPUT`.

```
void setup() {
  Serial.begin(9600);
  pinMode(pot, INPUT);   //Declare pot as INPUT port
  pinMode(LED, OUTPUT); //Set the LED port as OUTPUT.
}
```

`analogRead(pot)` will read the analog value from the potentiometer and stores it in the `potVal` variable. As the microcontroller used in Arduino Uno is a 10-bit microcontroller (ATmega328P), the analog values read from the potentiometer will be in the range of 0–1023 ($2^{10} - 1$). If you want to see the values read from the potentiometer, you can print the values on the serial using `Serial.println(potVal)`.

```
void loop() {
potVal = analogRead(pot);
```

In Arduino Uno, `analogWrite()` is used to write the values on a PWM pin in the range of 0–255. So, the LED's brightness will be highest at 255 and will be minimum at 0. In this part of the project, we'll control the brightness of the LED in five steps.

 Remember: Don't rotate the potentiometer such that its output analog value decreases below 200. If the analog value of the potentiometer decreases below 200, then its resistance value becomes closer to 0 which causes a short circuit. This will cause high current to pass through the potentiometer which may burn the potentiometer and spoil it (you can even see smoke and fire if left long).

 If the analog value of the potentiometer is below 200, then the LED will be in OFF state and a warning will be printed on the serial monitor. If the analog value of the potentiometer is between 200 and 400, then 75 is written on LED pin. If the analog value of the potentiometer is between 400 and 600, then 150 is written on LED pin. Similarly, if the analog value of the potentiometer is between 600 and 800, then 200 is written on LED pin. Finally, if the analog value of the potentiometer is between 800 and 1024, then the LED will be in ON state (at its maximum brightness).

```
// Control the brightness of LED in steps.
if(potVal <= 200) {
Serial.println("Don't further decrease the resistance of the Potentiometer!!")
analogWrite(LED, 0);
} else if(potVal <= 400)
    analogWrite(LED, 75);
```

```
  else if(potVal <= 600)
    analogWrite(LED, 150);
  else if(potVal <= 800)
    analogWrite(LED, 200);
  else
    analogWrite(LED, 255);
}
```

Remember: In C or C++ (in Arduino IDE as well), if only a single line of code needs to be written inside a control structure (`if-else`) or inside a loop (`for`, `while`), brackets need not be used. For example,

```
if(True){
Serial.println("Hello World");
} else {
Serial.println("Hi World");
}
```

is same as

```
if(True)
Serial.println("Hello World");
else
Serial.println("Hi World");
```

But

```
if(True){
Serial.println("Line 1");
Serial.println("Line 2");
} else {
Serial.println("Line 3");
Serial.println("Line 4");
}
```

is not the same as

```
if(True)
Serial.println("Line 1");
Serial.println("Line 2");
else
Serial.println("Line 3");
Serial.println("Line 4");
```

and might throw errors or cause problems.

After making the connections as mentioned in Table 3.4, upload the code to the Arduino Uno by selecting the appropriate board and COM port. After uploading the code, rotate the potentiometer to see the brightness of the LED vary in steps.

3.3.5 Try it

In this part of the project, you have controlled the brightness of the LED in steps. Try to write an Arduino code that can inversely control the brightness of the LED. (If the potentiometer value is 1024, then the LED should be at its minimum brightness and vice versa.)

3.4 SERVO MOTOR

The shaft of a commonly used AC/DC motor as depicted in Figure 3.5 rotates continuously when it is powered. It doesn't have any idea about its angular position at any given point in time. But a rotary servo motor has precise control of its shaft angular position and velocity. It has a closed-loop mechanism that incorporates positional feedback to control its rotational speed and position. For example, a regular motor's shaft cannot be moved to a particular position (132°) at a particular speed (10°/second), but the servo motor's shaft can be placed at a particular position at a particular speed by using positional feedback.

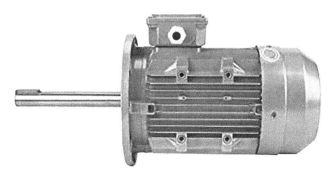

FIGURE 3.5   Shaft of an AC 3-phase motor

A servo motor is controlled with electric signals (either analog or digital) to move the shaft to a specified position at a definite speed. It consists of a suitable motor coupled with a sensor for positional feedback. In this project, we'll use a rotary DC servo motor as shown in Figure 3.6 that has precise control of its angular position and velocity. You can understand the pin configuration of a typical servo motor by referring to Figure 3.7.

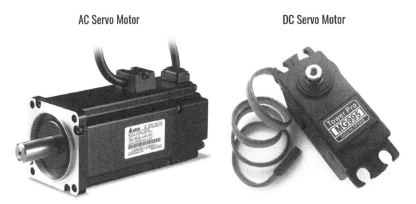

FIGURE 3.6   Types of servo motors

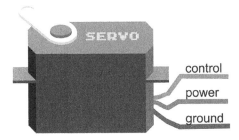

FIGURE 3.7   Servo motor pins

3.5 CONTROLLING A REMOTE ANTENNA USING POTENTIOMETER

In the previous two parts of the project, LED brightness is controlled using a potentiometer which is a simple application that can be used in our daily life. In this part of the project, let us simulate a real-time use case of a potentiometer.

Generally, large dish antennas (shown in Figure 3.8) used for scientific purposes or by broadcasting channels are not rotated by hand. These dishes are placed on a remotely controlled rotatable frame. The rotation of this frame is remotely controlled using a potentiometer. Depending on the angle of rotation of the potentiometer, the antenna also rotates proportionally.

FIGURE 3.8   Large dish antennas

3.5.1 Motive of the project

To rotate a remote antenna depending on the rotation of the potentiometer. In this part of the project, we'll use a servo motor instead of a rotatable frame. So, the objective of this part of the project is to rotate the servo motor depending on the rotation of the potentiometer.

3.5.2 Hardware required

TABLE 3.5   Hardware required for controlling a remote antenna using potentiometer

HARDWARE	QUANTITY
Arduino Uno	1
Potentiometer (10kΩ)	1
Servo motor	1
Breadboard	1
Jumper wires	As required

3.5.3 Connections

You can refer to the list of components required from Table 3.5, followed by Table 3.6, which presents the pin-wise connections. Also, the connection diagram is presented in Figure 3.9, which shall help get a visualization of the connections.

TABLE 3.6   Connection details for controlling a remote antenna using potentiometer

ARDUINO UNO	POTENTIOMETER	SERVO MOTOR
5 V	1st pin	VCC (red)
GND	3rd pin	GND (black)
6		OUT (yellow)
A5	2nd pin	

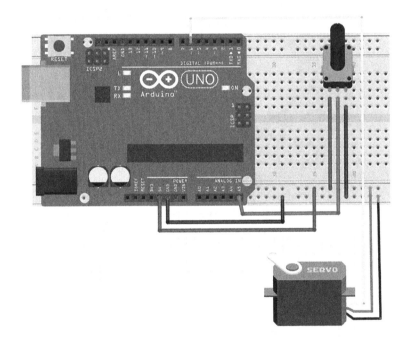

FIGURE 3.9   Connection layout for controlling a remote antenna using potentiometer

3.5.4 Arduino IDE code

For easy control of servo motors, `Servo.h` library is used which has functions to generate the PWM signals according to the given input. It is an inbuilt library that gets installed during the installation of Arduino IDE. In case, if it didn't get installed, then follow the below steps for installation. A similar procedure can be used for installing any Arduino library.

- Search for the library you want to install on the Internet. In this case, search *Download servo.h library.*
- Among the search results, open the relevant Github link of the library (in most of the cases, the first link will be the appropriate one), download the compressed file of the library and extract

it. In this case, download the `Servo.h` library compressed file at this link (https://github.com/
arduino-libraries/Servo) and extract it.

- Copy the extracted file at *C:\Program Files (x86)\Arduino\libraries* (Location at which
  Arduino is installed) or at *C:\Users\<USERNAME>\Documents\Arduino\libraries* - **but not
  at both the locations**. If you copy the extracted library file at both locations, then the library
  doesn't get installed properly.

Similar to C or C++, `#include <Library Name>` or `#include "Library Name"` is used for
importing libraries. So, `#include <Servo.h>` command is used for importing Servo.h library.

```
#include <Servo.h> //Import the servo library
```

`myServo` is a variable of `Servo` class used for easy access of `Servo.h` library functions. If you haven't
learnt about object-oriented programming (OOP) in the past, then it is recommended to read about it at
this link (https://www.w3schools.com/cpp/cpp_oop.asp) for having a better understanding of classes and
objects in C++ programming language. `servo` is used for defining the connection between servo motor
and Arduino Uno. For controlling a servo motor, it must be connected to a PWM pin. So, the servo motor
is connected to the digital pin 6 of Arduino Uno which is a PWM pin. `servoAngle` is used for storing
the angle to which the servo motor is rotated. `pot` is used for defining the connection between potentiometer and Arduino Uno. The potentiometer output value is stored in `potVal` and is initially set to 0.

```
Servo myServo;   //Servo class object
int servo = 6;   //Initialize the Servo motor pin
int servoAngle;

int pot = A5;    //Initialize the potentiometer pin
int potVal = 0;
```

Arduino's serial data transmission rate is initialized to 9600 baud rate using `Serial.begin()` command. As we need to read the values from the potentiometer, `pinMode` of the potentiometer is set as
`INPUT`. For accessing the functions in `Servo.h` library, the servo motor pin (`servo`) is attached to the
`myServo` using the `.attach()` function. Initially, the servo motor is set at 0° angle using the `.write()`
function.

```
void setup() {
  Serial.begin(9600);
  pinMode(pot, INPUT); //Declare pot as INPUT
  myServo.attach(servo); //Linking to the Arduino Uno's PWM pin
  myServo.write(0); //Initially the servo motor is set at 0°
}
```

`analogRead(pot)` will read the value from the potentiometer and stores it in the `potVal` variable.
`Serial.print()`and `Serial.println()` is used for printing the potentiometer values on the serial
monitor at a baud rate of 9600.

```
void loop() {
  potVal = analogRead(pot);   //Read the input from the potentiometer pin
  Serial.print("Potentiometer Value: ");
  Serial.println(potVal);
```

`map()`is an inbuilt mathematical function in Arduino IDE that will be used for mapping from one range
to another. Its syntax `map(var, fromLow, fromHigh, toLow, toHigh)`. The value `var` will be
mapped from the current range `fromLow – fromHigh` to target range `toLow – toHigh`. The `map()`
function doesn't constraint the mapping only in the specified range. Even if the value `var` is out of the

current range, the map() function uses integer math to give the mapped value. You can know more about the map() function at this Link (https://www.arduino.cc/reference/en/language/functions/math/map/).

Potentiometer outputs the analog values in the range of 0–1023, whereas the servo motor accepts the angular values in the range of 0–179. (If you have a servo motor that can rotate through 360°, then the range can be from 0 to 359.) So, the potentiometer values need to be linearly mapped from 0–1023 to 0–179. As the analog values of the potentiometer below 200 might damage it, we'll only consider the 200–1023 range of potentiometer values.

According to the potentiometer output value, the angle to which the servo motor is rotated is stored in the servoAngle variable. This is printed on the serial monitor and is written on the servo motor for turning it using the .write() function. Finally, a delay of 500 milliseconds is given for stabilizing the turn. You can even decrease or increase the time delay depending on the use case.

```
//Map the potentiometer to match the servo angle
servoAngle = map(potVal, 200, 1023, 0, 179);
Serial.print("Servo Angle: ");
Serial.println(servoAngle);
//Finally write the mapped value to the servo for turning.
myServo.write(servoAngle);
delay(500); //Give a delay of 500 milli seconds to stabilize the turn.
}
```

After making the connections as mentioned in Table 3.6, upload the code to the Arduino Uno by selecting the appropriate board and COM port. After uploading the code, as you rotate the knob of the potentiometer, the servo motor also rotates.

You can find all the resources of this project at this Github Link (https://github.com/anudeep-20/30IoTProjects/tree/main/Project%203).

In this project, you have learnt to control the brightness of an LED using a potentiometer and with and without using a microcontroller. You have also developed a smaller version of a real-time application of potentiometers where they are used to control larger appliances. In the next project, you'll build a door system that can be only opened using a password.

Let's build the next one!

Password-Authenticated Door Using Bluetooth

4

Safety and security are always paramount concerns for everyone. We do a lot of hard or smart work to earn wealth and to ensure the future is safe. The threat of theft is historically prevailing everywhere, and many solutions ranging from stronger locks to camera-based locks have been evolving. The market for locks is huge, and we are presenting one such product, password-authenticated door access using Bluetooth. The entire project is presented in a readable and understandable manner. Readers can clearly understand the importance and ease of doing the same.

4.1 MOTIVE OF THE PROJECT

In this project, we'll build a door system that can be opened only if the right password is entered. The door is operated by a microcontroller to which a Bluetooth module is connected. We need to connect our mobile phone to the Bluetooth module and enter the correct password for opening the door. If this project has to be implemented in real time, the normal hinge door has to be replaced with an electrically or electronically controlled motorized door. To make the project simple, we'll use a servo motor in the place of a door.

4.2 BLUETOOTH MODULE (HC-05)

Bluetooth is a wireless technology standard used for data exchange between devices over short distances operating at 2.45 GHz frequency. HC-05 is the most common Bluetooth module used in building IoT projects or prototypes. There are many types of HC-05 Bluetooth module available in the market like 4-pin HC-05 (with and without button), 6-pin HC-05 (with and without button) and many more. It is always recommended to purchase a 6-pin HC-05 Bluetooth module with a button as shown in Figure 4.1 as it will provide almost all the features.

The specifications and default settings of the HC-05 Bluetooth module are tabulated in Tables 4.1 and 4.2, respectively.

Other than HC-05, there are many other Bluetooth modules commonly used in IoT such as HC-04, HC-06, HC-12, HM-11 and many more. HC-12 is a special Bluetooth module that has a connectivity range of upto 1 km.

DOI: 10.1201/9781003147169-4

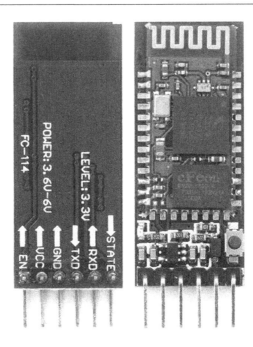

FIGURE 4.1   A 6-Pin HC-05 Bluetooth module (with Button)

TABLE 4.1   Specifications of HC-05 Bluetooth module

OPERATING VOLTAGE	*4–6 V (TYPICALLY +5 V)*
Operating current	30 mA
Range	Around 10 m
Communication	Serial communication (USART) and TTL compatible
Protocol standard	Follows IEEE 802.15.1 standardized protocol
Operation mode	Master, slave or master/slave mode
Supported baud rate	9600, 19200, 38400, 57600, 115200, 230400, 460800
Sensitivity	Typically −80 dBm

TABLE 4.2   Default settings of HC-05 Bluetooth module

BLUETOOTH NAME	*HC-05*
Password	1234 or 0000
Communication	Slave
Mode	Data mode
Data mode baud rate	9600, 8, N, 1
Command mode (AT mode) baud rate	38400, 8, N, 1
Firmware	LINVOR

4.3 BLUETOOTH TERMINAL APPLICATION

After pairing the mobile phone to the HC-05 Bluetooth module, we need to use a terminal application for sending or receiving data. For Android OS, you can download the *Bluetooth Terminal HC-05* application as shown in Figure 4.2 from the Play store (https://play.google.com/store/apps/details?id=project.

bluetoothterminal&hl=en_I). For iOS, you can download the *BLE Terminal HM-10* application as shown in Figure 4.3 from the App Store (https://apps.apple.com/us/app/ble-terminal-hm-10/id1398703795#?platform=iphone).

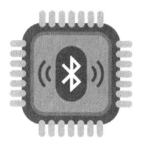

Bluetooth Terminal HC-05

mightyIT   Communication

3+

Contains Ads · Offers in-app purchases
ⓘ This app is compatible with all of your devices.

★ ★ ★ ★   709 ☻

🔖 Add to Wishlist

Install

FIGURE 4.2   Bluetooth Terminal HC-05

BLE Terminal HM-10   4+
BLE Terminal HM-10
Gopi Gadhiya
Designed for iPad

#169 in Productivity
★ ★ ★ ★ ★   4.1 • 16 Ratings

Offers In-App Purchases

FIGURE 4.3   BLE Terminal HM-10

4.4 HARDWARE REQUIRED

TABLE 4.3   Hardware required for the project

HARDWARE	QUANTITY
Arduino Uno	1
Servo motor	1
HC-05 Bluetooth module	1
LED	2 (1×green, 1×red)
Jumper wires	As required

4.5 CONNECTIONS

You can refer to the list of components required from Table 4.3, followed by Table 4.4, which presents the pin-wise connections. Also, the connection diagram is presented in Figure 4.4, which shall help get a visualization of the connections.

TABLE 4.4   Connection details

ARDUINO UNO	SERVO MOTOR	HC-05 BLUETOOTH MODULE	GREEN LED	RED LED
5 V		VCC		
3.3 V	VCC (red)			
GND	GND (black)	GND	−ve pin (short pin)	−ve in (short pin)
RXD		TXD		
TXD		RXD		
5	OUT (yellow)			
12			+ve pin (long pin)	
9				+ve pin (long pin)

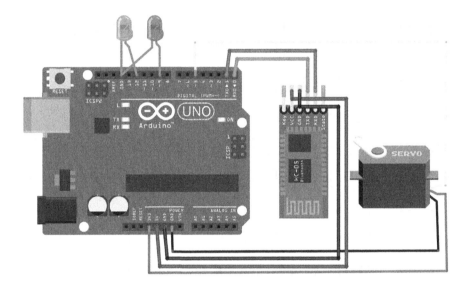

FIGURE 4.4   Connection layout

4.6 ARDUINO IDE CODE

As mentioned in the previous project, import the Servo library (`Servo.h`) for controlling the servo motor. In the previous project, a detailed explanation is provided regarding the usage of libraries in Arduino IDE. If you haven't followed the previous project, please refer to it for knowing about the handling of libraries in Arduino IDE.

```
#include <Servo.h> //Import the servo library
```

`myServo` is a variable of `Servo` class used for easy access of `Servo.h` library functions. `servoPin` variable is used for storing the digital pin number to which the servo motor is connected. For controlling a servo motor, it must be connected to a PWM pin. So, it is connected to the digital pin 5 of Arduino Uno which is a PWM pin. `GreenLED` and `RedLED` variables are used for storing the digital pin number to which the green and red LEDs are connected, respectively.

```
Servo myServo;     //Servo class object
int servoPin = 5;  //Initialize the Servo motor pin

int GreenLED = 12; //Initialize the Green LED to pin 12 of Arduino
int RedLED = 9; //Initialize the Red LED to pin 9 of Arduino
```

`pswd` is a `String` variable used to store your *Secret Password*. We have set the password as `MY SECRET PASSWORD`. It is generally preferred to set an alpha-numeric password but you can set it to any password of your wish. `recvPswd` is also a `String` variable used to store the received password obtained from your phone. Initially, `recvPswd` is stored as an empty string (`""`). `doorState` is a boolean variable (which can only store `true` or `false`) that is used for knowing the current state of the door (whether it is open or close). Initially, `doorState` is set to `false` which means the door is closed. `wrngPswd-Counter` is used as a counter for counting the number of wrong attempts by the used. Initially, the number of wrong attempts is set to 0.

```
String pswd = "MY SECRET PASSWORD"; //String to store your password
String recvPswd = ""; //String to store the received password

//Check whether door is open or close. Initially Closed.
bool doorState = false;
int wrngPswdCounter = 0; //Counter for number of wrong attempts
```

`void setup()` function is used for the initial setup of the Arduino program. The code in it will only be executed once. `Serial.begin(9600)` is used for initializing the Arduino's Serial at 9600 baud rate. `pinMode()` of `GreenLED`, and `RedLED` is set as `OUTPUT` to write voltages on the LED pins for controlling them. Initially, `GreenLED` and `RedLED` are set to LOW state (OFF state).

```
void setup() {
  Serial.begin(9600); //Switch on Serial communication
  pinMode(GreenLED,OUTPUT); //Set the mode of Green LED as OUTPUT
  pinMode(RedLED,OUTPUT); //Set the mode of Red LED as OUTPUT
  digitalWrite(GreenLED,LOW); //Let the Green LED stay off at the start
  digitalWrite(RedLED,LOW); //Let the Red LED stay off at the start
```

For accessing the functions in `Servo.h` library, the servo motor pin (`servo`) is linked to the `myServo` using the `.attach()` function. Initially, the servo motor is set at 0° angle using the `.write()` function.

```
  myServo.attach(servoPin); //Linking to the Arduino Uno's PWM pin
  myServo.write(0); //Initially Servo will be at 0° (Closed Door)
}
```

Unlike `void setup()` function, `void loop()` function is used for executing through the code written in it infinitely. `Serial.available()` is used for getting the number of bytes (characters) available for reading from the serial. If the number of bytes for reading is greater than 0, then `Serial.readString()` is used for reading the incoming String. As we know beforehand that the incoming data will be in the form of a string, we have used `Serial.readString()`. But if you don't know the type of incoming data, then you can use `Serial.read()` which reads byte by byte (character by character). `Serial.readStringUntil()` is used for reading the string until a particular character. For example, `Serial.readStringUntil("\n")` reads the incoming string until the \n character (line break).

```
void loop() {
  //Checks whether any information is available in Serial or not
  if(Serial.available() > 0){
    recvPswd = Serial.readString(); //Read the string
```

.trim() is used for removing any white space at the beginning or at the end of a string. Any string read from the serial will have \r (tab) and \n (line break) attached to it at the end. To determine the correct string entered by the user, we need to remove both \r and \n using the .trim() function. trim() modifies the string in place rather than returning a new modified string. To know more about the trim() function, visit this link (https://www.arduino.cc/reference/en/language/variables/data-types/string/functions/trim/). After removing all the white spaces at the end, the received password is printed on the serial monitor.

```
//Remove all the extra spaces at the beginning and at the end of the String.
It removes /r /n at the end of recvPswd
    recvPswd.trim();
    Serial.print("You have entered the Password: ");
    Serial.println(recvPswd);
  }
```

If there is no password received from the serial, then the code should not proceed further and again start from the beginning. In general, return is used for breaking out of a function which means no code gets executed after the return statement. But this doesn't exactly happen in the case of void loop() function of Arduino code. When the recvPswd is an empty string (""), the return statement is executed. So ideally, the code should exit from the void loop() function, and the Arduino should stop working as there is no code after the void loop() function. But this doesn't happen. Once a code is uploaded to the Arduino board, the void loop() function runs infinitely (i.e, until the board has a power supply to it). So, even after exiting from the void loop() function, the code again starts executing the void loop() function. If you are familiar with any other programming language, you can compare this with continue used in a for or while loop to start execution from the beginning.

```
//If No password is available then the loop() function starts from beginning.
if(recvPswd == ""){
    return;
}
```

If the received password (recvPswd) is the same as the assigned password (pswd), then you are authorized to open or close a door. The equality of two strings can be checked using == or .equals() function.

```
if(recvPswd == pswd) { //Compare the passwords
```

If the door is closed (i.e., doorState is false), then the door will open. First, the GreenLED is set to HIGH (switching on the green LED) to indicate that the entered password is correct and the door will open. Here, the opening of the door is indicated by rotating the servo motor from 0° to 90°. After the door is opened, the GreenLED is set to LOW (switching off the Green LED), and the doorState is changed to true (i.e., the door is open).

```
    if(doorState == false){ //If the door is closed then open it.
      Serial.println("You have entered CORRECT password !! Door is Opening.");
      digitalWrite(GreenLED, HIGH);
      delay(1000);
```

```
      for(int ang = 0; ang <= 90; ang++){
        myServo.write(ang);
        delay(20);
      }
      delay(1000);
      digitalWrite(GreenLED, LOW);
      doorState = true;
    }
```

If the door is open (i.e., doorState is true), then the door will close. First, the GreenLED is set to HIGH (switching on the Green LED) to indicate that the entered password is correct and the door will close. Here, the closing of the door is indicated by rotating the servo motor from 90° to 0°. After the door is closed, the GreenLED is set to LOW (switching off the green LED), and the doorState is changed to false (i.e, the door is closed). Finally, the data in the recvPswd is cleared (recvPswd = "").

If you don't understand why the data in recvPswd is cleared at the end, try removing the line and executing the code, you'll understand the reason. If the data in the recvPswd is not cleared, in the next iteration, the code will assume that the password typed by the user is the password stored in recvPswd. So, the code executes in the same way as executed before which leads to an uncontrolled infinite loop.

```
    else { //If the door is opened then close it.
      Serial.println("You have entered CORRECT password !! Door is Closing.");
      digitalWrite(GreenLED, HIGH);
      delay(1000);
      for(int ang = 90; ang >= 0; ang--){
        myServo.write(ang);
        delay(20);
      }
      delay(1000);
      digitalWrite(GreenLED, LOW);
      doorState = false;
    }
    recvPswd = ""; // Empty the recvPswd variable
  }
```

If the passwords don't match (i.e, recvPswd≠pswd), then the wrngPswdCounter will be incremented in steps of 1.

Remember: In most of the programming languages,

```
var = var + 100 is same as var += 100
var = var - 100 is same as var -= 100
var = var * 100 is same as var *= 100
var = var / 100 is same as var /= 100
```

```
  else { //If the passwords doesn't match
wrngPswdCounter += 1; // Increments the counter by 1
```

If wrong password is entered less than three times (wrngPswdCounter< 3), then for every wrong entry, a warning will be printed on the serial monitor and the Red LED will blink for 2 seconds.

```
//If the number of wrong attempts is less than 3 then another chance is
provided.
    if(wrngPswdCounter < 3) {
        Serial.print("You have entered WRONG password. You have more ");
        Serial.print(3 - wrngPswdCounter);
        Serial.println(" time(s) to enter the correct one.");
```

```
        digitalWrite(RedLED, HIGH);
        delay(2000);
        digitalWrite(RedLED, LOW);
    }
```

If a wrong password is entered more than three times, then the Arduino code pauses for 1 minute (60,000 milliseconds). If the user wants to re-enter the password, then the user needs to either reset the Arduino board or wait for 1 minute. The same instructions will be printed on the serial monitor. During the 1 minute pause, the red LED will be glowing indicating that the Arduino is paused.

Resetting the Arduino board can be done in two ways as shown in Figure 4.5:

• Using the onboard RESET pin.
• By pressing the onboard RESET button.

RESET Button

RESET Pin

FIGURE 4.5   Reset in Arduino Uno

```
    else { //If the number of wrong attempts equals 3
        Serial.println("You have exceeded the number of attempts. Please reset
the Arduino for re-entering the password or wait for 1 minute.");
        // Stops the whole process for 1 minute.
        digitalWrite(RedLED, HIGH);
        delay(60000);
        digitalWrite(RedLED, LOW);
        delay(1000);
```

After the 1 minute gets elapsed, both the LEDs together will blink for 2 seconds indicating that everything is reset and the user can now type the passwords again. After the 1 minute pause, the wrngPswd-Counter is again set to 0, and the recvPswd string variable is cleared (recvPswd = "").

```
      //Both the LEDs blink for 2 seconds indicating everything is reset.
      digitalWrite(RedLED, HIGH);
      digitalWrite(GreenLED, HIGH);
      delay(2000);
      digitalWrite(RedLED, LOW);
      digitalWrite(GreenLED, LOW);
      Serial.println("Now you can try to enter the passwords.");
      wrngPswdCounter = 0;
    }
    recvPswd = ""; // Empty the recvPswd variable
  }
}
```

Finally, upload the code to the Arduino Uno by selecting the appropriate board and COM port. After clicking the upload button, you will see the Arduino IDE *Uploading...* stuck or an error on the console as shown in Figures 4.6 and 4.7. This is because while uploading the code, the Rx and Tx pins of Arduino Uno are connected to the Bluetooth.

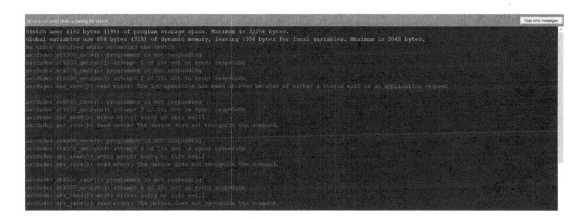

FIGURE 4.6   Uploading of code is stuck

FIGURE 4.7   Error uploading the code

Arduino Uno supports only half-duplex UART communication which means it can either send or receive data from a single device through UART at a time. While uploading the code to Arduino Uno from the PC, it is already connected to the HC-05 Bluetooth module over its default UART interface (Rx and Tx pins). So, the Arduino Uno cannot send or receive data from any other device. To upload code to Arduino Uno, we need to disconnect the Rx and Tx pins connected to the HC-05 Bluetooth module so that the code from the PC can be transferred.

So, whenever you upload code to Arduino Uno from the PC, all the connections to Rx and Tx pins need to be disconnected. After the code gets uploaded successfully (*Done Uploading* message will appear), re-connect the Rx and Tx pins. This entire process can be eliminated by using the `SoftwareSerial` library which will be introduced in future projects.

Note: For a point-to-point communication, there are three types of transmission modes.

- *Simplex Mode:* It is a unidirectional communication where the sender can send data to the receiver but cannot receive any data from the receiver.
- *Half-Duplex Mode:* It is a bidirectional communication but one way at a time. The sender can send data to the receiver or the sender can receive data from the receiver but not simultaneously.
- *Full-Duplex Mode:* It is a complete bidirectional communication where the sender and receiver can send and receive data at the same time.

4.7 USAGE OF BLUETOOTH TERMINAL APPLICATION

After the code is successfully uploaded, follow the below steps for entering the password to open or close the door.

- Switch on your mobile Bluetooth and Click *Pair a new device* as shown in Figure 4.8.

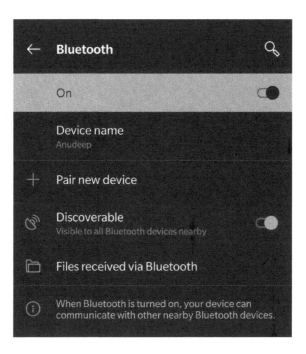

FIGURE 4.8   Pair new device

- Among the available devices, pair with your HC-05 Bluetooth module as shown in Figure 4.9.

FIGURE 4.9 Available devices

- Enter the password when prompted as shown in Figure 4.10. If you are using a new HC-05 Bluetooth module, then the password will be either *0000* or *1234*.

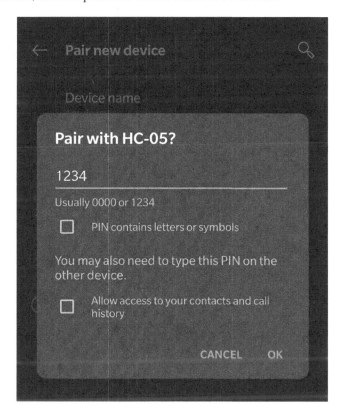

FIGURE 4.10 Enter the Bluetooth password

- After entering the correct password, *HC-05* gets stored in the previously connected devices as shown in Figure 4.11. For pairing the *HC-05*, Bluetooth Terminal App must be used.

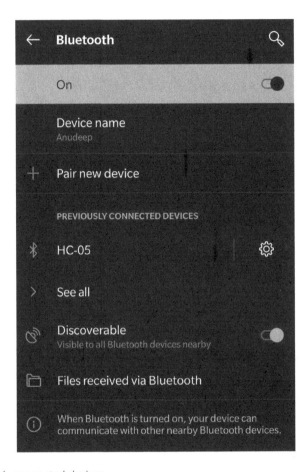

FIGURE 4.11   Previously connected devices

- Now, open the *HC-05 Bluetooth Terminal* or *BLE Terminal HM-10* App on your mobile phone. You can see all the previously paired Bluetooth devices listed in the App as shown in Figure 4.12. Among them will be HC-05 and clicking on it will pair the App with the HC-05 Bluetooth module.

FIGURE 4.12   Previously paired devices

- After the HC-05 Bluetooth module gets connected to the App, enter the password to open or close the door as shown in Figure 4.13.

FIGURE 4.13   Enter the door password

You can find all the resources of this project at this Github link (https://github.com/anudeep-20/30IoTProjects/tree/main/Project%204).

In this project, you have built a password-authenticated door which will open or close when the correct password is entered. In the process of building it, you have learnt to interface the HC-05 Bluetooth module and send data from your mobile phone to the Bluetooth module. In the next project, you'll learn to calculate or estimate the force exerted on a surface when touched or pressed.

Let's build the next one!

Calculate the Force Applied on a Surface Using FSR

5

Calculation or estimation of force is required at many places to solve important problems. This is an interesting project as it can be used as a core component in developing other applications. One of the prominent applications where impact estimation remains useful is during automobile accidents. In case of multiple accidents at a place that has limited medical resources, it would be difficult for the medical teams to decide which accident needs immediate medical attention. In case of mild impacts, the damage could be minimal and may not be requiring any medical attention or an emergency rescue team. But, if the impact is huge, the accident needs immediate medical attention. The impact estimations will also help the insurance companies in estimating the insurance claims.

In this project, we'll calculate or estimate the force applied on a surface using a resistor. Yes, it's a resistor but a force-sensitive resistor (similar to a light-dependent resistor) which is abbreviated as FSR. This project consists of three parts as detailed below:

- Visualize the estimated force applied on a surface using LED
- Estimate the force applied on a surface in steps
- Calculate the force applied on a surface

5.1 FORCE-SENSITIVE RESISTOR (FSR)

Force-sensitive resistor (FSR) is a variable resistor that changes its resistance based on the pressure applied on it. Some books or authors also refer to FSR as a *force sensor*, but in this book, we'll refer to force-sensitive resistor as FSR. These are sensors that will help in detecting the physical pressure, squeezing and weight applied to them. Figure 5.1 shows the image of a round-shaped FSR. The black region at the top of FSR is the sensitive area for physical pressure.

FSR is made of two layers separated by a spacer. As the physical pressure applied on the sensitive region increases, the number of active element dots touching the semiconductor also increases. The resistance of the FSR is inversely proportional to the number of dots touching the semiconductor. So, the resistance of FSR decreases as the number of active element dots touching the semiconductor increases and vice versa. All the layers of FSR are shown in Figure 5.2.

Similar to this round-shaped FSR, there are other types of FSRs as well used for different purposes like square-shaped FSR. You shouldn't completely depend on these sensors for weight calculation as these sensors cannot give the exact weight of the object in grams or pounds but can give an estimate. These sensors can be used for comparing the weights or force exerted by two objects or in any touch-sensitive based applications. Table 5.1 presents some basic specifications of round-shaped FSR.

DOI: 10.1201/9781003147169-5

FIGURE 5.1   Force-sensitive resistor

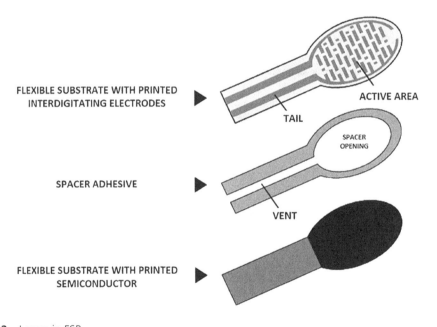

FIGURE 5.2   Layers in FSR

TABLE 5.1   Specifications of FSR

SIZE	12.7 MM (ACTIVE AREA)
Thickness	0.508 mm
Resistance Range	∞ Ω to 200 Ω (for light pressure, 100 kΩ)
Force Range	0–100 N
Current	<1 mA
Voltage	Depends on the pull-down resistor and supply voltage

5.2 MEASURING FORCE WITH FSR

We can measure the force applied on the FSR by using its property of decreasing resistance with the increase in force. When no force is applied, FSR acts as an open circuit, and when the force increases, the resistance decreases. Figure 5.3 shows the force vs resistance graph of FSR (obtained from the datasheet of FSR). Given the resistance output of FSR, we can interpret the force exerted by any object by using the graph shown in Figure 5.3. You can find complete details about FSR in its datasheet at this link (https://www.sparkfun.com/datasheets/Sensors/Pressure/fsrguide.pdf).

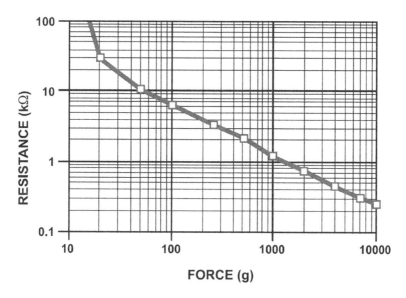

FIGURE 5.3 Force vs resistance graph of FSR

Note: This is not an exact mapping of force vs resistance of FSR but an approximate mapping of resistance at different force measurements. It is important to observe that the graph is a log/log graph (not a linear graph), and at very low force measurements, the resistance quickly goes to infinity. Force is never measured in grams or acceleration due to gravity but measured in Newtons (N). In Figure 5.3, the X-axis shows Force (g) which means the values indicated are in g-force. It is a type of force per unit mass measurement. To know more about g-force, refer to this link (https://en.wikipedia.org/wiki/G-force; last edited June 10, 2022)

5.3 VISUALIZE THE ESTIMATED FORCE APPLIED ON A SURFACE USING LED

5.3.1 Motive of the project

In this part of the project, you'll control the brightness of the LED depending on the force applied on the FSR. If maximum force is applied on the FSR, then the LED will glow with the maximum brightness and vice versa.

5.3.2 **Hardware required**

TABLE 5.2   Hardware required for estimating the force applied on a surface using LED

HARDWARE	QUANTITY
Arduino Uno	1
FSR	1
Resistor (10kΩ)	1
LED	1
Jumper wires	As required

5.3.3 **Connections**

Note: Similar to any other resistor, FSR also doesn't have polarity. So, Pin 1 and Pin 2 of FSR are identical. In the circuit shown in Figure 5.4, we haven't connected any series resistor between the LED −ve pin and GND. It is because the LED won't get damaged if it is connected between any digital pin or analog pin of Arduino Uno and GND. In case if you are not sure, it is always recommended to connect a resistor in series with the LED to prevent it from damage.

Remember: In the circuit shown in Figure 5.4, the resistor between the FSR and the GND is known as a pull-down resistor. The higher the value of pull-down resistance, the higher the accuracy of FSR values. Similarly, if a resistor is connected between a sensor and the VCC, then it is known as a pull-up resistor.

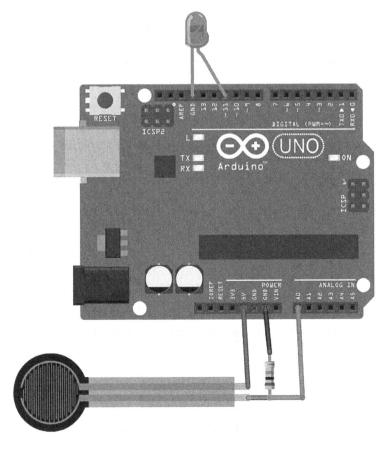

FIGURE 5.4   Connection layout for estimating the force applied on a surface using LED

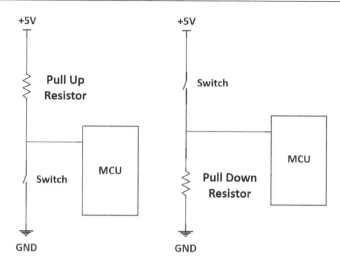

FIGURE 5.5 Pull-up and pull-down resistor

TABLE 5.3   Connection details for estimating the force applied on a surface using LED

ARDUINO UNO	FSR	RESISTOR (10 KΩ)	LED
5V	Pin 1		
A0	Pin 2	Pin 1	
GND		Pin 2	−ve pin (short pin)
11			+ve pin (long pin)

The lower the value of pull-up resistance, the higher the accuracy of the sensor values. So, a high-value resistor is used between the FSR and GND. If you have a higher value resistor than 10 kΩ, then you can use it as a pull-down resistor and this won't affect the performance. Refer to Figure 5.5 to better understand the difference between a pull-up and pull-down resistor.

You can refer to the list of components required from Table 5.2, followed by Table 5.3, which presents the pin-wise connections. Also, the connection diagram is presented in Figure 5.4, which shall help you to get a visualization of the connections.

5.3.4 Arduino IDE code

Similar to the other projects, initialize the variables for storing the Arduino Uno pin numbers to which the sensors are connected. FSRpin and LEDpin are used for storing the pin numbers to which FSR and LED are connected. FSRvalue is used to store the FSR reading, and the LEDbrightness is used to store the brightness of the LED.

```
int FSRpin = A0; // Initialize the FSR pin
int LEDpin = 11; // Initialize the LED pin

int FSRvalue; // Variable to store the FSR value
int LEDbrightness; // Variable to store the brightness of the LED
```

In void setup(), Serial.begin(9600) is used for initializing the Arduino's Serial at 9600 baud rate. As we are reading the FSR values, we'll declare the direction of FSRpin as INPUT. Depending on the FSRvalue, we'll write the brightness (PWM signals) on to the LEDpin, so the direction of LEDpin is set as OUTPUT. Initially, LEDpin is set to OFF.

```
void setup() {
  Serial.begin(9600); //Start the Serial communication
  pinMode(FSRpin,INPUT); //Set the mode of FSR as INPUT
  pinMode(LEDpin,OUTPUT); //Set the mode of LED as OUTPUT
  digitalWrite(LEDpin,LOW); //Set the LED OFF at the start
}
```

analogRead(FSRpin) reads the FSR value and will store it in the FSRvalue. The FSRvalue will be printed on the serial monitor using Serial.println(). FSRvalue ranges from 0 to 1023, but the analogWrite() values range from 0 to 255. To map the range from 0-1023 to 0-255, map() function is used. The mapped value is stored in the LEDbrightness and analogWrite() is used to write this value on the LED. The last step is similar to the controlling brightness of an LED which you have seen in two of the previous projects.

```
void loop() {
  FSRvalue = analogRead(FSRpin); //Read the FSR values
  Serial.print("FSR Value: "); //Print the FSR values
  Serial.println(FSRvalue);

  // Map the FSR values from 0-1023 to 0-255
  LEDbrightness = map(FSRvalue, 0, 1023, 0, 255);
  analogWrite(LEDpin, LEDbrightness); //Write the brightness value on LED

  delay(100);
}
```

Upload the code to Arduino Uno after selecting the appropriate board and COM port. After successfully uploading the code, open the serial monitor for seeing the FSR values. Don't forget to select the correct baud rate in the serial monitor. Now, press the FSR and then see the brightness of the LED change.

5.4 ESTIMATE THE FORCE APPLIED ON A SURFACE IN STEPS

5.4.1 Motive of the project

This is a very simple part of the project in which we'll just read the FSR values and estimate the force/pressure exerted on the FSR.

5.4.2 Hardware required

TABLE 5.4   Hardware required for estimating the force applied on a surface in steps

HARDWARE	QUANTITY
Arduino Uno	1
FSR	1
Resistor (10 kΩ)	1
Jumper wires	As required

5.4.3 Connections

You can refer to the list of components required from Table 5.4, followed by Table 5.5, which presents the pin-wise connections. Also, the connection diagram is presented in Figure 5.6, which shall help get a visualization of the connections.

TABLE 5.5 Connection details for estimating the force applied on a surface in steps

ARDUINO UNO	FSR	RESISTOR (10 KΩ)
5V	Pin 1	
A0	Pin 2	Pin 1
GND		Pin 2

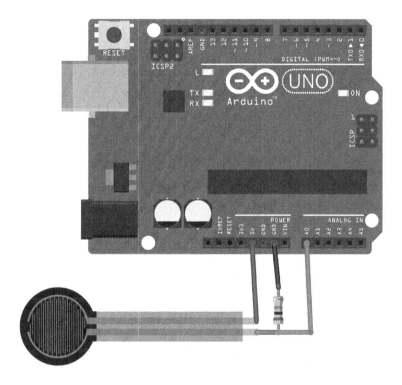

FIGURE 5.6 Connection layout for estimating the force applied on a surface in steps

5.4.4 Arduino IDE code

Initially, declare the required variables. FSRpin is used to store the analog pin number to which the FSR is connected to Arduino Uno, and the FSRvalue variable is used to store the FSR values.

```
int FSRpin = A0; //Initialize the FSR pin
int FSRvalue; //Variable to store the FSR value
```

Arduino's Serial is initialized at a baud rate of 9600 using Serial.begin(). As we are reading the values from the FSR, the direction of FSRpin is set as INPUT.

```
void setup() {
  Serial.begin(9600); //Start the Serial communication
  pinMode(FSRpin,INPUT); //Set the mode of FSR as INPUT
}
```

analogRead() is used for reading the analog values from the FSRpin. These values are printed on the serial monitor using Serial.println(). Depending on the FSRvalue, the amount of pressure exerted on the FSR is classified into five types (soft touch, hard touch, slight press, little press and hard press). Using stacked if-else, the corresponding interpretation for the FSRvalue is printed on the serial monitor.

```
void loop() {
  FSRvalue = analogRead(FSRpin); //Read the FSR values
  Serial.print("FSR Value: "); //Print the FSR values
  Serial.println(FSRvalue);

  // Interpret the FSR values
  if(FSRvalue == 0)
    Serial.println("There is no pressure. \n");
  else if(FSRvalue < 200)
    Serial.println("Very little force is exerted on FSR. It can be considered
as a soft touch. \n");
  else if(FSRvalue < 400)
    Serial.println("Little force is exerted on FSR. It can be considered as a
hard touch. \n");
  else if(FSRvalue < 600)
    Serial.println("High force is exerted on FSR. It can be considered as a
slight press. \n");
  else if(FSRvalue < 800)
    Serial.println("Very High force is exerted on FSR. It can be considered
as a little press. \n");
  else
    Serial.println("Maximum force is exerted on FSR. It can be considered as
a hard press. \n");

  delay(500);
}
```

Upload the code to Arduino Uno after selecting the appropriate board and COM port. After successfully uploading the code, open the serial monitor for seeing the FSR values and the force interpretation. Don't forget to select the correct baud rate in the serial monitor.

5.5 CALCULATE THE FORCE APPLIED ON A SURFACE

In almost all the cases while using FSR, you'll be using the previous two parts for interpreting pressure exerted on a surface.

5.5.1 Motive of the project

This part of the project is a little advanced where you'll be calculating the approximate force exerted on the FSR in Newton(s). In this part of the project, the force will be calculated using the graphs in the FSR datasheet.

5.5.2 Hardware required

TABLE 5.6   Hardware required for calculating the force applied on a surface

HARDWARE	QUANTITY
Arduino Uno	1
FSR	1
Resistor (10 kΩ)	1
Jumper wires	As required

Note: In this part of the project, use only a 10 kΩ resistor and don't use any other resistor as results may vary.

5.5.3 Connections

You can refer to the list of components required from Table 5.6, followed by Table 5.7, which presents the pin-wise connections. Also, the connection diagram is presented in Figure 5.7, which shall help get a visualization of the connections.

TABLE 5.7   Connection details for calculating the force applied on a surface

ARDUINO UNO	FSR	RESISTOR (10 KΩ)
5V	Pin 1	
A0	Pin 2	Pin 1
GND		Pin 2

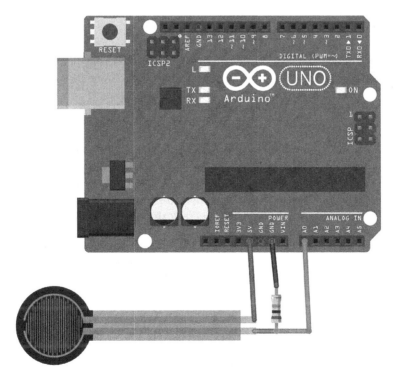

FIGURE 5.7   Connection layout for calculating the force applied on a surface

In this example, we are using a 5 V power supply (from Arduino Uno's Vcc) and a 10 kΩ pull-down resistor (between FSR and Arduino Uno). When force is applied on the FSR, its resistance decreases which means the total resistance (FSR + pull-down resistor) decreases. This causes the current flowing through both the resistors to increase which will, in turn, increase the voltage across the fixed pull-down resistor. The total resistance of FSR and the pull-down resistor will vary from ∞ Ω (open circuit) to 10 kΩ.

TABLE 5.8   Voltage based on force applied on FSR

FORCE (LB)	FORCE (N)	FSR RESISTANCE	TOTAL RESISTANCE (FSR + PULL-DOWN)	CURRENT THROUGH (FSR + PULL-DOWN) (MA)	VOLTAGE ACROSS PULL-DOWN RESISTOR (V)
No Force	No Force	∞ Ω (open circuit)	∞ Ω (open circuit)	0	0
0.04	0.2	30 kΩ	40 kΩ	0.13	1.3
0.22	1	6 kΩ	16 kΩ	0.31	3.1
2.2	10	1 kΩ	11 kΩ	0.45	4.5
22	100	250 Ω	10.25 kΩ	0.49	4.9

Table 5.8 shows the change in current and voltage across the pull-down resistor when variable force is applied on the FSR. One thing to keep in mind is that these values are not perfectly accurate and are approximated. So, by using this data, we can approximately calculate the force applied on the FSR in Newtons (N).

5.5.4 Arduino IDE code

Initially, declare the required variables. `FSRpin` is used to store the analog pin number to which the FSR is connected to Arduino Uno and the `FSRvalue` variable is used to store the FSR values.

```
int FSRpin = A0;   //Initialize the FSR pin
int FSRvalue;      //Variable to store FSR analog values
```

Next, declare the required variables for calculating the force applied on the FSR. `voltage` is used to store the voltage values converted from `FSRvalue`. From the `voltage` value, we need to calculate the resistance and conductance across the FSR. These values will be very large and a normal `int` may not be able to store it and may cause integer overflow. To know more about Integer overflow, visit this link (http://www.cplusplus.com/articles/DE18T05o/#:~:text=Overflow%20is%20a%20phenomenon%20 where,be%20larger%20than%20the%20range).

When we declare `int var;` then the variable `var` can store both positive and negative values. For recognizing a number stored inside the variable as positive or negative, it uses an extra bit at the start which decreases the overall positive numbers stored. If you know that the value you are storing in the variable `var` is always positive, then you can use `unsigned int var;` which will increase the number of positive integers stored in it.

Note: The minimum value of a signed integer is −2147483648 (-2^{31}), and the maximum value is 2147483647 ($2^{31}-1$). Any value stored above or below these limits in `int` will cause integer overflow.

Similarly, the minimum value of an unsigned integer is 0, and the maximum value is 4294967295 ($2^{32} - 1$). Even for unsigned int, if any value is stored above or below these limits will cause integer overflow.

Any value (positive or negative) exceeds the limits of int, then we use long or long long for the declaration of a variable. Similar to unsigned int, unsigned long also doesn't store negative values. As Arduino language is similar to C++ (in terms of syntax), the space occupied by data types in Arduino will be the same as C++. If you want to know the maximum and minimum values of different data types in C++, then go to this link (https://docs.microsoft.com/en-us/cpp/cpp/integer-limits?view=msvc-160).

int is used for voltage as it stores small values. unsigned long is used for storing resistance, conductance and force values.

```
int voltage; //Variable to store FSR values to Voltage
//Variable to store the resistance calculated from the Voltage
unsigned long resistance;
//Variable to store the conductance calculated from the Resistance
unsigned long conductance;
//Variable to store the force on FSR calculated from the Conductance
unsigned long force;
```

Arduino's Serial is initialized at 9600 baud rate using Serial.begin(). As we are reading the values from the FSR, the direction of FSRpin is set as INPUT.

```
void setup() {
  Serial.begin(9600);      //Initialize the Arduino's serial
  pinMode(FSRpin, INPUT); //Set the FSR mode as INPUT
}
```

FSRvalue stores the FSR values by reading from the FSRpin using analogRead(). These values are printed on the serial monitor using Serial.println(). The FSRvalue which will be in the range of 0–1023 is converted to voltage (in millivolts) by using the map() function. As the FSR is connected to 5 V (5,000 mV), the FSRvalue will be mapped from 0–1023 to 0–5000 and will be stored in a voltage variable.

```
void loop() {
  FSRvalue = analogRead(FSRpin); //Read the FSR values
  Serial.print("FSR Value: "); //Print the FSR values
  Serial.println(FSRvalue);

  //Analog FSR value is converted to Voltage
  voltage = map(FSRvalue, 0, 1023, 0, 5000);
  Serial.print("FSR Voltage (in mV): "); //Print voltage
  Serial.println(voltage);
```

When there is no voltage across the FSR, then no pressure/force is exerted on it. If there is a non zero voltage across the FSR, then there is a non-zero force being exerted on it.

Now, we need to calculate the resistance across FSR for calculating the force exerted on it. From Figure 5.8, we can obtain the resistance across FSR in known terms,

$$V_{FSR} = Vcc * \frac{R_{pull-down}}{R_{pull-down} + R_{FSR}} \tag{5.1}$$

$$R_{FSR} = \frac{(Vcc - V_{FSR}) * R_{pull-down}}{V_{FSR}} \tag{5.2}$$

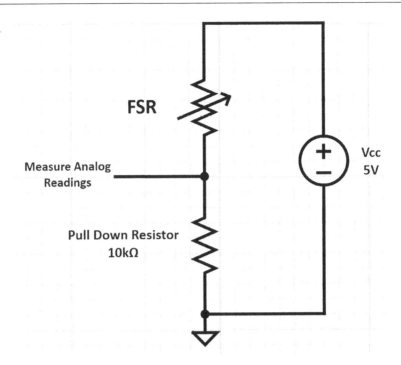

FIGURE 5.8 Circuit diagram of FSR and pull-down resistor

Here, Vcc is 5 V, $R_{pull\text{-}down}$ is 10,000 Ω, and V_{FSR} is the voltage which is calculated in the previous step.

    Note: You need to observe in the code that the resistance across FSR is calculated in steps i.e., first $Vcc - V_{FSR}$ is calculated then it is multiplied by the $R_{pull\text{-}down}$ and at last divided by V_{FSR}. If all the three operations are done in a single step, during the calculation, there is a possibility of overflow, to avoid it we calculate step by step.

```
if (voltage == 0) { // If voltage is 0 then No pressure on FSR
Serial.println("No Pressure (ON)");
} else {
//Calculate the resistance from the non zero voltage
resistance = (5000 - voltage);
resistance *= 10000;
resistance /= voltage;
Serial.print("FSR Resistance (in ohms): "); //Print resistance in Ohms
Serial.println(resistance);
```

After obtaining the resistance value, we need to calculate the conductance. Conductance is nothing but the inverse of resistance. The unit of conductance is Mho (reverse of Ohm). It is denoted by various symbols like S, Ω^{-1} or $\mho$.

$$R_{FSR} = \frac{1}{C_{FSR}} \tag{5.3}$$

$$R_{FSR} = \frac{10^6}{C_{FSR}}\mu \tag{5.4}$$

Figures 5.9 and 5.10 show the graphs of conductance vs force on FSR obtained from the datasheet of FSR. By using these graphs and conductance values, we can calculate the force exerted on FSR.

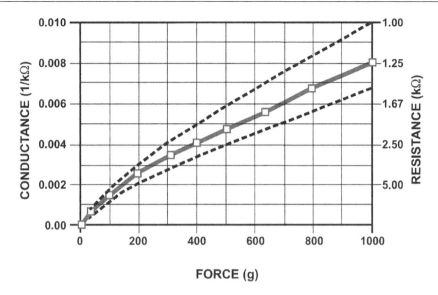

FIGURE 5.9 Conductance vs force (0–1 kg) low force range

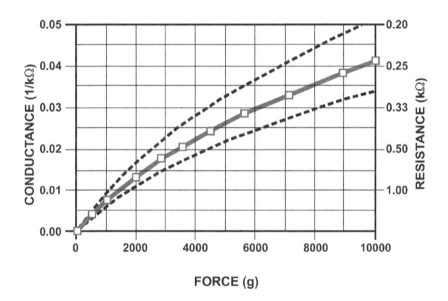

FIGURE 5.10 Conductance vs force (0–10 kg)

If the conductance of FSR is less than 1,000 µ, then

$$F_{FSR}\left(\text{in } N\right) = \frac{C_{FSR}\ \left(\text{in } \mu\mho\right)}{80}$$

else

$$F_{FSR}\left(\text{in } N\right) = \frac{C_{FSR}\ \left(\text{in } \mu\mho\right) - 1000}{30}$$

After calculation of force in Newton(s) using the Equations 5.5 and 5.6, it is printed on the serial monitor using `Serial.println()`.

```
        //Calculate conductance in microMho
  conductance = 1000000/resistance;

  if (conductance <= 1000) { //If conductance is less than 1000 (Low force
Range)
      force = conductance / 80;
      Serial.print("Force (in Newtons): "); //Print force in Newton
      Serial.println(force);
  } else { //If conductance is greater than 1000 (High force Range)
      force = (conductance - 1000)/30;
      Serial.print("Force (in Newtons): "); //Print force in Newton
      Serial.println(force);
  }
}
Serial.println("");
delay(1000);
}
```

Finally, upload the code to Arduino Uno after selecting the appropriate board and COM port. After successfully uploading the code, open the serial monitor to see the force applied on the FSR in Newton(s). Don't forget to select the correct baud rate in the serial monitor.

Now, you can start touching or pressing the FSR and observe the change in force exerted on it. As you increase the pressure on the FSR, the force displayed on the serial monitor also decreases and vice versa. One point to remember is that these values are not accurate and cannot be used for accurate calculation but can be used for an estimation of actual values.

You can find all the resources of this project at this Github link (https://github.com/ anudeep-20/30IoTProjects/tree/main/Project%205).

In this project, you have used FSR for calculation and estimation of force exerted on it. You have also learnt most of the features of FSR and ways of using it in IoT. In the next project, you'll build an automatic sanitizer or soap dispenser without the need to touch it using IoT.

Let's go to the next one!

Automatic Sanitizer or Touch-Free Soap Dispenser

6

The COVID-19 pandemic has shown the importance of cleanliness and personal hygiene to the entire world. Cleanliness and hygiene are something that could save anyone from any dangerous contagious disease. During a pandemic or an epidemic, all the government organizations and the World Health Organization (WHO) request people to be clean and hygienic to reduce disease transmission. For maintaining cleanliness and hygiene, people are requested to wash their hands frequently. If people are at home or offices, they use soap for cleaning their hands, and at public places, they'll be using sanitizers as carrying soap and water to all the places is not possible. But, if everyone has to touch the sanitizer or soap dispenser manually to get their hands sanitized, it could prove dangerous for others as well, as even infected persons use the same sanitizer or soap dispensers. How do to handle the situation? Automate it. That's what we have done and this project shall give a clear idea to the readers about building an automated sanitizer or soap dispenser.

6.1 MOTIVE OF THE PROJECT

In this project, you'll use sensors for automatically dispensing sanitizer or soap from a bottle without touching it. This is similar to the automatic dispensing of water from sensor-based faucets (used in most airports or shopping malls). These faucets don't have any handle for opening or closing the water outlet instead they have sensors which will recognize whether any hand is placed near them or not. Depending on the presence of hand, these faucets automatically open or close the water outlet. Sensor-based faucets are useful in reducing water wastage and the number of touch contacts in the public washrooms which will reduce disease transmission.

Generally, sensor-based faucets as shown in Figure 6.1 use IR sensors for recognizing the presence of hands. Similarly, in this project, we'll use an IR sensor to determine whether a hand is placed in front of the sanitizer or soap bottle for dispensing it.

DOI: 10.1201/9781003147169-6

FIGURE 6.1   Sensor-based faucet

6.2 INFRARED (IR)

Infrared is the acronym for IR. Infrared rays are also a form of electromagnetic radiation but cannot be seen using human eyes because the wavelength of infrared rays is longer than those of visible light. IR radiation is used in the field of industrial, scientific, military and medicine for developing many applications. Figure 6.2 shows the energy, frequency and wavelength relation of all the rays in the electromagnetic spectrum.

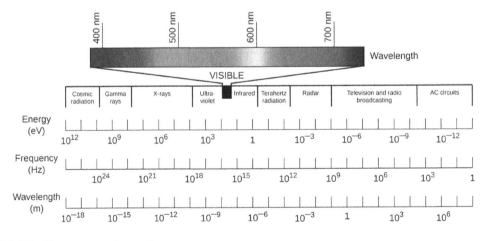

FIGURE 6.2   Electromagnetic spectrum

6.3 IR SENSOR

There are different types of IR sensors available in the market for diverse applications and for varied accuracy requirements. In this project, we'll use an IR sensor similar to the one shown in Figure 6.3. This IR sensor has two LEDs which are IR emitter (transparent LED) and IR receiver (black LED). It also has

FIGURE 6.3 IR sensor

an onboard potentiometer which will be used for adjusting its sensitivity. An IR emitter LED is similar to a normal RGB LED but the difference is that the IR emitter LED emits IR rays whereas the normal RGB LED emits rays in the visible spectrum.

Not only in IR sensors, but we also use the combination of IR emitter and IR receiver in our day-to-day life as well. The remote which we use for operating the television uses this combination. The television remote has an IR emitter, and the television has an IR receiver. For each button on the television remote, the IR emitter emits IR rays in a unique binary format which will be received by the IR receiver. This information is decoded, and the corresponding action is performed by the television.

Fun Fact: Even though you cannot see the IR rays with your naked eye, cameras can see it. Don't believe it? Then, power your IR sensor and notice its LEDs through any of your cameras. You'll be able to see a violet light (IR rays) emitting from the IR emitter as shown in Figure 6.4.

FIGURE 6.4   IR rays emitting from IR emitter as seen from a camera

6.4 WORKING OF IR SENSOR

IR sensor is a digital sensor. It outputs only a digital value (HIGH or LOW) depending on whether an object is present in front of it or not. After the IR sensor is powered ON, the IR emitter LED continuously emits IR rays. If an object is placed in front of the IR sensor as shown in Figure 6.5, then these IR rays are reflected and are received by the IR receiver LED. If the IR receiver LED on the IR sensor receives the IR rays, then the IR sensor detects that an object is present in front of it and outputs a digital signal.

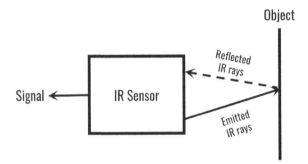

FIGURE 6.5   Working of IR sensor

6.5 HARDWARE REQUIRED

TABLE 6.1   Hardware required for the project

HARDWARE	QUANTITY
Arduino Uno	1
Servo Motor	1
IR Sensor	1
LED	1
Power Source	1
Thread	As required
Double Tape	As required
Jumper wires	As required

6.6 CONNECTIONS

You can refer to the list of components required from Table 6.1, followed by Table 6.2, which presents the pin-wise connections. Also, the connection diagram is presented in Figure 6.6, which shall help you to get a visualization of the connections.

TABLE 6.2 Connection details

ARDUINO UNO	SERVO MOTOR	IR SENSOR	GREEN LED
5V		VCC	
3.3V	VCC (red)		
GND	GND (black)	GND	−ve pin (short pin)
5		OUT (yellow)	
2		OUT	
11			+ve pin (long pin)

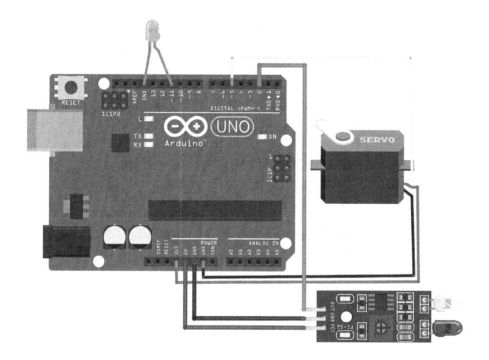

FIGURE 6.6 Connection layout

6.7 ARRANGEMENT OF THE HARDWARE

In all the previous projects, the arrangement of hardware doesn't matter. But, in this project for the automatic sanitizer or soap dispenser to work, the hardware needs to be arranged in a particular format. The arrangement of the hardware is explained in the below steps and can also be visualized in Figure 6.7.

- Tie one end of the thread to the servo motor's rotating leg and stick the other end on a plain surface by passing it over the sanitizer or soap bottle.
- Next, make all the hardware connections as shown in Figure 6.6 and connect the Arduino to a power source.
- Place the IR sensor on the sanitizer or soap bottle such that when a hand is placed in front of the bottle then the IR rays are interrupted.
- When the IR rays get interrupted by a hand, the IR sensor sends a signal to the Arduino that a hand is detected. Now, the Arduino rotates the servo motor causing a pressing effect on top of the sanitizer or soap bottle which leads to the dispensing of soap or sanitizer.

FIGURE 6.7   Arrangement of hardware

6.8 ARDUINO IDE CODE

Initially, import the Servo.h library for easy control of the servo motor. Similar to the previous projects, declare myServo variable of Servo class for easy access of Servo.h library functions. servoPin variable is used for storing the digital pin number to which the servo motor is connected.

```
#include <Servo.h> //Import the servo library

Servo myServo;    //Servo class object
int servoPin = 5; //Initialize servo motor pin
```

The OUT pin of the IR sensor is connected to Digital Pin 2 of Arduino Uno which is stored in the IRpin variable. The IRval variable stores the digital output value from the IR sensor.

An LED is also used for indicating whether the sanitizer or soap is being dispensed from the bottle. The +ve pin (longer pin) of the LED is connected to the digital pin 11 of Arduino Uno which is stored in the LEDpin variable.

```
int IRpin = 2;   //Initialize IR sensor pin
int IRval;       //Used for storing the IR sensor value

int LEDpin = 11; //Initialize LED pin
```

Arduino's serial is initialized at 9600 baud rate using Serial.begin(9600). To use the functions in the Servo.h library, we need to link the myServo variable with the servoPin. Initially, the Servo is set at 0° angle.

As we are taking input from the IR sensor, the direction of the IRpin is set to INPUT. Similarly, as we are writing voltage on the LEDpin, its direction is set as OUTPUT.

```
void setup() {
  Serial.begin(9600); //Begin the Serial communication
  myServo.attach(servoPin); //Linking to the Arduino Uno's PWM pin
  myServo.write(0); //Initially Servo will be at 0°
  pinMode(IRpin, INPUT); //Define the direction of IR sensor
  pinMode(LEDpin, OUTPUT); //Define the direction of LED
}
```

digitalRead() is used for reading from a digital pin of Arduino Uno. IRval is used for storing the digital output read from the IRpin and is printed on the serial monitor.

There are two types of IR sensors depending on their digital output:

- Output 1 (HIGH) when an object is present in front of it and Output 0 (LOW) when there is no object.
- Output 0 (LOW) when an object is present in front of it and Output 1 (HIGH) when there is no object.

In this project, we have used an IR sensor of the second type which will output 0 (LOW) when an object is present in front of it. Before setting up the hardware and uploading the entire code, check which type of IR sensor you are using and then change the code accordingly.

When the IR sensor detects an object (in this case *hand*), the LED is switched ON and the servo motor is rotated to 150° to dispense sanitizer or soap from the bottle. It is not a thumb rule that rotating the servo motor for 150° will generate a pressing effect on the sanitizer or soap bottle. We have tested for different angles and 150° worked for us perfectly in generating a pressing effect and dispensing sanitizer or soap from the bottle. You can also try with different angles and arrive at an angle at which the sanitizer or soap is getting dispensed perfectly. This angle can be less than or greater than 150° and also depends upon the orientation in which the servo motor is placed. Before uploading, change this angle accordingly in the code.

After the servo motor rotates to 150° (or the angle which you determined), keep it at that angle for at least 5 seconds so that the sanitizer or soap gets dispensed properly and then rotate it back to 0° and switch OFF the LED.

```
void loop() {
  IRval = digitalRead(IRpin); //Store the IR sensor output
  Serial.print("IR sensor: ");
  Serial.println(IRval); //Print the IR sensor output

  if(IRval == 0) { //If anything is infront of IR sensor
    Serial.println("Hand detected. Releasing Soap / Sanitizer.");
    digitalWrite(LEDpin, HIGH); //Turn ON LED
    myServo.write(150); //Rotate the servo motor to 150°
    delay(5000); //Keep the servo motor at 150° for 5 seconds
    digitalWrite(LEDpin, LOW); //Turn OFF LED
    myServo.write(0); //Again bring back the servo motor to 0°
  }

  delay(2000); //Check the IR sensor output for every 2 seconds
}
```

Note: In this project, we have used a low torque servo motor (SG90) as shown in Figure 6.8 and a normal thread for generating a pressing effect on the soap bottle which we have used. If the servo motor or the thread is not able to generate a pressing effect on the sanitizer or soap bottle which you are using, then use a high torque servo motor (MG995) as shown in Figure 6.9 or high tensile wire for generating a pressing effect.

FIGURE 6.8   SG90 servo motor

FIGURE 6.9   MG995 servo motor

At 4.8 V, the SG90 servo motor generates a torque of 1.2 kg-cm, whereas the MG995 servo motor generates a torque of 13 kg-cm. So, if the SG90 servo motor used in your project is not able to generate the required torque, then the MG995 servo motor will provide you with enough torque.

Finally, upload the code to Arduino Uno after selecting the appropriate board and COM port. After successfully uploading the code, connect the Arduino Uno to a power source. Now, you can place your hand in front of the soap or sanitizer bottle for automatic dispensing of soap or sanitizer without touching the bottle.

You can find all the resources of this project at this Github link (https://github.com/ anudeep-20/30IoTProjects/tree/main/Project%206).

In this project, you have used an IR sensor and servo motor for automatically dispensing the soap or sanitizer from a bottle without touching it. You have also learnt about infrared rays, IR sensors and the usage of IR sensor in IoT. In the next project, you'll remotely control an LED using NodeMCU and Internet. In the process of building this project, you'll also learn about building basic web pages using HTML.

Let's build the next one!

Controlling Devices over the Internet Using a Webpage

7

The way the Internet has revolutionized the modern world is appreciable, and the growth is unimaginable. The usage of the Internet in the coming years is going to multiply in many folds due to the increase in the usage of smart things and IoT products.

Instead of it being Internet of Things, it has become Internet of Everything. Yes, everything is going to be connected to the Internet and will be controlled also through the Internet in the near future. Having conveyed this, controlling the devices and equipment over the Internet is possible, and plenty of innovations have been happening in this sector. To let the readers understand how to control devices over the Internet using a webpage, this project has been built.

7.1 MOTIVE OF THE PROJECT

In this project, you'll control an LED over the Internet using a webpage. You can control any number of devices or sensors using the same procedure shown in this project. Until the previous project, we haven't used Internet in building the projects. But from this project, we'll be using the Internet and cloud as part of the development. If you need Internet in your project, then you cannot use Arduino Uno and have to use an ESP-based evaluation board. One of the most accurate, convenient and cheap ESP-based evaluation boards available in the market is NodeMCU. A detailed introduction of NodeMCU is explained in Section 1.10 of the 1st project. If you haven't configured your NodeMCU yet, refer to Section 1.11.

7.2 BASICS OF HTML

In this project, you will build a small webpage for which you need to know the basics of HTML. Don't worry, this won't be a computer science chapter, and we'll keep it short and sweet.

In any HTML page, there will be a <head> tag and a <body> tag. <head> tag is useful in the initial setup, and <body> tag contains all the main contents displayed on a webpage. HTML pages might also contain CSS (used for styling) and JavaScript (used for controlling elements through coding).

Every tag in HTML has an opening tag and a closing tag. If the opening tag is <X>, then its closing tag will be </X>. All the functionality-related information must be written in between these two tags. But there are some self-closing tags as well like <br/>, <meta/>, etc. You can write these self-closing tags as either <Y> or as <Y/>. All the tags used in this project are tabulated in Table 7.1.

Note: HTML tags are not case sensitive, but it is always preferred to write the tags in a single case (lower or upper). We generally prefer to write in lowercase.

DOI: 10.1201/9781003147169-7

TABLE 7.1   HTML tags and their usage

TAG	USAGE
<html>	All the HTML code must be written in this tag
<head>	Used for initial setup
<body>	Contains all the information displayed on a webpage
<meta>	Used for defining charset
<title>	Used for naming the title of the webpage tab
<h1>	Used for writing a heading of level 1
<a>	Used for inserting a link that redirects to another webpage or a section in the same webpage
<div>	Used as a container for a section of webpage
 	Line break in HTML (similar to \n in C)

7.3 HARDWARE REQUIRED

TABLE 7.2   Hardware required for the project

HARDWARE	QUANTITY
NodeMCU	1
Resistor (220 Ω)	1
LED	1
Jumper wires	As required

7.4 CONNECTIONS

You can refer to the list of components required from Table 7.2, followed by Table 7.3, which presents the pin-wise connections. Also, the connection diagram is presented in Figure 7.1, which shall help get a visualization of the connections.

TABLE 7.3   Connection details

NODEMCU	RESISTOR (220 Ω)	LED
GND	Pin 1	
	Pin 2	−ve pin (short pin)
D1		+ve pin (long pin)

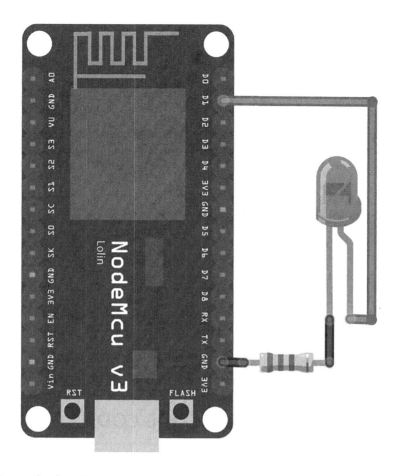

FIGURE 7.1   Connection layout

7.5 HTML CODE

Before developing any web-based project in IoT, write the HTML code in an editor and confirm that you are getting your desired HTML page. For writing HTML code, you need not download any separate editor from the Internet. You can use your default text editor.

```
<html>

<head>
    <meta charset="UTF-8" />
    <title> LED Control </title>
</head>

<body>
    <h2> Controlling an LED using a Webpage </h2>

    <div>
        <a href="/ledon"> <button> Switch ON LED </button> </a>
    </div>
```

```
    <br />
    <div>
        <a href="/ledoff"> <button> Switch OFF LED </button> </a>
    </div>
</body>

</html>
```

After writing the HTML code in a text editor, save it as a .html file (not .txt file). You can run this HTML file using any one of the available browsers on your PC like Microsoft Edge, Google Chrome, Mozilla Firefox or any other browser. After running the above HTML code using a Google Chrome browser, you'll see a webpage as shown in Figure 7.2.

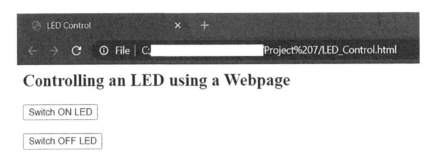

FIGURE 7.2   Webpage

7.6 ARDUINO IDE CODE

For connecting NodeMCU (ESP8266) to the WiFi, we need to use `ESP8266WiFi.h` library. You can download this library at this link (https://github.com/ekstrand/ESP8266wifi). After downloading the .zip file of the library, extract it and paste it into the libraries folder.

In C or C++, `#define` is a preprocessor directive and can be used for declaring a constant variable. The data stored in these variables is fixed and cannot be altered throughout the execution. `const` can also be used to declare constant variables. `const int var = 0;` is same as `#define var 0`.

```
#include <ESP8266WiFi.h>

#define LED D1 //Initialize LED pin
```

`Ssid` and `password` are of `String` type variables used for storing the name and password of your WiFi. These are useful for connecting the NodeMCU to your WiFi network. Write your WiFi name and password in the place of YOUR WIFI NAME and YOUR WIFI PASSWORD, respectively.

NodeMCU is configured as a server at port 80 using `WiFiServer` class which listens for any incoming client requests. Visit this link (https://www.arduino.cc/en/Reference/WiFiServer) for knowing more about `WiFiServer` class.

```
String ssid = "YOUR WIFI NAME"; //Write your WiFi name
String password = "YOUR WIFI PASSWORD"; //Write your WiFi Password

WiFiServer server(80); //Initialize a server at port 80
```

`void setup()` contains all the code required for WiFi connection and initial setup of NodeMCU. NodeMCU's serial is initialized at 115200 baud rate using `Serial.begin()`. Generally, many people in

IoT prefer to have 9600 baud rate for Arduino-based boards and 115200 baud rate for ESP-based boards. But any valid baud rate can be chosen for the serial communication. As we are writing the voltage on to the LED, we'll declare the direction of LED as OUTPUT. Initially, the LED is set to OFF state.

```
void setup() {
  Serial.begin(115200); //Initialization of serial communication
  pinMode(LED, OUTPUT); //Define the direction of LED pin
  digitalWrite(LED,LOW); //Initially LED is set to OFF state
```

WiFi.begin(SSID, PASSWORD) is used for connecting the NodeMCU to the WiFi network whose WiFi name is SSID and password is PASSWORD. Until the NodeMCU gets connected to a WiFi network, the WiFi.status() command returns WL_IDLE_STATUS, and as soon as the NodeMCU gets connected to a WiFi network, it returns WL_CONNECTED. So, until the NodeMCU gets connected to a WiFi network, we'll print dots(.) on the serial monitor indicating NodeMCU is in the process of connecting to a WiFi network. As soon as the NodeMCU gets connected to a WiFi network, WiFi connected message is printed on the serial monitor.

```
  //Connecting the NodeMCU to WiFi
  Serial.println("");
  Serial.println("");
  Serial.print("Connecting to ");
  Serial.println(ssid);
  WiFi.begin(ssid, password);
  while (WiFi.status() != WL_CONNECTED) {
    delay(500);
    Serial.print(".");
  }
  Serial.println("");
  Serial.println("WiFi connected");
```

After the NodeMCU gets connected to a WiFi network, a server is started using the server.begin() command. Any device connected to a WiFi network will be assigned with an IP (Internet Protocol) address by the router. The IP address of the NodeMCU is required for accessing the webpage hosted on its server. WiFi.localIP() command is used for knowing the local IP address of NodeMCU. Visit this link (*https://en.wikipedia.org/wiki/IP_address*; last edited June 8, 2022) for knowing more about IP address, types of IP address and their usage in the Internet world.

```
//After the WiFi gets connected, start the server
  server.begin();
  Serial.println("Server started");
  //Print the local IP address of NodeMCU
  Serial.println(WiFi.localIP());
}
```

server.available() is used for getting the client information connected to the server. It returns a client object when a client is available or else will return false. WiFiClient class is used for creating a client object which can store the returned client object from the server.available().

If the client (variable created by WiFiClient) contains a client object, then new client is printed on the serial monitor or else the execution doesn't go forward. client.available() is used for getting any information (request) the new client sends to the server.

```
void loop() {
  //Check whether a new client is available
  WiFiClient client = server.available();
```

```
if (!client) {
  return;
}
Serial.println("new client");
while(!client.available()) {
  delay(1);
}
```

If a new request is available from the client, the request is read as a String until tab (\r) using the .readStringUntil('\r'). The read request is stored in the req variable and also printed on the serial monitor. client.flush() is used for waiting until all the outgoing characters in the buffer are sent so that no information from the client is missed.

```
//If new request is available, read the request from the client
  String req = client.readStringUntil('\r');
  Serial.println(req); //Print the request
  //Waits until all outgoing characters in the buffer are sent
  client.flush();
```

STR.indexOf("SUBSTR") is used for determining the index of the substring SUBSTR in the string STR. If the substring SUBSTR is not present in the string STR, then it returns −1 or else will return a positive number (index of the substring).

If the request from the client (req) contains /ledoff, then the LED is set to LOW state. Similarly, if the request from the client (req) contains /ledon, then LED is set to HIGH state. Any change in the state of LED is printed on the serial monitor.

```
if (req.indexOf("/ledoff") != -1) { //If the request is ledoff
  digitalWrite(LED,LOW); //Then switch off the LED
  Serial.println("LED OFF\n");
} else if(req.indexOf("/ledon") != -1) { //If the request is ledon
  digitalWrite(LED,HIGH); //Then switch on the LED
  Serial.println("LED ON\n");
}
```

For the client to visualize the webpage, the entire HTML code of the webpage must be printed on the client side by the server. The entire HTML code must be printed at once in a single String on the client side.

web is a variable of String type which will be used to store the entire HTML code. If we write the entire HTML code in a single line, it will look clumsy and cannot be understood. So, the entire HTML code of the webpage is concatenated line by line to the web String. Finally, client.print() is used for printing the entire HTML code on the client.

```
// Attach the HTML code to a String
String web;
web += "HTTP/1.1 200 OK\r\nContent-Type: text/html\r\n\r\n";
web += "<html>\r\n";

web += "<head>\r\n";
web += "<meta charset=\"UTF-8\"/>\r\n";
web += "<title> LED Control </title>\r\n";
web += "</head>\r\n";

web += "<body>\r\n";
web += "<h1>Controlling LED using Webpage</h1>\r\n";

web += "<div>\r\n";
web += "<a href=\"/ledon\"><button> Switch ON LED </button></a>\r\n";
```

```
web += "</div>\r\n";
web += "<br/>\r\n";

web += "<div>\r\n";
web += "<a href=\"/ledoff\"><button> Switch OFF LED </button></a>\r\n";
web += "</div>\r\n";

web += "</body>\r\n";
web += "</html>\r\n";

client.print(web); //Show the webpage to the client
}
```

Now, upload the code to NodeMCU after selecting the appropriate board and COM port. Immediately after uploading the code, open the serial monitor and change the baud rate to 115200 to see the serial data.

Until the NodeMCU gets connected to the WiFi, it prints dots on the serial monitor. As soon as it gets connected to the WiFi, its local IP address will be printed on the serial monitor as shown in Figure 7.3. Note down this IP address (very important).

Note: If no information as shown in Figure 7.3 is printed on your serial monitor, click the *RST* button on the NodeMCU by keeping the serial monitor open. Now, the entire process starts again, and the information as shown in Figure 7.3 will be printed on the serial monitor.

If the local IP address of NodeMCU is not printed even after waiting for 5 minutes and only the dots are being printed on the serial monitor, then check your WiFi network or the WiFi credentials given in the code.

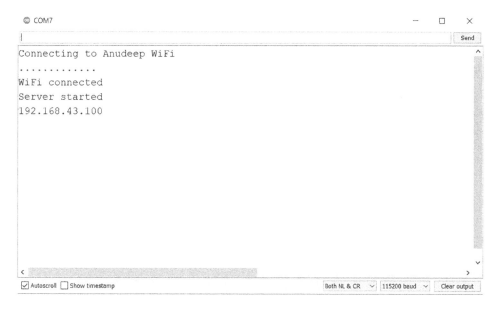

FIGURE 7.3   Local IP address of NodeMCU

After obtaining the local IP address of NodeMCU, open any one of your favourite browsers and enter the IP address in the search bar. It has to be remembered that the device (PC or Mobile) in which you are opening the webpage must be connected to the same WiFi network as the NodeMCU.

This IP address will host the webpage to control the LED. It will be the same website as shown in Figure 7.2. Figure 7.4 shows the hosted webpage in chrome browser after entering the NodeMCU's IP address in the search bar.

Controlling LED using Webpage

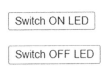

FIGURE 7.4   Webpage for controlling the LED

After opening the webpage in the browser, go to the serial monitor. You can see the data printed on the serial monitor telling that a new client is available as shown in Figure 7.5.

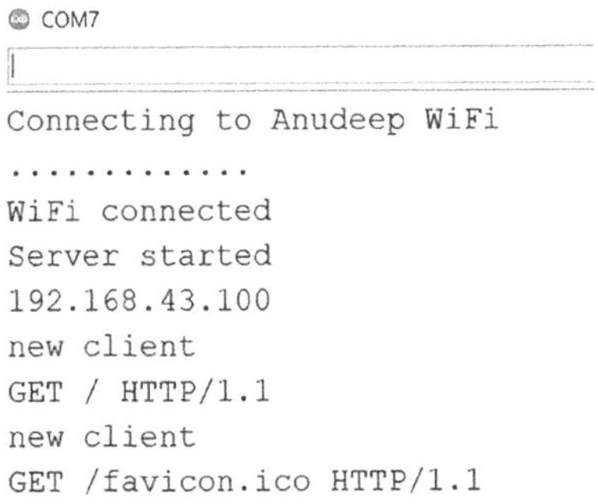

```
COM7

Connecting to Anudeep WiFi
.............
WiFi connected
Server started
192.168.43.100
new client
GET / HTTP/1.1
new client
GET /favicon.ico HTTP/1.1
```

FIGURE 7.5   Printed when a new client is available

Now, click on the *Switch ON LED* button, you'll see the LED switch ON similarly if you click *Switch OFF LED* button, the LED will be switched OFF.

After Switching ON and OFF LED once, go to the serial monitor. You can see the requests sent by the client to the NodeMCU server printed on the serial monitor as shown in Figure 7.6.

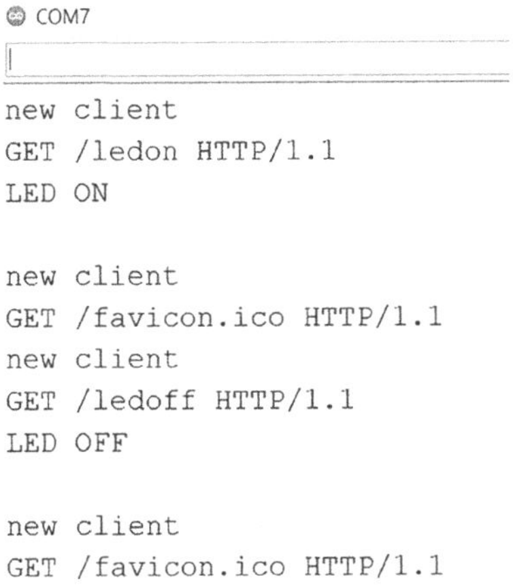

```
COM7

new client
GET /ledon HTTP/1.1
LED ON

new client
GET /favicon.ico HTTP/1.1
new client
GET /ledoff HTTP/1.1
LED OFF

new client
GET /favicon.ico HTTP/1.1
```

FIGURE 7.6   Request data from the client

Now, you can use the webpage to remotely Switch ON and Switch OFF the LED. You can also control the LED from your mobile phone by opening the same webpage on a browser on your mobile phone.

7.7 TRY IT

This is a small exercise that is an extension of the main project. In this exercise, you need to build a webpage that can control the LED using a single button instead of two buttons as shown in Figure 7.7. This is a simple exercise to test your skills. Try to do it on your own. You might fail many times but try to complete it. As an addition, you can also add text at the bottom which tells the state of the LED whether it is presently ON or OFF.

Note: For adding normal text in HTML, you can use <p> tag. It can be used as <p> TEXT TO BE WRITTEN </p>.

If you are unable to solve it, don't worry!! The below code is the solution for it. Explanation of the code is not required as it is almost similar to the previous code.

7.7.1 Arduino IDE code

```
#include <ESP8266WiFi.h>

#define LED D1

String ssid = "YOUR WIFI NAME";
String password = "YOUR WIFI PASSWORD";
String LEDstate; // State variable for LED to track its state

WiFiServer server(80);

void setup() {
  Serial.begin(115200);
  pinMode(LED, OUTPUT);
  digitalWrite(LED,LOW);
  LEDstate = "OFF";

  Serial.println("");
  Serial.println("");
  Serial.print("Connecting to ");
  Serial.println(ssid);
  WiFi.begin(ssid, password);
  while (WiFi.status() != WL_CONNECTED) {
    delay(500);
    Serial.print(".");
  }
  Serial.println("");
  Serial.println("WiFi connected");
  server.begin();
  Serial.println("Server started");
  Serial.println(WiFi.localIP());
}
void loop() {
  WiFiClient client = server.available();
```

```
  if (!client) {
    return;
  }

  Serial.println("new client");
  while(!client.available()) {
    delay(1);
  }
  String req = client.readStringUntil('\r');
  Serial.println(req);
  client.flush();

  // Depending on the present state of LED, toggle its state
  if (req.indexOf("/ledtoggle") != -1) {
    if(LEDstate == "ON") {
      digitalWrite(LED,LOW);
      LEDstate = "OFF";
      Serial.println("LED OFF\n");
    } else {
      digitalWrite(LED,HIGH);
      LEDstate = "ON";
      Serial.println("LED ON\n");
    }
  }

  String web;
  web += "HTTP/1.1 200 OK\r\nContent-Type: text/html\r\n\r\n";
  web += "<html>\r\n";

  web += "<head>\r\n";
  web += "<meta charset=\"UTF-8\"/>\r\n";
  web += "<title> LED Control </title>\r\n";
  web += "</head>\r\n";

  web += "<body>\r\n";
  web += "<h1>Controlling LED using Webpage & Single Button</h1>\r\n";

  web += "<div>\r\n";
  // Use a single variable i.e, /ledtoggle instead of two
  web += "<a href=\"/ledtoggle\"> <button> Toggle LED state </button> </a>\r\n";
  web += "</div>\r\n";

  web += "<br/>\r\n";

  web += "<p>\r\n";
  // You can concatenate the state of the LED with normal text
  web += "LED State: "+ LEDstate +"\r\n";
  web += "</p>\r\n";

  web += "</body>\r\n";
  web += "</html>\r\n";

  client.print(web);
}
```

Change the WiFi SSID, password appropriately and upload the code. Don't forget to select the appropriate board and COM port. After successfully uploading the code, open the serial monitor and select the appropriate baud rate.

7.7.2 Result

After the NodeMCU's local IP address gets printed, note it down and enter it in a browser to see the single button webpage as shown in Figure 7.7. Now, you are all set to control the LED using a single button.

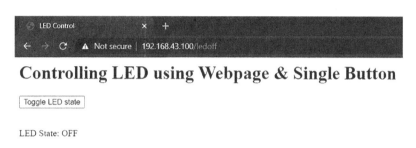

FIGURE 7.7   Webpage with a single button

You can find all the resources of this project at this Github link (**https://github.com/anudeep-20/30IoTProjects/tree/main/Project%207**).

In this project, you have learnt to remotely control an LED using NodeMCU and a webpage. You can extend this similar concept in controlling various sensors, modules and appliances in IoT. You have also learnt the basics of HTML which will help build basic webpages for IoT applications. In the next project, you'll build a logistics tracker using GPS and Adafruit cloud for tracking the logistics in the transit.

Let's move to the next project!

Logistics Tracker Using GPS

8

It is time for us to get better and stronger in learning. This time, the reader shall learn to build a logistics tracker. It is always very important to track the goods and consignments for safe delivery. It is very essential not only in the courier or freight transfer but also in the supply chain management. Apart from logistics, the same setup can also be used in the tracking of kids, elders or anyone who needs to be tracked. It can also be used in live train tracking or live bus tracking which is useful for the travelling passengers to track the location of their train or bus. This is an interesting project and would need a little more time than in the past to complete the task.

8.1 MOTIVE OF THE PROJECT

In this project, you'll be building a logistics tracker using cloud services and GPS which can be used in tracking goods during their transit.

8.2 GPS

If you are using smartphone or PC in your daily life, then there is a high chance that you might be knowing about GPS and its uses. GPS is the acronym for *Global Positioning System*. As the name suggests, GPS is used for determining the position of the objects in the world. Nowadays, every smartphone, PC and even most smartwatches are equipped with GPS modules. These are useful in tracking our movement, suggesting routes, determining our location and many more.

Do you know? Many present-day innovations are once developed by the military for internal usage and later made available to the general public due to their wide range of use cases. One among them is GPS which is a great innovation that has revolutionized many sectors like supply chain, consumer electronics and many more. It was first started as a prototype by the United States Department of Defense in 1973, and the prototype was launched in 1978 for internal usage. It was only made available to the general public in the late 1980s.

GPS used to be very inaccurate in its initial days due to a smaller number of GPS satellites. Later in 1993, a full constellation of 24 operational satellites was launched which improved the accuracy of GPS. So, what is the connection between a GPS module and a satellite? How does GPS work?

Before discussing the working of GPS, you need to know that until the mid-2000s there is only one accurate GPS provider in the world which is provided by the United States government. So, the government can selectively decide whether to provide its GPS to other countries or not. As a result, several other countries have developed and are in the process of developing their own version of GPS and satellite navigation systems. Some of the other GPS systems developed by other countries are

DOI: 10.1201/9781003147169-8

- Global Navigation Satellite System (GLONASS) – Russia
- BeiDou Navigation Satellite System – China
- Galileo Positioning System – European Union
- Indian Regional Navigation Satellite System (NavIC) – India
- Quasi-Zenith Satellite System (QZSS) – Japan

You can find more details about these positioning and navigation systems and their working from the Internet. As we cannot explain the working of all the positioning and navigation systems, we'll explain the working of the first positioning and navigation system developed by the United States (GPS). Don't worry! Working of all these positioning and navigation systems will be almost similar. So, understanding a positioning and navigation system will help you in understanding the remaining ones as well.

GPS is a satellite-based navigation system that is used to provide time and location details of the GPS receiver depending on their position in the world. It works in all weather conditions provided the receiver has no obstruction to at least four GPS satellites. For many, there is a general misconception that GPS doesn't work without Internet or cellular network. This is wrong. GPS operates independent of Internet and cellular network, but the presence of Internet or cellular network will increase its effectiveness. Presently, there are 31 satellites continuously orbiting our earth for providing accurate position details irrespective of the place in the world. The greater the number of satellites in the line of sight to the GPS receiver, the greater the accuracy in determining the position.

The GPS satellites which are in the line of sight of the GPS receiver will send information about the position and current time of the GPS receiver in the form of NMEA (National Marine Electronics Association) data at regular intervals of time. There should be at least four satellites in the line of sight of the GPS receiver to determine the 3D (latitude, longitude, altitude) location of the GPS receiver. If there are only three satellites in the line of sight of the GPS receiver, then only the 2D (latitude, longitude) position of the receiver can be calculated and the GPS receiver assumes that to be at sea level.

By now you might be getting a doubt, if every satellite sends the position information to the GPS receiver then why is it necessary for four satellites to be in the line of sight? The calculation of position by the GPS receiver based on the information provided by the satellite is not straightforward. GPS receivers use a method known as *Trilateration* for determining their accurate position. Trilateration is the process of determining a position based on the intersection of spheres. When a GPS receiver receives a signal from one of the satellites, it calculates its distance from the satellite and constructs a 3D sphere with the satellite located at the centre. After performing the same with three other GPS satellites, the GPS receiver finds the intersection point of the spheres to calculate its location. This is a simple explanation of the working of GPS, but there are many other components and challenges faced by GPS which are not discussed and can easily be found on the Internet.

8.3 U-BLOX NEO-6M GPS MODULE

In this project, we will use a GPS receiver module developed by u-blox. A Neo-6M GPS chip from u-blox is placed at the centre of the module which is less than the size of a postal stamp. The 50 channel u-blox 6 positioning engine can track up to 22 satellites and achieves the industry's highest level of sensitivity i.e. −161 dB tracking while consuming only 45 mA supply current. The u-blox 6 positioning engine also boasts a Time-To-First-Fix (TTFF) of under 1 second. This GPS module can do up to five location updates per second and can be used even for indoor localization applications as it has a high horizontal position accuracy of 2.5 m.

One of the best features the chip provides is the Power Save Mode (PSM) which allows a reduction in system power consumption by selectively switching parts of the receiver ON and OFF. This dramatically reduces the power consumption of the module to just 11 mA making it much suitable for power-sensitive applications like smartwatches with inbuilt GPS. Some of the important specifications of the Neo-6M GPS module are tabulated in Table 8.1.

TABLE 8.1   Specifications of u-blox Neo-6M GPS module

RECEIVER TYPE	50 CHANNELS, GPS L1(1575.42 MHZ)
Horizontal position accuracy	2.5 m
Navigation update rate	1 Hz (5 Hz maximum)
Navigation sensitivity	−161 dBm
Communication protocol	NMEA, UBX Binary, RTCM
Serial baud rate	4800–230400 (default 9600)
Operating temperature	−40°C–85°C
Operating voltage	2.7–3.6 V
Operating current	45 mA
TXD/RXD impedance	510 Ω

The GPS receiver module must have at least four GPS satellites in its line of sight for a position fix. The greater the number of satellites in the line of sight of the GPS receiver, the greater is the GPS location accuracy. For knowing the status of the position fix of the Neo-6M GPS module, there is an LED in the module that has two different rates of blink depending on the state of the module.

- No blinking: The module is searching for satellites.
- Blink rate of 1 second: Position fix is found. (The module can see enough satellites.)

From Table 8.1, you can see that the operating voltage of the Neo-6M GPS module is between 2.7 V and 3.6 V. But as this GPS module is mostly used along with the microcontrollers which have more than 3.6 V output, it is equipped with an on-board ultralow dropout 3.3 V voltage regulator. So, you can use this GPS module along with any microcontroller without using any logic level converter.

The Neo-6M GPS module contains a rechargeable button battery which acts as a super-capacitor. It is also equipped with an HK24C32 two-wire serial EEPROM (Electrically Erasable Programmable Read-Only Memory) which is 4 KB in size and connected to the chip via I²C. An EEPROM along with a battery helps in retaining the battery-backed RAM (BBR). The BBR contains clock data, latest position data (GNSS orbit data) and module configuration. But it is not meant for permanent data storage. As the battery retains the clock and last position, the TTFF significantly reduces to 1s which allows for a much faster position fix. Without the battery, the GPS module will always cold-start so the initial position fix takes more time. The battery is automatically charged when power is applied and maintains data for up to two weeks without power. Figure 8.1 shows the u-blox Neo-6M GPS module along with the naming of all the important parts discussed.

For any kind of wireless communication, an antenna is required. So, the Neo-6M GPS module comes with a patch antenna having −161 dBm sensitivity. Figure 8.2 shows the antenna attached to the GPS module using of U.FL connector.

You are not only limited to the u-blox Neo-6M GPS module for use in your IoT projects, there are other GPS receiver modules as well like Skylab SKG13BL, SIM808 GSM+GPS module and many more which can be used in your IoT applications. But the u-blox Neo-6M GPS module is widely used in IoT applications due to its compact packaging and high precision.

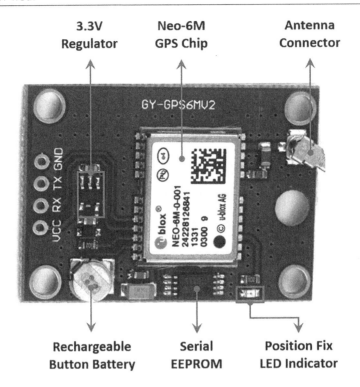

3.3V Regulator   **Neo-6M GPS Chip**   **Antenna Connector**

Rechargeable Button Battery   **Serial EEPROM**   **Position Fix LED Indicator**

FIGURE 8.1   U-blox Neo-6M GPS module

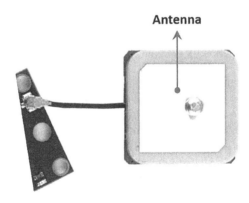

Antenna

FIGURE 8.2   Antenna for GPS module

8.4 NMEA DATA

The raw data received from the GPS receiver module is known as NMEA data which is the standard data format for almost all GPS receivers. It cannot be used directly and need to be processed before usage. Figure 8.3 shows the NMEA data from the GPS receiver module when there is a position fix, and Figure 8.4 shows the NMEA data from the same GPS receiver when there is no position fix.

The NMEA data can be formatted easily by using most of the programming languages. There are also libraries in these programming languages to easily format the NMEA data making our job much more easier. For formatting the NMEA data using the Arduino IDE, we'll use a library known as *TinyGPS++*. The rate at which the NMEA data received by the GPS receiver is known as the *update rate*. The update

```
$GPRMC,144513.000,A,1054.1074,N,07653.7804,E,0.23,162.25,050418,,,A*61
$GPVTG,162.25,T,,M,0.23,N,0.43,K,A*39
$GPGGA,144914.000,1054.1074,N,07653.7801,E,1,4,1.77,234.5,M,-93.2,M,,*7A
$GPGLL,1054.1074,N,07653.7801,E,144914.000,A,A*5F
$GPGSA,A,3,24,02,05,12,,,,,,,,,2.03,1.77,1.00*01
$GPGSV,2,1,05,05,61,161,41,02,50,349,24,12,41,336,26,24,33,251,29*7E
$GPGSV,2,2,05,193,,,*47
$GPRMC,144914.000,A,1054.1074,N,07653.7801,E,0.22,150.86,050418,,,A*6A
$GPVTG,150.86,T,,M,0.22,N,0.41,K,A*32
$GPGGA,144915.000,1054.1073,N,07653.7801,E,1,4,1.77,234.5,M,-93.2,M,,*7C
$GPGLL,1054.1073,N,07653.7801,E,144915.000,A,A*59
$GPGSA,A,3,24,02,05,12,,,,,,,,,2.03,1.77,1.00*01
$GPGSV,2,1,05,05,61,161,41,02,50,349,22,12,41,336,26,24,33,251,29*78
$GPGSV,2,2,05,153,,,*47
```

FIGURE 8.3   NMEA data (with position fix)

```
$GPRMC,,V,,,,,,,,,,,N*53
$GPVTG,,,,,,,,,,N*30
$GPGGA,,,,,,0,00,99.99,,,,,,*48
$GPGSA,A,1,,,,,,,,,,,,,99.99,99.99,99.99*30
$GPGSV,4,1,14,01,,,26,02,,,23,03,,,24,04,,,25*79
$GPGSV,4,2,14,05,,,25,06,,,24,07,,,25,08,,,26*74
$GPGSV,4,3,14,09,,,24,10,,,24,11,,,25,13,,,25*71
$GPGSV,4,4,14,14,,,25,15,,,25*7D
$GPGLL,,,,,,V,N*64
$GPRMC,,V,,,,,,,,,,,N*53
$GPVTG,,,,,,,,,,N*30
$GPGGA,,,,,,0,00,99.99,,,,,,*48
$GPGSA,A,1,,,,,,,,,,,,,99.99,99.99,99.99*30
$GPGSV,4,1,13,02,,,20,03,,,21,04,,,22,05,,,21*7C
$GPGSV,4,2,13,06,,,21,07,,,22,08,,,09,09,,,21*74
$GPGSV,4,3,13,10,,,21,11,,,22,13,,,21,14,,,09*73
$GPGSV,4,4,13,15,,,21*7C
$GPGLL,,,,,,V,N*64
```

FIGURE 8.4   NMEA data (without position fix)

rate of the Neo-6M GPS receiver is 1 Hz (one update per second) and can be increased to 5 Hz. From Figures 8.3 and 8.4, you can see that there are many sentences in the NMEA data. All the sentences in the NMEA data are not useful in obtaining the location data (latitude and longitude). Only the two sentences starting with $GPRMC (Global Positioning Recommended Minimum Coordinates) and $GPGGA (Global Positioning System Fix Data) which are highlighted in Figure 8.3 are useful. The usage and interpretation of each of the two sentences are listed below:

$GPRMC – This sentence provides the time, date, latitude, longitude, altitude and estimated velocity information of the GPS receiver. This information is written in a particular format in the $GPMRC sentence and can be extracted easily. For example, consider the first $GPRMC sentence in Figure 8.3.

$GPRMC,144513.000,A,1054.1074,N,07653.7804,E,0.23,162.25,050418,,,A*61

$GPGGA – This sentence provides essential fix data which presents the 3D location and accurate data of the GPS receiver. This information is written in a particular format in the $GPGGA sentence and can be extracted easily. For example, consider the first $GPGGA sentence in Figure 8.3:

$GPGGA,144914.000,1054.1074,N,07653.7801,E,1,4,1.77,234.5,M,-93.2,M,,*7A

Information in sentences starting with $GPRMC in NMEA data can be extracted by following Table 8.2. Similarly, information in sentences starting with $GPGGA in NMEA data can be extracted from Tables 8.3 and 8.4.

TABLE 8.2 $GPRMC data format

NAME	EXAMPLE	UNITS	DESCRIPTION
Message ID	$GPRMC		RMC protocol header
UTC Time	144513.000		hhmmss.sss
Status	A		A = data valid or V = data not valid
Latitude	1054.1074		ddmm.mmmm
N/S indicator	N		N = nort h or S = south
Longitude	07653.7804		dddmm.mmmm
E/W indicator	E		E = east or W = west
Speed over ground	0.23	knots	
Course over ground	162.25	degrees	True
Date	050418		ddmmyy
Magnetic variation'		degrees	E = east or W = west
Mode	*A*		*A = Autonomous, D = DGPS, E = DR*
Checksum	*61		
<CR><LF>			End of message termination

TABLE 8.3 $GPGGA data format

NAME	EXAMPLE	UNITS	DESCRIPTION
Message ID	$GPGGA		GGA protocol header
UTC time	144914.000		hhmmss.sss
Latitude	1054.1074		ddmm.mmmm
N/S indicator	N		N = north or S = south
Longitude	07653.7801		dddmm.mmmm
E/W indicator	E		E = east or W = west
Position fix indicator	1		See Table 8.4
Satellites used	4		Range 0–12
HOOP	1.77		Horizontal Dilution of Precision
MSL altitude	234.5	meters	
Units	M	meters	
Geoid separation	−93.2	meters	
Units	M	meters	
Age of diff. corr.		second	Null fields when DGPS is not used
Diff. ref. station ID			
Checksum	*7A		
<CR><LF>			End of message termination

TABLE 8.4   Position fix indicator

VALUE	DESCRIPTION
0	Fix not available or invalid
1	GPS SPS mode, fix valid
2	Differential GPS, SPS mode, fix valid
3–5	Not supported
6	*Dead Reckoning Mode, fix valid*

Note: In Table 8.2, Tables 8.3 and 8.4, the *rows written in Italics* are only applicable to NMEA version 2.3 (and later).

More information about the NMEA data can be found from the NMEA reference manual at this link (https://www.sparkfun.com/datasheets/GPS/NMEA%20Reference%20Manual1.pdf).

8.5 USB TO TTL SERIAL CONVERTER

Generally, for controlling the devices based on the data obtained from a UART module (like HC-05 Bluetooth, Neo-6M GPS, etc.), we use a microcontroller and Arduino IDE. But, just to view the serial data from these UART modules, you can use a USB to TTL serial converter. There are many manufacturers of USB to TTL serial converters in the market. But the most reliable and commonly used ones in IoT are CP2102 USB to TTL serial converter and PL2303 USB to TTL serial converter (shown in Figure 8.5). The PL2303 USB to TTL serial converter is also called prolific USB to TTL serial converter. Both the serial converters are almost the same cost and also work almost the same way. You can use either of them depending on their availability. If both serial converters are available, then it is recommended to use the CP2102 USB to TTL serial converter.

Similar to a microcontroller which has a UART interface (Rx, Tx pins), a USB to TTL serial converter also has Rx and Tx pins along with 5V and GND pins for powering the UART module. Some USB to TTL serial converters has extra pins like DTR, 3.3V for other purposes.

Usually, the drivers of the CP2102 USB to TTL serial converter are pre-installed in most PCs. If they are not installed, they get automatically installed as soon as the CP2102 USB to TTL serial converter is connected to the PC. Even if the drivers don't get installed automatically, they can be easily downloaded and installed from this link (https://www.silabs.com/developers/usb-to-uart-bridge-vcp-drivers). But this is not the case with the drivers of PL2303 USB to TTL serial converter. The automatically installed drivers of the PL2303 USB to TTL serial converter by your PC often get corrupted or outdated. The drivers don't have proper updates or online support. It is very difficult to resolve the errors of PL2303 drivers, and it is also difficult to find the updated drivers from the Internet.

New PL2303 USB to TTL serial converters might not cause many issues but old PL2303 USB to TTL serial converters may cause several errors. Most of the users mainly come across the below two errors while using the PL2303 USB to TTL serial converter:

- Missing driver error
- Phased out error

After connecting the PL2303 USB to TTL serial converter to your PC, open the Device Manager and see the assigned COM ports to know whether the PL2303 drivers installed in your PC are corrupted or

CP2102 USB to TTL Serial Converter

PL2303 USB to TTL Serial Converter

FIGURE 8.5 USB to TTL serial converters

not up to date. If a COM port (communication port) is assigned to the PL2303 USB to TTL serial converter without any exclamation mark or warning symbol or error symbol as shown in Figure 8.15, then the PL2303 drivers installed on your PC are up to date and the PL2303 USB to TTL serial converter is ready to use.

8.5.1 Resolving missing driver error

This error is mainly caused if the installed PL2303 drivers are missing or corrupted. It is often seen not only in old PL2303 USB to TTL serial converters but also in new ones. After connecting the PL2303 USB to TTL serial converter to your PC, open the Device Manager and see the assigned COM ports. If a COM port is assigned to the PL2303 USB to TTL serial converter with an exclamation mark as shown in Figure 8.6, then the PL2303 drivers are not properly installed on your PC.

You can also find this error by seeing the properties of the COM port assigned to your PL2303 USB to TTL serial converter. Right click on the COM port which will show a list of options as shown in Figure 8.7. Select the *Properties* option which will open the Properties window as shown in Figure 8.8.

In the properties window, read the text in the *Device status* column. If you see *The device cannot start. (Code 10)*, then the PL2303 drivers are not properly installed on your PC.

To solve this error, right click on the COM port assigned to the PL2303 USB to TTL serial converter and click *Update driver*. This will open a window as shown in Figure 8.9. Among the listed options, click *Search automatically for drivers*. This will search for the best drivers online and will install on your PC.

If your PC cannot install the required drivers or even after installing the best drivers you are not able to solve the error, then right click on the COM port assigned to the PL2303 USB to TTL serial converter and click *Uninstall device*. This will open a new window as shown in Figure 8.10. Check the box beside *Delete the driver software for this device* and then click *Uninstall*. This will uninstall the previously installed PL2303 drivers in your PC.

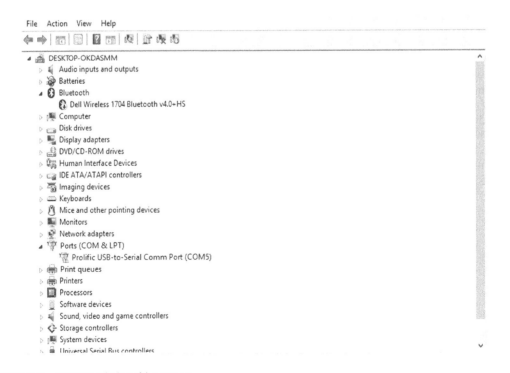

FIGURE 8.6   PL2303 missing driver error

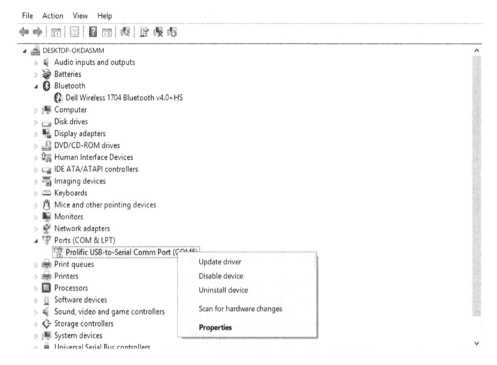

FIGURE 8.7 Right click on COM port

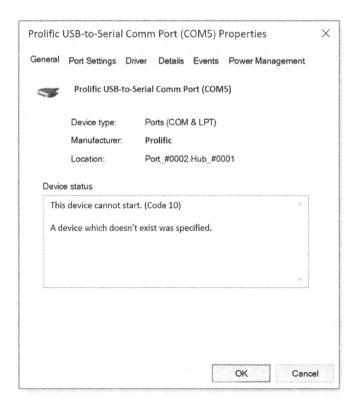

FIGURE 8.8 Properties of COM port

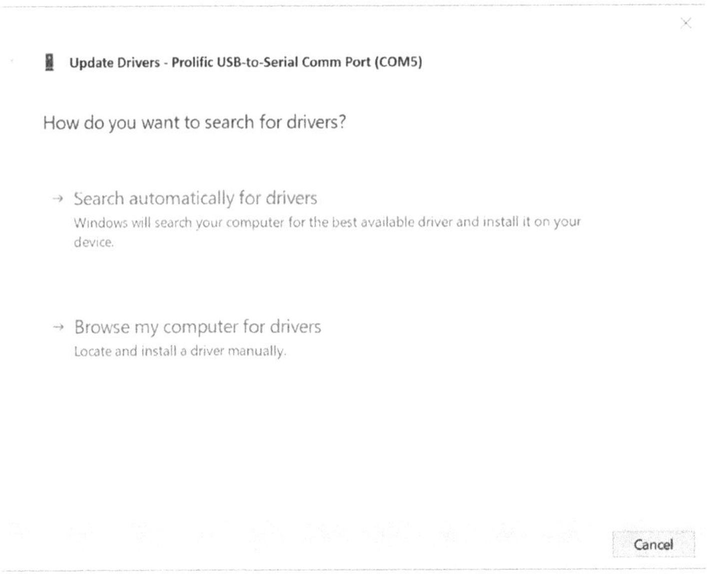

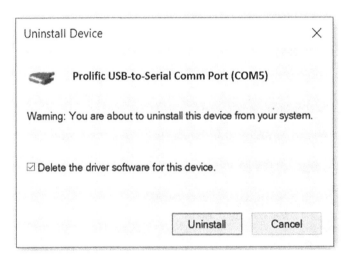

×

■ Update Drivers - Prolific USB-to-Serial Comm Port (COM5)

How do you want to search for drivers?

→ Search automatically for drivers
Windows will search your computer for the best available driver and install it on your device.

→ Browse my computer for drivers
Locate and install a driver manually.

Cancel

FIGURE 8.9   Update drivers

Uninstall Device   ✕

Prolific USB-to-Serial Comm Port (COM5)

Warning: You are about to uninstall this device from your system.

☑ Delete the driver software for this device.

Uninstall   Cancel

FIGURE 8.10   Uninstall device

Now, download the *PL2303_64bit_Installer.exe* executable file from this link (https://github.com/anudeep-20/30IoTProjects/tree/main/Project%208) and run it. If the .NET framework is not installed on your PC, then you'll see a window as shown in Figure 8.11. Among the listed options, click *Download and install this feature*. This will install the .NET Framework 3.5. After you see the successfully installed window as shown in Figure 8.12, restart your PC for changes to come into effect.

After restarting your PC, connect the PL2303 USB to TTL serial converter and uninstall the driver in Device Manager again. Now, run the *PL2303_64bit_Installer.exe* executable file. This will open a window as shown in Figure 8.13. If you want to donate to the developer of the software and support their project, click *Donate*. If you want to proceed with fixing the error, click *Continue*. If .NET Framework 3.5 is previously installed on your PC, then you'll directly see a window as shown in Figure 8.13.

After successfully finishing the process, you'll see a window as shown in Figure 8.14. Click *Close* and remove the PL2303 USB to TTL serial converter. Now, restart your PC for the changes to take effect. After restarting your PC, open the Device Manager and you should see no exclamation mark beside the COM port assigned to the PL2303 USB to TTL serial converter as shown in Figure 8.15.

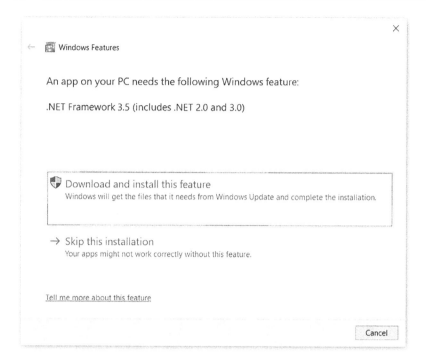

FIGURE 8.11 Installing .NET framework

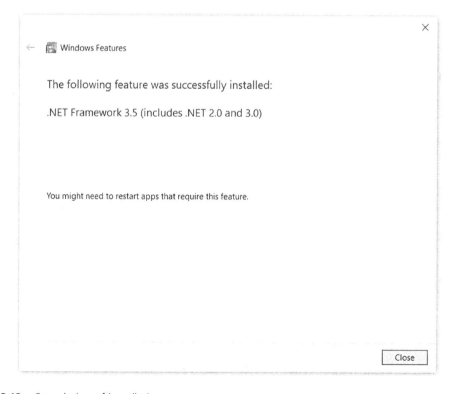

FIGURE 8.12 Completion of installation

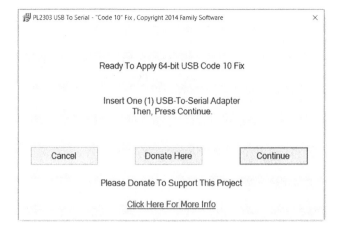

FIGURE 8.13   "Code 10" fix

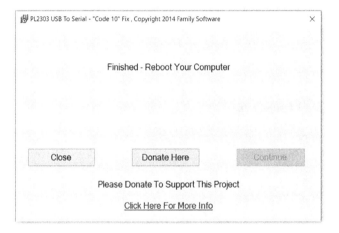

FIGURE 8.14   Error fix completion

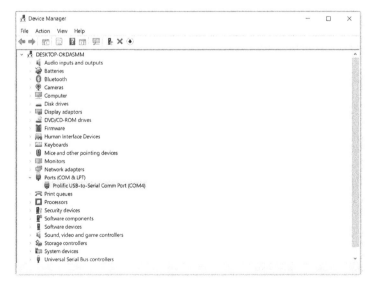

FIGURE 8.15   COM port without any error

8.5.2 Resolving phased out error

This error is mainly caused if the PL2303 drivers installed on your PC are not compatible with the PL2303 USB to TTL serial converter you are using. For example, the PL2303 USB to TTL serial converter which you are using has an old chip but the drivers installed on your PC are the latest ones. This error is often seen in old PL2303 USB to TTL serial converters.

After connecting the PL2303 USB to TTL serial converter to your PC, open the Device Manager and see the *Ports* section. If you see *PL2303HXA PHASED OUT SINCE 2012. PLEASE CONTACT YOUR SUPPLIER.* as shown in Figure 8.16, then the correct PL2303 drivers are not installed on your PC.

To solve this error, you need to download and install old PL2303 drivers. To do so, download the *PL2303_Prolific_GPS_1013_20090319.exe* executable file from this Link (https://github.com/anudeep-20/30IoTProjects/tree/main/Project%208) and run it. After running the executable file, you'll see a window as shown in Figure 8.17. Click *Next >*, and you'll see a license agreement as shown in Figure 8.18. Read the License Agreement completely, select *I accept the terms of the license agreement* and click *Next >*. Now, wait until the installation of PL2303 drivers gets completed. After completing the installation, you'll see a window as shown in Figure 8.19 and click *Finish*. Now, connect the PL2303 USB to TTL serial converter to your PC. If you have inserted the PL2303 USB to TTL serial converter before running the installation setup, remove and connect the PL2303 USB to TTL serial converter again.

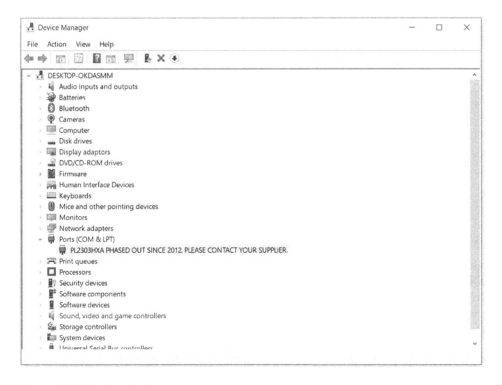

FIGURE 8.16 PL2303 phased out error

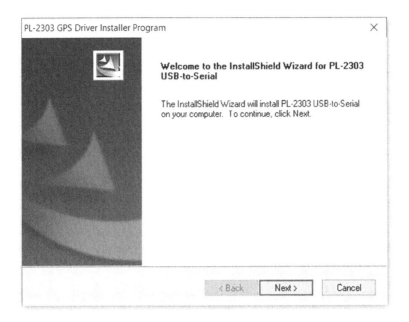

FIGURE 8.17   InstallShield wizard

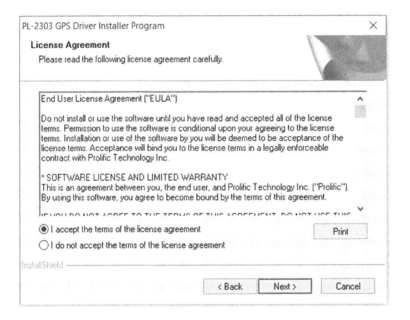

FIGURE 8.18   License agreement

Now, open the Device Manager and expand the *Ports* section. Then right click on the phased out COM port (*PL2303HXA PHASED OUT SINCE 2012. PLEASE CONTACT YOUR SUPPLIER.*) and then click *Update driver* as shown in Figure 8.20. This will open a new window as shown in Figure 8.21. Among the listed options, click on *Browse my computer for drivers* which will open another window as shown in Figure 8.22. Among the newly listed options, click on *Let me pick from a list of available drivers on my computer* which will open a new window to select the correct driver for installation as shown in Figure 8.23.

In the *model* column in Figure 8.23, you can see all the available PL2303 drivers on your PC ready for installation. As your PL2303 USB to TTL serial converter has an older chip and has a PHASED OUT

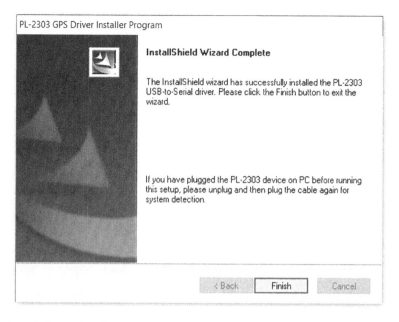

FIGURE 8.19   Completion of installation

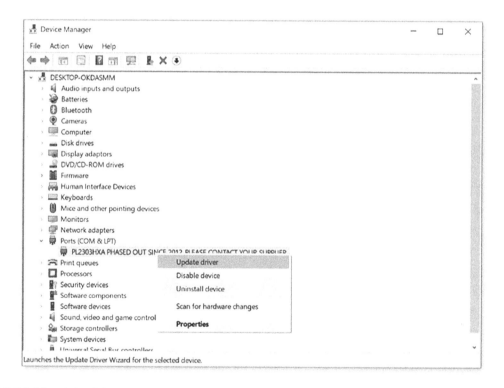

FIGURE 8.20   Right click on COM port

SINCE 2012 error, select the driver model before 2012. In our case, among the listed drivers, the first one is a 2008 model of the PL2303 driver and the second one is a 2020 model of the PL2303 driver. So, select the first model (*Prolific USB-to-Serial Comm Port Version: 3.3.2.105 [27-10-2008]*) and click *Next*. After completing the installation, you'll see a window as shown in Figure 8.24. Now, once again open the

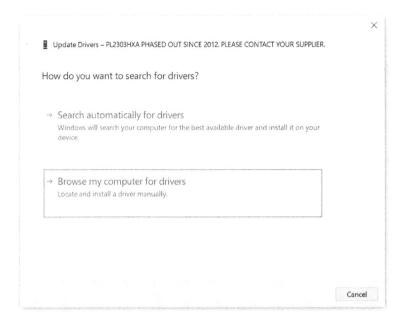

FIGURE 8.21   Update drivers

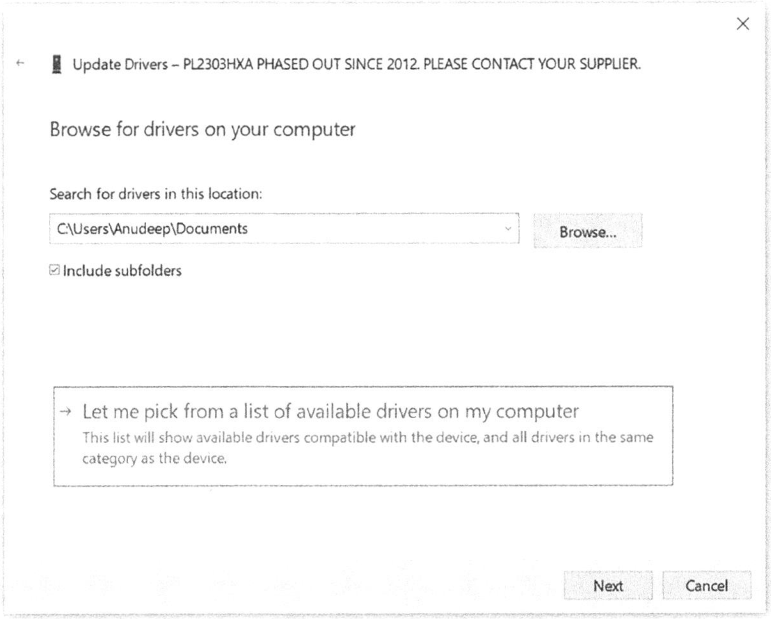

FIGURE 8.22   Browse for drivers

Device Manager and expand the *Ports* section to see an error less COM port assigned to the PL2303 USB to TTL serial converter as shown in Figure 8.15.

Note: The executable files which were used in Sections 8.5.1 and 8.5.2 are not created by us. These are obtained from some third-party developers, and they may not be associated with the original company.

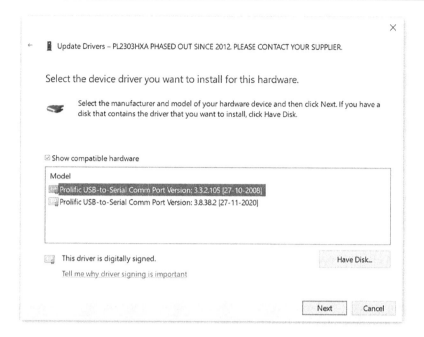

FIGURE 8.23   Selection of driver to be installed

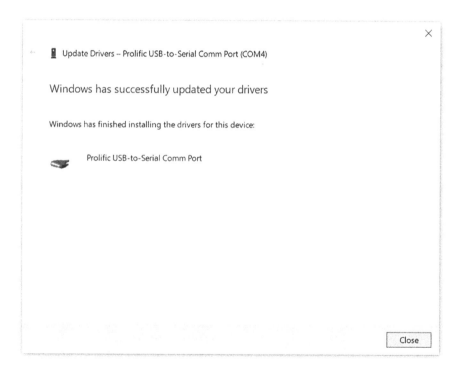

FIGURE 8.24   Completion of Installation

These executable files *may* contain harmful virus or malware. So, it is completely under your sole discretion to use these executable files. But there is no need to worry too much as these executable files are used by most of the people in the community so there might be very little chance of a virus or malware attack.

8.6 TERA TERM

Usage of USB to TTL serial converter is very simple, just connect it to one of the USB drives on your PC. But how to visualize the data received by it or send data from it?

It can be done in two ways:

- Using an open-source software like Tera Term (recommended method).
- Using Arduino IDE's Serial Monitor.

Tera Term is an open-source, free terminal emulator software that can be used for reading the serial data and sending data through the serial ports. Installation and usage of Tera Term is very simple and straightforward.

Download the executable file of Tera Term from this link (https://osdn.net/projects/ttssh2/releases/) and run it. You can also find previous releases of Tera Term in the same link. Read the complete license agreement, terms and conditions and then select the *I accept the agreement* radio button then click *Next>* as shown in Figure 8.25.

Now, the setup process will ask the file location at which Tera Term must be installed. Browse the location and click *Next >* as shown in Figure 8.26.

Next, select the type of installation among *Standard Installation* (Recommended), *Full Installation*, *Compact Installation* or *Custom Installation* and then click *Next >* as shown in Figure 8.27.

After selecting the type of installation, select the language in which you want to use the application as shown in Figure 8.28 and then select the start menu folder as shown in Figure 8.29.

Finally, select the Additional Tasks to be performed by the setup process by checking the required tasks as shown in Figure 8.30 and then click Install as shown in Figure 8.31.

After waiting for 1–2 minutes depending on your PC speed, the installation of Tera Term gets completed and you'll see a window as shown in Figure 8.32.

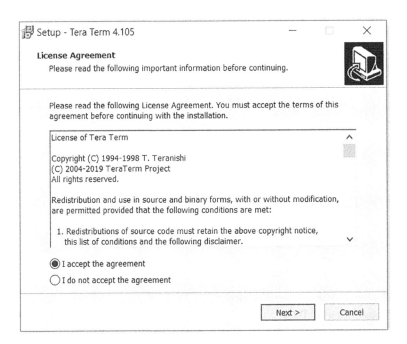

FIGURE 8.25 License agreement

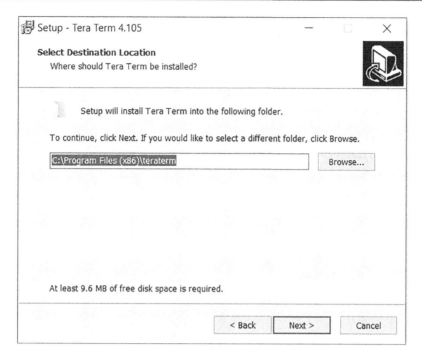

FIGURE 8.26   Selection of installation location

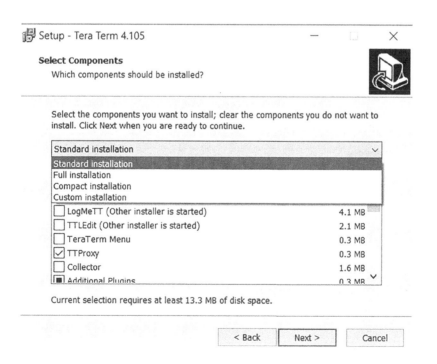

FIGURE 8.27   Selection of installation components

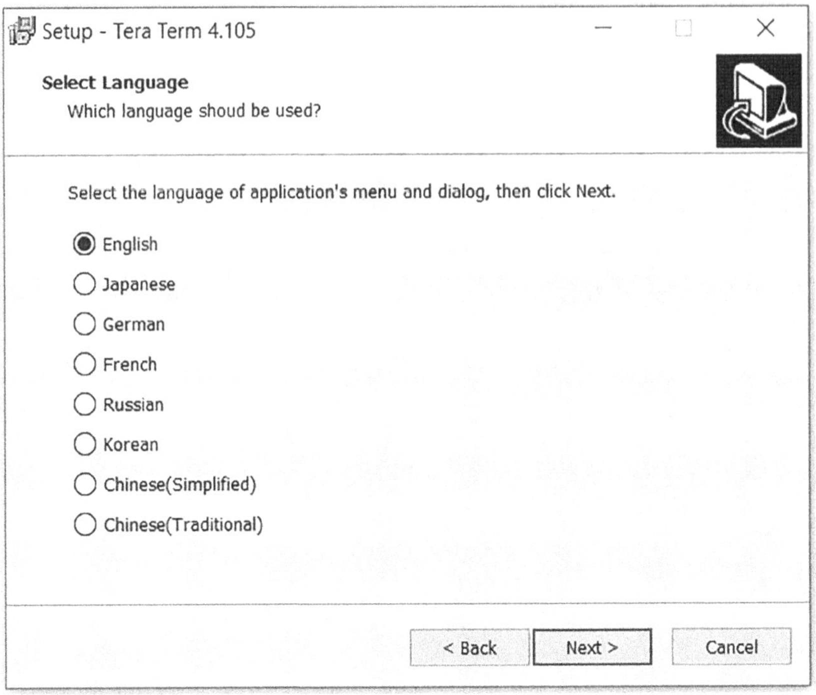

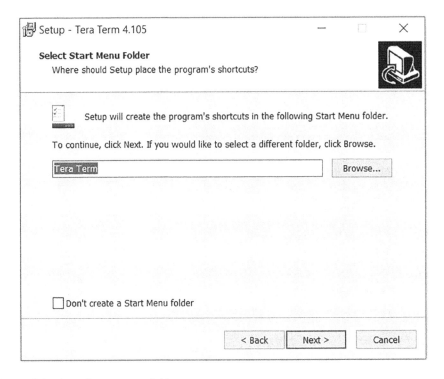

FIGURE 8.28   Application language selection

FIGURE 8.29   Selection of start menu folder

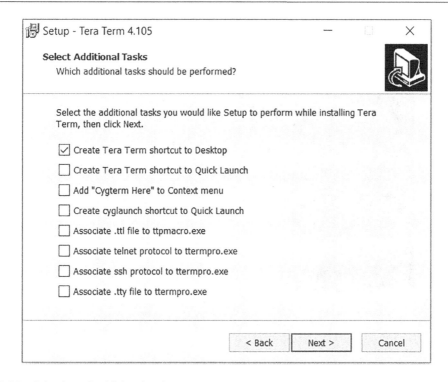

FIGURE 8.30   Selection of additional tasks

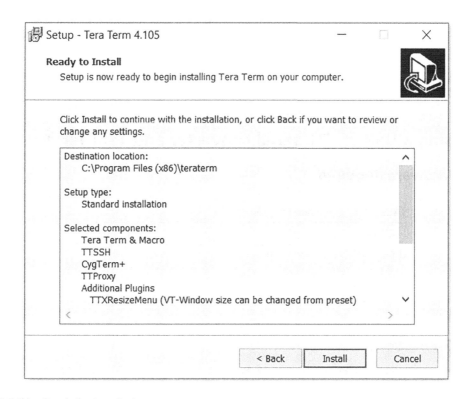

FIGURE 8.31   Ready for installation

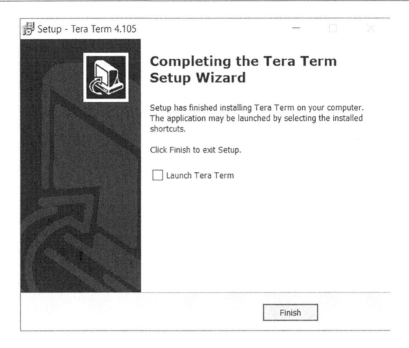

FIGURE 8.32   Completion of installation

8.7 VISUALIZING THE NMEA DATA

There are many cases in which you want to visualize the raw NMEA data shown in Figure 8.3 and Figure 8.4. It can be done in two ways.

8.7.1 Using a USB to TTL serial converter and Tera Term

8.7.1.1 Hardware Required

TABLE 8.5   Hardware required

HARDWARE	QUANTITY
U-blox Neo-6M GPS module	1
USB to TTL serial converter	1
Jumper wires	As required

8.7.1.2 Connections

After connecting the required hardware (Table 8.5) as shown in Figure 8.33 and Table 8.6, connect the USB to TTL serial converter to one of the USB drives on your PC. Next, find the COM port assigned to the connected USB drive as explained in the 1st project.

FIGURE 8.33   Connection layout

TABLE 8.6   Connection details

U-BLOX NEO-6M GPS MODULE	USB TO TTL SERIAL CONVERTER
VCC	5V
RxD	TxD
TxD	RxD
GND	GND

Now, open the Tera Term software which will open a window as shown in Figure 8.34. Select *Serial* among the two radio buttons at the left and then from the drop-down, select the appropriate COM port assigned to the USB to TTL serial converter connected to your PC. Then click *OK*. Now, you'll be able to see the NMEA data as shown in Figure 8.3 or 8.4.

FIGURE 8.34   Selection of COM port

8.7.2 Using a microcontroller and Arduino IDE

You can refer to Table 8.7 for the list of required hardware, followed by Table 8.8, which presents the pin-wise connections. Also, the connection diagram is presented in Figure 8.32, which shall help get a visualization of the connections. Next, open the Arduino IDE and upload the below code after selecting the appropriate board and COM port. No explanation is needed for the below code as it is very simple and easy to understand.

```
#include <SoftwareSerial.h>

SoftwareSerial gpsSerial(D5, D6); //Rx Pin, Tx Pin of NodeMCU

void setup() {
```

```
  Serial.begin(115200); //Initialisation of NodeMCU's Serial
  gpsSerial.begin(9600); //Initialisation of GPS's Serial
}

void loop() {
  //If data is available on GPS's Serial
  while (gpsSerial.available() > 0)
    Serial.write(gpsSerial.read()); // then write on NodeMCU's Serial
}
```

Note: In the last line of the code, we have used Serial.write() instead of Serial.print() or Serial.println(). There is a reason for using it, and you can understand the subtle difference between Serial.write() and Serial.print() by referring to this Stack Exchange Question (https://arduino.stackexchange.com/questions/10088/what-is-the-difference-between-serial-write-and-serial-print-and-when-are-they).

After successfully uploading the code, open the serial monitor and select the baud rate as 115200 to view the raw NMEA data from the GPS module as shown in Figure 8.3 or Figure 8.4.

8.8 U-CENTER SOFTWARE

U-center is a very useful software tool developed by u-blox for evaluation, performance analysis and configuration of GPS receivers including u-blox Neo-6M. It is a very important tool to be known by anyone dealing with GPS receivers as it can display real-time structured and graphical data visualization of any GPS receiver.

Installation and usage of u-center software are easy and straightforward. U-center software for windows can be downloaded from the official website of u-blox at this link (https://www.u-blox.com/en/product/u-center). Unfortunately, u-center is not available for iOS and can only be downloaded and used on Windows OS.

Download the zip file of the software from the website and extract the u-center executable file. Next, select the language of the Installer and then click *OK* as shown in Figure 8.35.

FIGURE 8.35   Installer language selection

Now, read the complete License Agreement and click I Agree as shown in Figure 8.36. Next, choose the components for installation and location of installation as shown in Figures 8.37 and 8.38. Finally, click *Install* for completing the installation procedure.

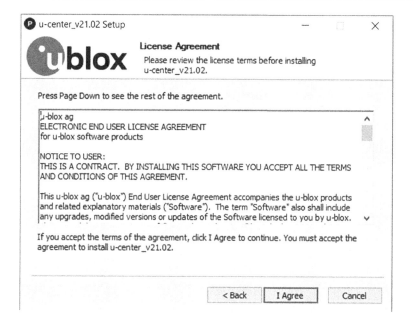

FIGURE 8.36   License agreement

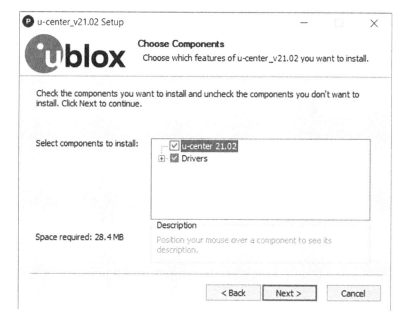

FIGURE 8.37   Components for installation

After successfully finishing the installation, click *Finish* as shown in Figure 8.39. Now, open the u-center software installed on your PC. Figure 8.40 shows the startup page of the u-center software.

U-center provides a wide range of features for users of GPS receivers. It can display real-time structured and graphical data visualization from any GPS receiver such as satellite summary view, navigation summary view, data recording, playback functionality and many more. You can find all the functionalities

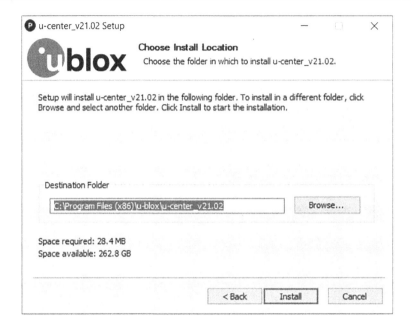

FIGURE 8.38   Installation location

FIGURE 8.39   Completion of installation

of u-center by referring to the user guide (https://www.u-blox.com/sites/default/files/u-center_Userguide_UBX-13005250.pdf).

You can also view the raw NMEA data using the u-center software. Use the hardware mentioned in Table 8.5 and connect them using Figure 8.33 and Table 8.6. Now, select the COM port associated with the USB to TTL serial converter from *Receiver > Connection* as shown in Figure 8.41. Next, select the baud rate of the u-blox Neo-6M GPS module (default is 9600) from *Receiver >Baudrate* as shown in Figure 8.42.

FIGURE 8.40 Startup page of u-center

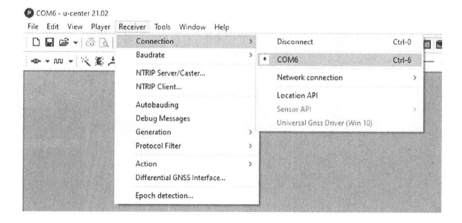

FIGURE 8.41 COM port selection

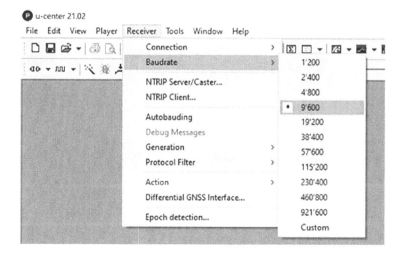

FIGURE 8.42 Baud rate selection

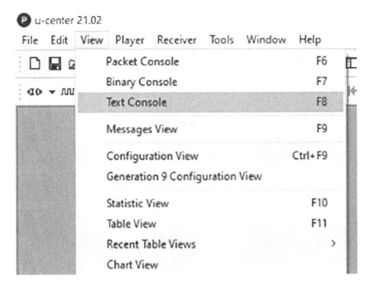

FIGURE 8.43 Selection of text console

Finally, for viewing the NMEA data, select *View > Text Console* as shown in Figure 8.43. A Text Console window opens as shown in Figure 8.44 which will show the NMEA data.

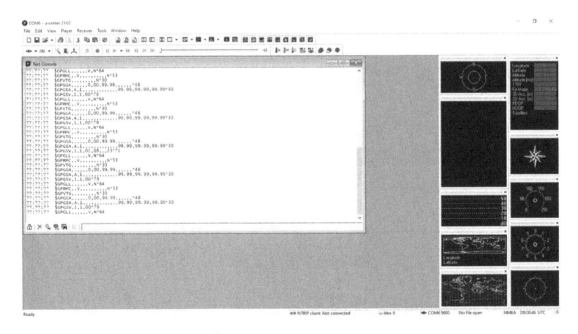

FIGURE 8.44 Visualization of NMEA data

8.9 ADAFRUIT CLOUD

Cloud is nothing but someone's computer at a remote location that can be accessed across the Internet. Companies that provide these computers for usage are known as cloud service providers and Adafruit is one such company. Most of these companies provide cloud services for free for limited usage and charge a nominal subscription fee or a one-time payment fee for complete usage.

Not only Adafruit there are many other cloud service providers in IoT such as DigitalOcean, Thingspeak, Ubidots and many more. Usage of all these cloud services is almost similar with minor feature changes. In this book, we'll be using Adafruit cloud services wherever cloud is required and you can try other services as well by making necessary changes in the code.

In the present world, we are adding more and more IoT devices to our home which are generating a huge amount of data. Cloud services help in the storage and analysis of this data so that we can get the maximum benefit of an IoT infrastructure. With the help of IoT and cloud services, we can also control and monitor a place from a remote location.

For using the services of Adafruit cloud, you need to create an account in their website (https://io.adafruit.com/). Open their website and on the home page click *Get Started for Free* if you don't have an account in Adafruit IO or click *Sign In* if you already have an account in Adafruit IO as shown in Figure 8.45.

FIGURE 8.45   *Getting started*

For creating an account in Adafruit IO, you need to provide some necessary details like First name, Last name, Email ID, Username and Password as shown in Figure 8.46. After filling in the details click *CREATE ACCOUNT* button and your account will be created. Now, you can directly sign in by providing the email address/username and password as shown in Figure 8.47.

After signing in, you'll be redirected to a page as shown in Figure 8.48. This page shows the dashboards you have built in your account. This is an *IO Free* account, and all the features available for an *IO Free* account can be seen in Profile Section as shown in Figure 8.49. If you need more services, you can update your account to *IO+* from the Profile Section.

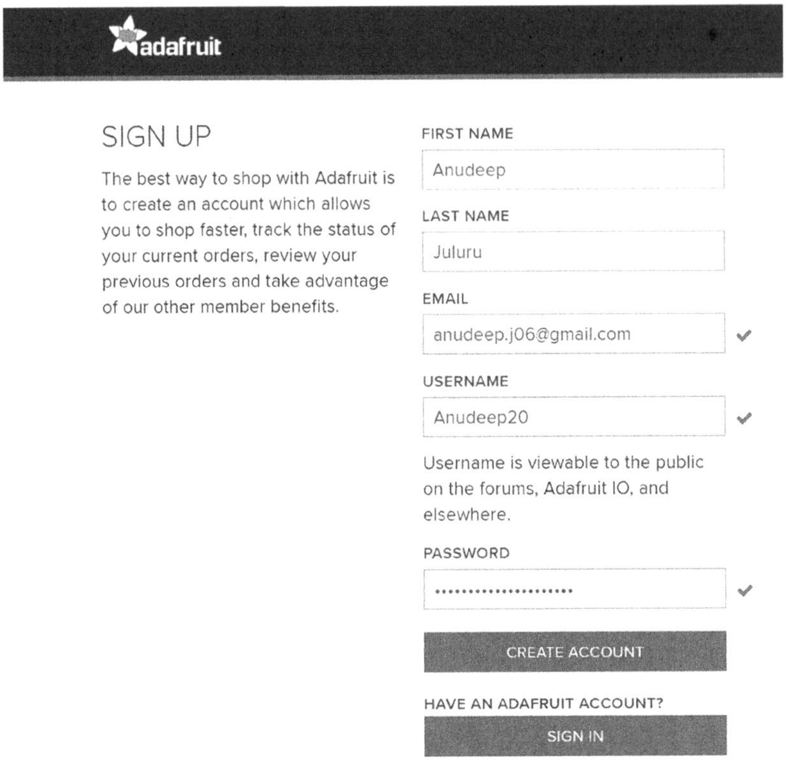

FIGURE 8.46   Registration of an account

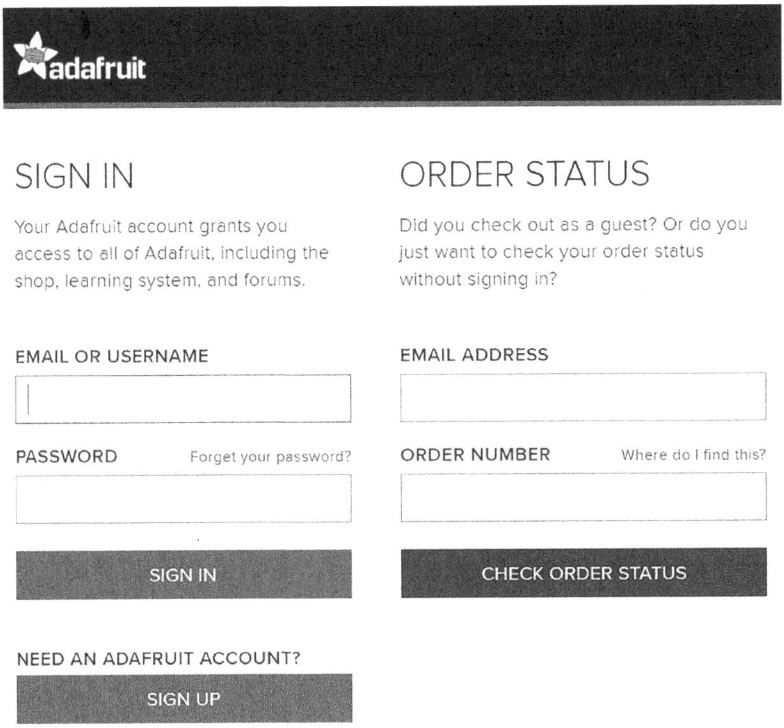

FIGURE 8.47   Sign in screen of Adafruit IO

FIGURE 8.48   Adafruit dashboards

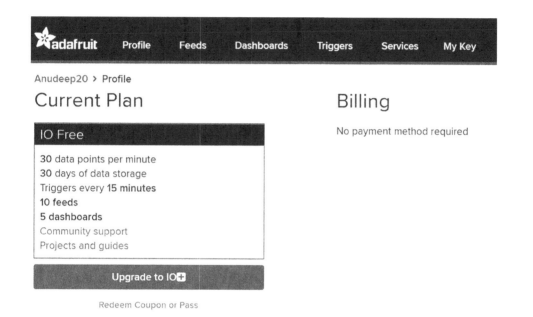

FIGURE 8.49   Adafruit profile

8.10 MQTT PROTOCOL

A protocol is a set of rules used to either transmit or receive data between devices. These rules include parameters like the size of the packet to be transmitted, the timing gap between two packets, connectivity type (Broadcast, P2P connection, etc.), etc. Anyone can design their own protocol with simple parameters and use it in their desired applications. This makes the device more secure as it is proprietary and not available to the public. Custom protocols are only used by some individuals and some companies where security is the utmost priority.

In this book, we'll not design a custom protocol but use widely used protocols in IoT. There are many protocols in IoT that can be used to publish or transmit data like Constrained Application Protocol (CoAP), Message Queue Telemetry Transport Protocol (MQTT), Advanced Message Queuing Protocol (AMQP), Data Distribution Service (DDS) and many more.

In this project, we'll use the MQTT protocol to publish and subscribe data from Adafruit. MQTT is an OASIS standard messaging protocol for Internet of Things (IoT). It is designed as an extremely lightweight publish/subscribe messaging transport that is ideal for connecting remote devices with small code and minimal network bandwidth. MQTT is used in a wide variety of industries, such as automotive, manufacturing, telecommunications, oil and gas, etc.

MQTT protocol defines two types of network entities: Broker and Client. MQTT broker is a server that receives all messages from the clients and then routes the messages to the appropriate destination clients. MQTT client can be any device (from a microcontroller up to a fully fledged server) that runs an MQTT library and connects to an MQTT broker over a network.

The MQTT broker acts as a post office. It doesn't use the address of the intended recipient but uses the subject line called *Topic* and anyone who wants a copy of that message will subscribe to that *Topic*. Multiple clients can receive the message from a single broker (one to many capability). Similarly, multiple publishers can publish topics to a single subscriber (many to one). Each MQTT client can both receive and produce data by subscribing and publishing a topic (MQTT is a bidirectional communication protocol). For example, the client devices can publish sensor data and at the same time can receive the configuration information or control commands. This helps in both sharing data, managing and controlling devices. The architecture of the MQTT protocol is shown in Figure 8.50.

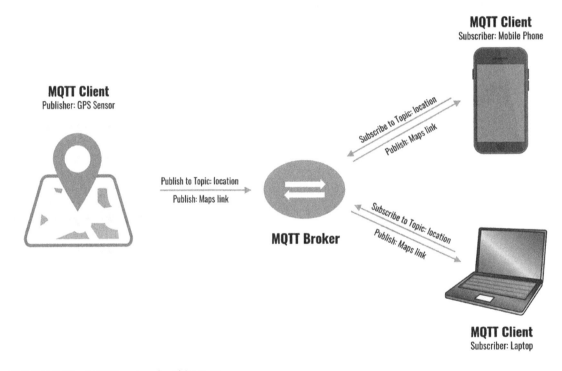

FIGURE 8.50   MQTT protocol architecture

8.11 HARDWARE REQUIRED

TABLE 8.7   Hardware required for the project

HARDWARE	QUANTITY
NodeMCU	1
U-blox Neo-6M GPS Module	1
Jumper wires	As required

8.12 CONNECTIONS

You can refer to the list of components required from Table 8.7, followed by Table 8.8, which presents the pin-wise connections. Also, the connection diagram is presented in Figure 8.51, which shall help get a visualization of the connections.

TABLE 8.8   Connection details

NODEMCU	U-BLOX NEO-6M GPS MODULE
3V3	VCC
D5	TxD
D6	RxD
GND	GND

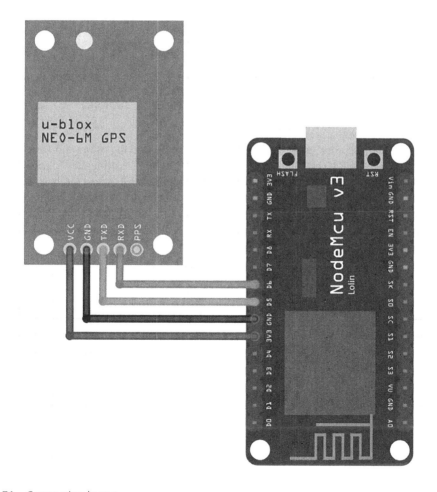

FIGURE 8.51   Connection layout

8.13 ADAFRUIT FEED CREATION

For storing data in the Adafruit IO cloud, we need to create a feed. Feeds in Adafruit are storage containers that can be used for storing incoming data. In an IO Free Account, you can create 10 Feeds and can upload 30 data points per minute. Each data point that needs to be uploaded cannot be greater than 1 kb. Also, the uploaded data into the feed will not stay for more than 30 days in an IO Free Account.

To create a new feed, first sign in to your Adafruit IO Account. After you see the landing page as shown in Figure 8.48, click on *Feeds > view all* in the menu bar at the top. Now, you'll be able to see all the feeds created in your account. By default, you should have no feeds created in your account as shown in Figure 8.52.

FIGURE 8.52   Adafruit feeds

For creating a new feed, click on the *+ New Feed* button and fill in the *Name* and *Description* details of the feed and then click the *Create* button as shown in Figure 8.53. Filling *Description* details in the creation of a feed is optional. Adafruit allows using all the alphabets and almost all the special characters in naming a feed. But, after you create the feed, a *Key* will be generated corresponding to the *Feed Name* using only the alphabets in the *Feed Name* and the – special character. After a feed is successfully created by default, it'll be added to the *Default* group as shown in Figure 8.54. You can also create multiple feeds and group them into separate groups by using the *+ New Group* button.

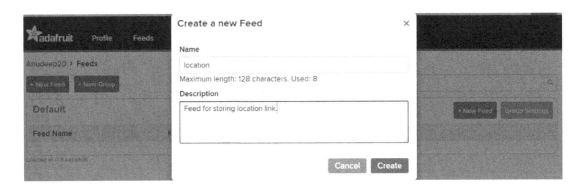

FIGURE 8.53   New feed creation

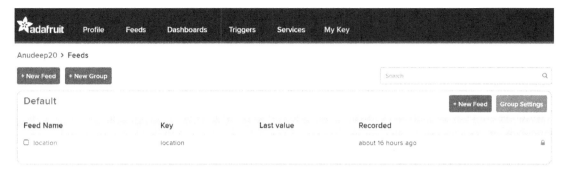

FIGURE 8.54 *location* feed created

Click on the newly created feed *location* which will open a new page as shown in Figure 8.55. Now, you can access all the settings of this feed like Feed Info, Privacy, Sharing and many more. If you add numerical data to this feed, you can see a graphical representation of the data. You can also perform many tasks on this feed like manually adding new data, downloading all the stored data, filtering the stored data and many more.

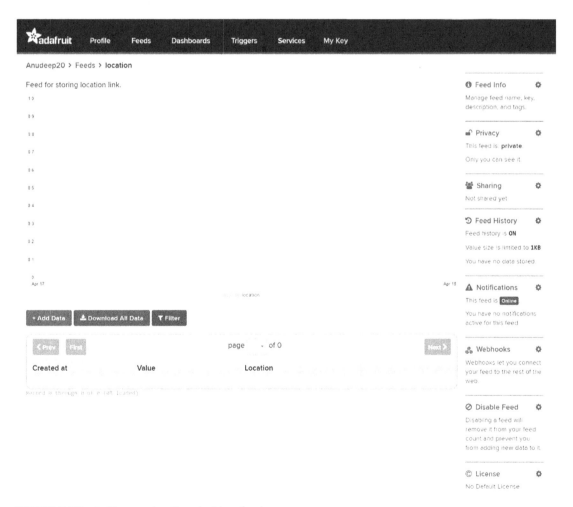

FIGURE 8.55 Settings and options inside a feed

For accessing feeds in an Adafruit account, you need to have an Adafruit IO key. You can view your Adafruit IO key by clicking on *My Key* in the menu bar at the top (written in yellow) which will open a window as shown in Figure 8.56. Copy the first code snippet written below Arduino.

```
#define IO_USERNAME  "Anudeep20"
#define IO_KEY       "aio_iEsC88tR8nwXsI6ZNXBzxRio6ptz"
```

The Adafruit IO key is equivalent to your Adafruit Password. You need to keep it safe and do not share it with anyone as they can access and modify data in your feeds. If you find that your Adafruit IO key is leaked, don't worry! You can change it using the *REGENERATE KEY* button. But remember to change the old key with the new one in all your code manually after regenerating it.

YOUR ADAFRUIT IO KEY   ✕

Your Adafruit IO Key should be kept in a safe place and treated with the same care as your Adafruit username and password. People who have access to your Adafruit IO Key can view all of your data, create new feeds for your account, and manipulate your active feeds.

If you need to regenerate a new Adafruit IO Key, all of your existing programs and scripts will need to be manually changed to the new key.

Username | Anudeep20 |

Active Key | aio_iEsC88tR8nwXsI6ZNXBzxRio6ptz | [REGENERATE KEY]

Hide Code Samples

Arduino

```
#define IO_USERNAME  "Anudeep20"
#define IO_KEY       "aio_iEsC88tR8nwXsI6ZNXBzxRio6ptz"
```

Linux Shell

```
export IO_USERNAME="Anudeep20"
export IO_KEY="aio_iEsC88tR8nwXsI6ZNXBzxRio6ptz"
```

Scripting

```
ADAFRUIT_IO_USERNAME = "Anudeep20"
ADAFRUIT_IO_KEY = "aio_iEsC88tR8nwXsI6ZNXBzxRio6ptz"
```

FIGURE 8.56   Adafruit IO key

8.14 ARDUINO IDE CODE

TinyGPS++.h library is used for parsing the raw NMEA data from the GPS module into useful data. It can be downloaded at this link (https://github.com/mikalhart/TinyGPSPlus). Whenever a module with UART lines (Rx, Tx) is interfaced with a microcontroller's Tx and Rx pins, the UART connections must be removed before uploading the code and to be reconnected after uploading the code. To avoid this procedure every time while uploading the code, SoftwareSerial.h library is used. It can replicate the serial communication functionality on other digital pins of the microcontroller with the help of software. SoftwareSerial.h library will be installed by default during the installation of Arduino IDE. You can also download this library at this link (https://github.com/PaulStoffregen/SoftwareSerial). For connecting NodeMCU (ESP8266) to the WiFi, ESP8266WiFi.h library is used. Finally, Adafruit _ MQTT _ Client.h library is used for publishing or subscribing data from Adafruit. You can download Adafruit _ MQTT _ Client.h library at this link (https://github.com/adafruit/Adafruit_MQTT_Library). After downloading the files of all new libraries, extract and copy them in the libraries folder.

```
#include <TinyGPS++.h> //Library for GPS module
#include <SoftwareSerial.h> //Library for replicating Rx, Tx
#include <ESP8266WiFi.h> //Library for WiFi functionalities
#include <Adafruit_MQTT_Client.h> //Library for Adafruit MQTT Client
```

server, port, IO _ USERNAME and IO _ KEY are constant variables that don't change in the entire program execution. These are the details required for connecting to Adafruit (MQTT Broker). In the place of AIO _ USERNAME and AIO _ KEY, paste the credentials (shown in Figure 8.56) which you have copied from the Adafruit website.

```
//Adafruit Details
#define server "io.adafruit.com"
#define port 1883
#define IO_USERNAME  "AIO_USERNAME"
#define IO_KEY       "AIO_KEY"
```

gps is a TinyGPSPlus object which is used for parsing the raw NMEA data. gpsSerial is a SoftwareSerial object used for replicating the default serial functionality on other digital GPIO pins. Here, gpsSerial(D5,D6) means D5 andD6 are the new Rx and Tx pins of the microcontroller respectively. ssid and password are of String type variables used for storing the name and password of your WiFi. These are useful for connecting the NodeMCU to your WiFi network. Write your WiFi name and password in the place of YOUR WIFI NAME and YOUR WIFI PASSWORD, respectively. Similarly, lat _ str, lng _ str and loc _ link are also String type variables used for storing the decoded location details. lat _ str is used for storing the latitude of the location, lng _ str is used for storing the longitude of the location and loc _ link is used for storing the Google maps link of the location.

```
TinyGPSPlus gps;  //TinyGPSPlus object
SoftwareSerial gpsSerial(D5, D6); //Rx Pin, Tx Pin
String ssid = "YOUR WIFI NAME"; //SSID of your WiFi
String password = "YOUR WIFI PASSWORD"; //Password of your WiFi
String lat_str, lng_str, loc_link; //Storing location details
```

esp is a WiFiClient object used for creating a client that can connect to a specific IP address and port. mqtt is an Adafruit _ MQTT _ Client object used for connecting to Adafruit. feed is an Adafruit _ MQTT _ Publish object used for publishing data to a particular feed in Adafruit. The second argument of the Adafruit _ MQTT _ Publish() is a combination of your Adafruit IO username and the key name of the feed. The key name of a feed can be obtained from the second column of the feed list shown in Figure 8.54. In this case, the feed name is location and the key name is also location.

If you are new to programming or C language, you might not understand the usage of * and &. They are nothing but pointer and address, respectively. To know more about this concept, we recommend you to refer to this tutorial (https://www.javatpoint.com/c-pointers).

```
WiFiClient esp; //Create a WiFi Client

//Creation of Adafruit MQTT Client
Adafruit_MQTT_Client mqtt(&esp, server, port, IO_USERNAME, IO_KEY);
//Variable for publishing data to Adafruit Feed
Adafruit_MQTT_Publish feed = Adafruit_MQTT_Publish(&mqtt, IO_USERNAME"/feeds/
location");
```

void setup() contains all the code required for the initial setup of NodeMCU, WiFi connection and Adafruit connection. NodeMCU's serial is initialized at 115200 baud rate using Serial.begin(), and GPS module's serial is initialized at 9600 baud rate using gpsSerial.begin(). The code used for connecting NodeMCU to WiFi is the same code used in the previous project.

```
void setup() {
  Serial.begin(115200); //Initialisation of NodeMCU Serial
  gpsSerial.begin(9600); //Initialisation of GPS Serial

  //Connecting NodeMCU to WiFi
  Serial.println("");
  Serial.print("Connecting to ");
  Serial.println(ssid);
  WiFi.begin(ssid, password);
  while (WiFi.status() != WL_CONNECTED) {
    delay(500);
    Serial.print(".");
  }
  Serial.println("");
  Serial.println("WiFi connected");
  //Print the local IP address of NodeMCU
  Serial.println(WiFi.localIP());
```

mqtt.connect() is used for connecting the NodeMCU (MQTT Client) to the Adafruit (MQTT Broker)..'s will be printed until there is a successful MQTT connection between NodeMCU and Adafruit. Data stored in String variable cannot be uploaded to the Adafruit feed. So, the data stored in loc _ link is converted to a character array and stored in charLocation.

```
//Connecting to MQTT
  Serial.print("Connecting to MQTT");
  while (mqtt.connect()) {
    Serial.print(".");
  }
}
char charLocation[100];
```

gpsSerial.available() > 0 checks whether there is any incoming data from the GPS module. If there is any incoming data, gpsSerial.read() will read the incoming data from the GPS module.

gps.encode() will encode the raw NMEA data that is read from the GPS module. If the encoded NMEA data has a valid location and the NodeMCU is connected to Adafruit, then latitude and longitude are extracted from the encoded NMEA data.

```
void loop() {
  while (gpsSerial.available() > 0) //Data available in GPS Serial
    //Read and encode the NMEA data
    if (gps.encode(gpsSerial.read())) {
      //Check whether GPS location is valid & MQTT is connected
      if (gps.location.isValid() && mqtt.connected()) {
        //Latitude, Longtitude of the location is extracted
        lat_str = String(gps.location.lat() , 6);
        lng_str = String(gps.location.lng() , 6);
```

For tracking any logistics, the google maps link would be useful than providing latitude and longitude details. If you have latitude and longitude details, then the google maps link can be easily generated using https://www.google.com/maps/search/?api=1&query=<latitude>, <longitude> where <latitude> and <longitude> are to be replaced with latitude and longitude details, respectively. So, a google maps link is generated and stored in the loc _ link variable using the latitude and longitude details stored in the lng _ str and lng _ str variables. Next, the google maps link stored in the loc _ link variable is converted to a character array and stored in the charLocation variable.

.publish() is used for publishing the google maps link stored in the charLocation to Adafruit feed. If the publishing is successful, then it returns True; else it returns False. If the publishing is successful, then the program execution is paused for 2,000 milliseconds as we cannot upload more than 30 times per minute to a feed in Adafruit IO Free account. You can decrease the delay time if you are having an Adafruit IO+ account as it allows you to upload 60 data points per minute. If you exceed the upload speed limit to a feed, then you'll be seeing a warning as shown in Figure 8.57. So, give proper delay after uploading data to a feed.

FIGURE 8.57 Throttle warning

```
      //Google Maps Location link using the latitude and longitude
      loc_link = "https://www.google.com/maps/
search/?api=1&query="+lat_str+","+lng_str;

      Serial.println(loc_link); //Print the Google Maps link
      loc_link.toCharArray(charLocation, 100); //Convert string Google maps
link to character array

      //Publish the location link to the Adafruit
      if (feed.publish(charLocation)) {
        Serial.println("Maps Link Published Successfully");
        delay(2000);
```

```
        } else
          Serial.println("Maps Link Publishing Failed");
      }
    }
}
```

Finally, upload the code to NodeMCU after selecting the appropriate board and COM port. After successfully uploading the code, open the serial monitor and change the baud rate to 115200 to see the location updates and acknowledgements of publishing the data to Adafruit.

Next, open the *location* feed in your Adafruit IO Account and see the google maps link getting updated every 2,000 milliseconds as shown in Figure 8.58. Complete information regarding a data point can be seen by clicking the *View* next to that data point which will open a window as shown in Figure 8.59.

For tracking a logistic, connect the entire NodeMCU and GPS module setup to a portable power source and attach it to that logistic. Now, you can see the live location updates of that logistic in the Adafruit feed remotely. You can also build a mobile application that shows the live location of the logistic on a GUI (Graphical User Interface) using the data in the Adafruit feed.

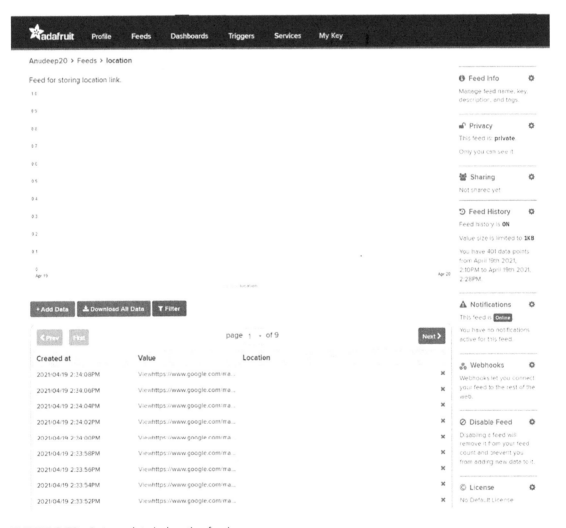

FIGURE 8.58   Data update in *location* feed

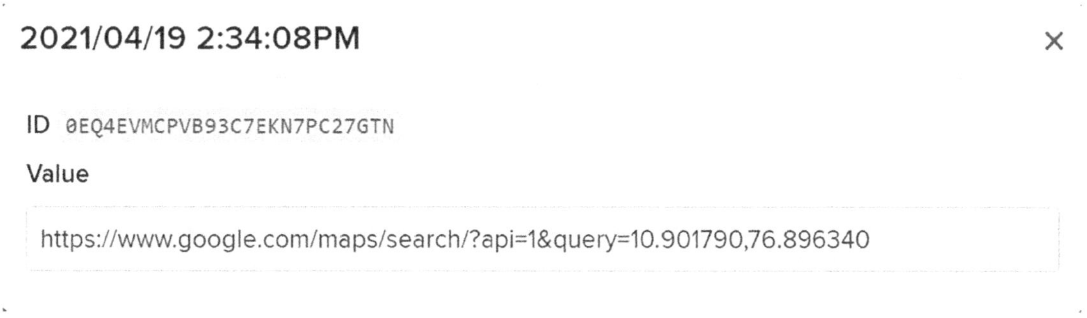

2021/04/19 2:34:08PM   ✕

ID `0EQ4EVMCPVB93C7EKN7PC27GTN`

Value

> https://www.google.com/maps/search/?api=1&query=10.901790,76.896340

FIGURE 8.59   Information in a datapoint

8.15 TRY IT

This is a small exercise to test your skills and understanding. In this exercise, you'll build the same logistics tracker without using MQTT and cloud but using a local webpage. By using the knowledge of webpages from the previous project, build a local webpage containing the latitude, longitude, google maps link, date and time as shown in Figure 8.60. The details have to be updated automatically according to the NMEA data from the GPS module. Try to do it on your own. If you are unable to solve it, don't worry! The below code is the solution for it. Explanation of the code is not required as it is just similar to the previous code.

8.15.1 Arduino IDE code

```
#include <TinyGPS++.h>
#include <SoftwareSerial.h>
#include <ESP8266WiFi.h>
TinyGPSPlus gps;
SoftwareSerial gpsSerial(D5, D6);
String ssid = "YOUR WIFI NAME";
String password = "YOUR WIFI PASSWORD";
int year , month , date, hour , minute , second;
String date_str , time_str , lat_str , lng_str, loc_link;
String web;
int isPM;

WiFiServer server(80);

void setup() {
  Serial.begin(115200);
  gpsSerial.begin(9600);
  Serial.println();
  Serial.print("Connecting to ");
  Serial.println(ssid);
  WiFi.begin(ssid, password);
  while (WiFi.status() != WL_CONNECTED) {
    delay(500);
```

```
    Serial.print(".");
  }
  Serial.println("");
  Serial.println("WiFi connected");
  server.begin();
  Serial.println("Server started");
  Serial.println(WiFi.localIP());
}

void loop() {
  while (gpsSerial.available() > 0)
    if (gps.encode(gpsSerial.read())) {
      if (gps.location.isValid()) {
        lat_str = String(gps.location.lat() , 6);
        lng_str = String(gps.location.lng() , 6);
        loc_link = "https://www.google.com/maps/
search/?api=1&query="+lat_str+","+lng_str;
      }
      if (gps.date.isValid()) {
        date_str = "";
        date = gps.date.day();
        month = gps.date.month();
        year = gps.date.year();
        if (date < 10)
          date_str = '0';
        date_str += String(date);
        date_str += " / ";

        if (month < 10)
          date_str += '0';
        date_str += String(month);
        date_str += " / ";
        if (year < 10)
          date_str += '0';
        date_str += String(year);
      }
      if (gps.time.isValid()) {
        time_str = "";
        hour = gps.time.hour();
        minute = gps.time.minute();
        second = gps.time.second();
        minute = (minute + 30);
        if (minute > 59) {
          minute = minute - 60;
          hour = hour + 1;
        }
        hour = (hour + 5) ;
        if (hour > 23)
          hour = hour - 24;
        if (hour >= 12)
          isPM = 1;
        else
          isPM = 0;
        hour = hour % 12;
        if (hour < 10)
          time_str = '0';
```

```
          time_str += String(hour);
          time_str += " : ";
          if (minute < 10)
            time_str += '0';
          time_str += String(minute);
          time_str += " : ";
          if (second < 10)
            time_str += '0';
          time_str += String(second);
          if (isPM == 1)
            time_str += " PM ";
          else
            time_str += " AM ";
      }
   }

  WiFiClient client = server.available();
  if (!client) {
    return;
  }

  web = "HTTP/1.1 200 OK\r\nContent-Type: text/html\r\n\r\n <!DOCTYPE html>
<html>";
  web += "<head> <title>Logistics Tracker</title> </head>";
  web += "<body> <h1> LOGISTICS TRACKER </h1>";
  web += "<h3> Location Details </h3>";
  web += "<table> <tr> <th>Latitude:</th>";
  web += "<td>";
  web += lat_str;
  web += "</td> </tr> <tr> <th>Longitude:</th> <td>";
  web += lng_str;
  web += "</td> </tr> <tr> <th>Maps Link:</th> <td>";
  web += "<a href=\""+loc_link+"\">Google Maps Link</a>";
  web += "</td> </tr> <tr> <th>Date:</th> <td>";
  web += date_str;
  web += "</td></tr> <tr> <th>Time:</th> <td>";
  web += time_str;
  web += "</td>   </tr> </table> ";

  web += "</body> </html>";

  client.print(web);
  delay(100);
}
```

Don't forget to change the WIFI _ SSID and WIFI _ PASSWORD with your WiFi SSID and password respectively before uploading the code. Finally, upload the code after selecting the appropriate board and COM port. After successfully uploading the code, open the serial monitor and select the suitable baud rate to see the data printed on the serial monitor.

8.15.2 Result

After the NodeMCU's local IP address gets printed, note it down and enter it in a browser to see the logistics tracker webpage as shown in Figure 8.60.

LOGISTICS TRACKER

Location Details

Latitude:	10.901790
Longitude:	76.896340
Maps Link:	Google Maps Link
Date:	19/04/21
Time:	04:48:23 PM

FIGURE 8.60   Logistics tracker using a webpage

You can find all the resources of this project at this Github link (https://github.com/anudeep-20/30IoTProjects/tree/main/Project%208).

In this project, you have learnt to build a logistics tracker from scratch using NodeMCU, GPS module and Adafruit which can be used to track logistics remotely. The same concept can be extended in building live train tracking or live bus tracking systems that can be useful for the passengers to track the location of their train or bus. You have also learnt about GPS, NMEA data and the usage of u-center software in this project. In the next project, you'll be building a system using PIR sensor and Adafruit cloud that can remotely monitor the movement inside a room or in a constrained space.

Let's GO to the next project!

Motion Detector Using PIR Sensor

9

This will be an interesting project and each of us can easily connect to an application scenario through this project. PIR sensor is one of the most useful sensors, and its application possibilities are enormous. One of its major applications is to detect the motion of a human and control the electrical equipment accordingly. The fans or lights can be switched on only when the human presence is detected. This would certainly make the home intelligent as well as cost effective. It is time that the reader can go through the project to understand things in detail. Happy learning!

9.1 MOTIVE OF THE PROJECT

In this project, you'll use a PIR sensor to detect movement inside a room or in a constrained space and log the time in an Adafruit feed. Motion detection can be employed in a lot of places like automatic control of electronic appliances (fan, light) inside a room depending on the human presence, remote monitoring of a room or a constrained space without using a camera and much more.

9.2 PIR SENSOR

PIR sensor is the acronym of passive infrared sensor and also regarded as pyroelectric sensor or infrared motion sensor. They are mainly used to sense motion and detect whether a human has moved in or out of the sensor's range. PIR sensors are more complicated compared to the sensors used in the previous projects as multiple factors affect their input and output.

All objects with a temperature above absolute zero (−273.15°C or 0 K) emit heat energy in the form of infrared radiation (IR). The hotter an object is, the higher the radiation it emits. PIR sensors are basically made of pyroelectric sensor which can detect levels of infrared radiation. The pyroelectric sensor itself has two slots in it, and each slot is made up of a special material that is sensitive to IR. When the sensor is idle or when there is no movement around the sensor, both slots of the pyroelectric sensor detect the same amount of IR emitting from the room or walls or outdoors. But, when a warm body like a human or an animal passes by the sensor, it first intercepts one half of the sensor which causes a positive differential change between the two halves, and when the warm body leaves the sensing area, the reverse happens thereby the sensor generates a negative differential change as shown in Figure 9.1. These change pulses are detected as motion by the PIR sensor.

In almost all the PIR sensors, the pyroelectric sensor is covered with a specially designed lens known as the Fresnel lens. It is a convex lens that is used to increase the detection area of the pyroelectric sensor by converging the IR rays around the sensor on to it.

DOI: 10.1201/9781003147169-9

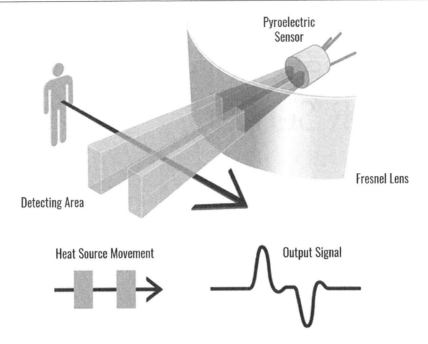

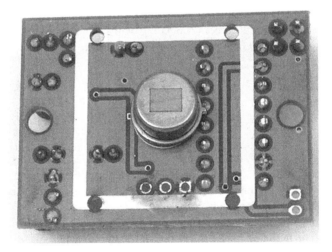

FIGURE 9.1   Working of PIR sensor

FIGURE 9.2   Pyroelectric sensor

For most of the IoT projects that need to detect the motion of people or to detect whether a person has entered an area or not, the HC-SR501 PIR sensor is used. The HC-SR501 PIR sensor is made up of a pyroelectric sensor, a Fresnel lens and a bunch of supporting electronic components. Figure 9.2 shows the pyroelectric sensor beneath the Fresnel lens, and Figure 9.3 shows the HC-SR501 PIR sensor with a Fresnel lens on top of it.

All the supporting circuitry of the HC-SR501 PIR sensor will be at the backside of the pyroelectric sensor as shown in Figure 9.4. Similar to most of the sensors, the HC-SR501 PIR sensor also has three pins Vcc, Output and Ground that can be interfaced with a microcontroller. It also has a voltage regulator so that it can be powered by any DC voltage between 4.5 V and 12 V, but typically 5 V is used. Also, it works normally even when powered with a 3.3 V DC voltage.

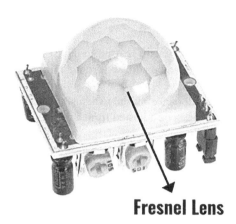

Fresnel Lens

FIGURE 9.3   HC-SR501 PIR sensor

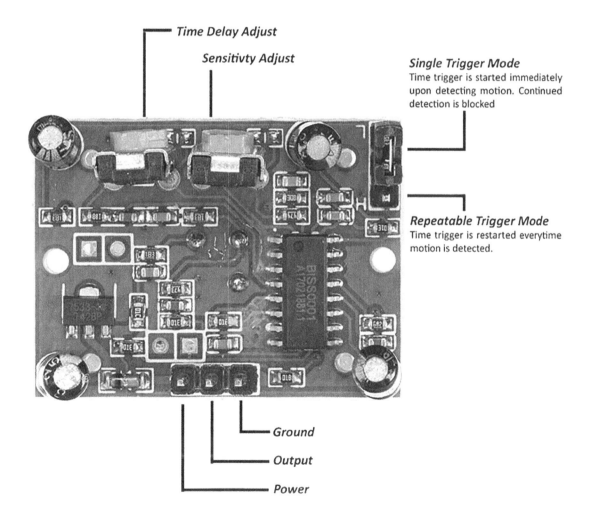

Time Delay Adjust

Sensitivty Adjust

Single Trigger Mode
Time trigger is started immediately
upon detecting motion. Continued
detection is blocked

Repeatable Trigger Mode
Time trigger is restarted everytime
motion is detected.

Ground

Output

Power

FIGURE 9.4   Supporting circuitry of HC-SR501 PIR sensor

On the HC-SR501 PIR sensor, there are two onboard potentiometers that can be used to adjust sensitivity and time.

- Time Delay Adjust – This potentiometer is used to adjust the time for which the output will remain HIGH after detection. It can be adjusted from a minimum of 3 seconds to a maximum of 300 seconds.
- Sensitivity Adjust – This potentiometer is used to adjust the maximum distance up to which the sensor can detect the motion. It can be adjusted from a minimum distance of 3 m to a maximum distance of approximately 7 m.

The HC-SR501 PIR sensor has three additional pins along with a jumper which can be seen at the top right corner in Figure 9.4. According to the placement of the jumper (used for shorting two pins) on these pins, it offers two different modes.

- L (Single Trigger Mode) – In this position, the HC-SR501 will output HIGH only when movement is detected and will stay HIGH for the period set by the Time adjust potentiometer. Figure 9.5 shows the jumper arrangement for the single trigger mode, and Figure 9.6 shows the input and output timing diagram of the PIR sensor in the single trigger mode.

FIGURE 9.5   Single trigger mode (L)

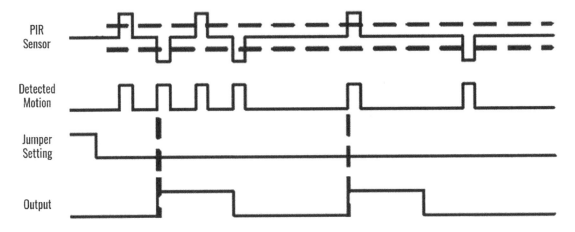

FIGURE 9.6   Timing diagram of single trigger mode

- H (Repeatable Trigger Mode) – In this position, the HC-SR501 will continue to output HIGH as long as the sensor continues to detect movement. Figure 9.7 shows the jumper arrangement for the repeatable trigger mode, and Figure 9.8 shows the input and output timing diagram of the PIR sensor in the repeatable trigger mode.

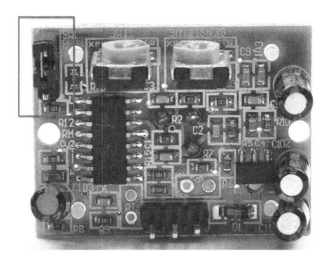

FIGURE 9.7   Repeatable trigger mode (H)

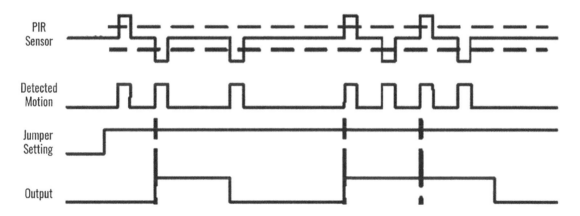

FIGURE 9.8   Timing diagram of repeatable trigger mode

For most real-time IoT applications, repeatable trigger mode (jumper in H position) is a better choice. But always try to test both the modes and choose a mode suitable for your application.

9.3 HARDWARE REQUIRED

Note: In this project, the usage of an LED is optional. It is only used to let the user know that movement is detected by the PIR sensor.

TABLE 9.1   Hardware required for the project

HARDWARE	QUANTITY
NodeMCU	1
HC-SR501 PIR sensor	1
Resistor (220 Ω)	1
LED	1
Jumper wires	As required

9.4 CONNECTIONS

For implementing this project, you can refer to the list of components required from Table 9.1, followed by Table 9.2, which presents the pin-wise connections. Also, the connection diagram is presented in Figure 9.9, which shall help you to get a visualization of the connections.

TABLE 9.2   Connection details

NODEMCU	HC-SR501 PIR SENSOR	RESISTOR (220 Ω)	LED
3V3	VCC		
D6	OUT		
D4			+ve pin (long pin)
		Pin 1	−ve pin (short pin)
GND	GND	Pin 2	

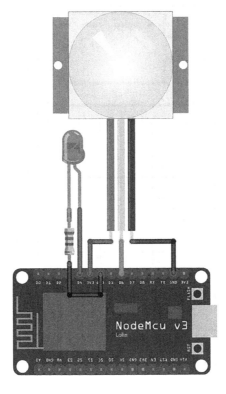

FIGURE 9.9   Connection layout

9.5 ADAFRUIT FEED CREATION

After successfully connecting the components, create a new feed *pir* in your Adafruit account as explained in the previous project. While creating the new feed in Adafruit, fill in the name and description details as shown in Figure 9.10. After successfully creating the *pir* feed, it will be added to the list of feeds in your account as shown in Figure 9.11.

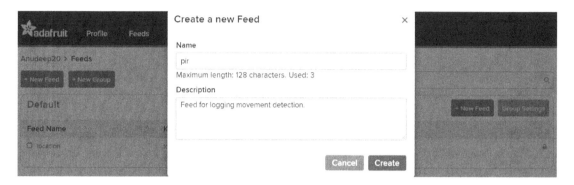

FIGURE 9.10   New feed creation

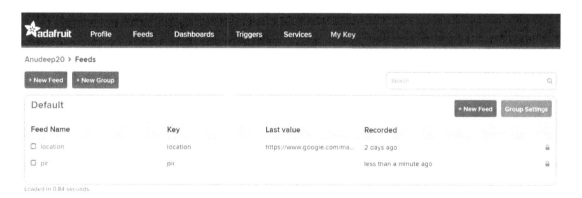

FIGURE 9.11   *pir* feed created

9.6 ARDUINO IDE CODE

ESP8266WiFi.h library is used for connecting the NodeMCU to the WiFi and the Adafruit _ MQTT _ Client.h library is used for connecting the NodeMCU to Adafruit. For connecting the NodeMCU to WiFi and Adafruit, the same code used in the previous project is used. The only change is that in the previous project we have uploaded the data to the *location* feed but in this project, we are uploading the data to *pir* feed. So, change it accordingly in the second argument of Adafruit _ MQTT _ Publish(). In the previous code, IO _ USERNAME"/feeds/location" is used which has to be changed to IO _ USERNAME"/feeds/pir".

```
#include <ESP8266WiFi.h> //Library for WiFi functionalities
#include <Adafruit_MQTT_Client.h> //Library for Adafruit MQTT Client

//Adafruit Details
#define server "io.adafruit.com"
#define port 1883
#define IO_USERNAME "AIO_USERNAME"
#define IO_KEY       "AIO_KEY"

String ssid = "YOUR WIFI NAME"; //SSID of your WiFi
String password = "YOUR WIFI PASSWORD"; //Password of your WiFi

WiFiClient esp; //Create a WiFi Client

//Creation of Adafruit MQTT Client
Adafruit_MQTT_Client mqtt(&esp, server, port, IO_USERNAME, IO_KEY);
//Feed variable for accessing data in Adafruit Feed
Adafruit_MQTT_Publish feed = Adafruit_MQTT_Publish(&mqtt, IO_USERNAME"/feeds/
pir");
```

PIR and LED are two variables used to store the digital pins to which the PIR sensor and LED are connected to the NodeMCU respectively. PIR _ val is another variable used to store the output from the PIR sensor. NodeMCU's serial is initialized at 115200 baud rate using Serial.begin(). As we are reading the values from the PIR sensor, the direction of PIR is set as INPUT. Similarly, as we are writing voltage on the LED, its direction is set as OUTPUT.

```
int PIR = D6; //Initialize the PIR output pin
int LED = D4; //Initialize the LED pin
int PIR_val = 0; //Used for storing PIR output value

void setup()
{
  Serial.begin(115200); //Initialization of NodeMCU Serial
  pinMode(PIR, INPUT); //Defining the direction of PIR sensor
  pinMode(LED, OUTPUT); //Defining the direction of LED

  //Connecting NodeMCU to WiFi
  Serial.println("");
  Serial.print("Connecting to ");
  Serial.println(ssid);
  WiFi.begin(ssid, password);
  while (WiFi.status() != WL_CONNECTED)
  {
    delay(500);
    Serial.print(".");
  }
  Serial.println("");
  Serial.println("WiFi connected");
  //Print the local IP address of NodeMCU
  Serial.println(WiFi.localIP());

  //Connecting to MQTT
  Serial.print("Connecting to MQTT");
  while (mqtt.connect())
```

```
  {
    Serial.print(".");
  }
}
```

`mqtt.connected()` is used to check whether the NodeMCU (MQTT client) is connected to Adafruit (MQTT broker) or not. If the NodeMCU is connected to Adafruit, then the output from PIR is read using `digitalRead()` and stored in the `PIR_val`. If the PIR sensor detects any movement, then it'll output HIGH. This will switch ON the LED and publish `Movement Detected` to the Adafruit feed. If no movement is detected, then the LED will be switched OFF.

One of the main objectives of this project is to know the time at which movement is detected. But we are not uploading the timestamp to the Adafruit feed when movement is detected. This is because Adafruit automatically logs time for every published data point as shown in Figure 9.12; hence it is redundant to upload the time stamp again.

Created at	Value	Location	
2021/04/21 11:01:42PM	Movement Detected		✕

FIGURE 9.12 Data point along with the timestamp

```
void loop()
{
  if(mqtt.connected()) {
    PIR_val = digitalRead(PIR); //Read the PIR output

    if(PIR_val) { //If movement is detected
      Serial.println("Movement Detected.");
      digitalWrite(LED, HIGH);

      //Publish data to the Adafruit feed
      if(feed.publish("Movement Detected")){
        Serial.println("Successfully data is Uploaded to the Adafruit.\n");
        delay(2000);
      } else
        Serial.println("Sorry, data cannot be Uploaded.\n");

    } else { //If movement is not detected
      Serial.println("No Movement Detected.");
      digitalWrite(LED, LOW);
    }
  }
}
```

Before uploading the code, don't forget to change your Adafruit username, Adafruit IO key in the place of `AIO_USERNAME` and `AIO_KEY`, respectively. Also, change the `WIFI_SSID` and `WIFI_PASSWORD` with your WiFi name and password, respectively.

Finally, upload the code to NodeMCU after selecting the appropriate board and COM port. After successfully uploading the code, open the serial monitor and change the baud rate to 115200 to see the acknowledgements of publishing the data to Adafruit.

Did you know? Even if you don't create the feed (skip Section 9.5) and directly upload the code, the publishing of data will not fail. If the feed to which you are trying to upload is not created in your account, then the Adafruit IO automatically creates it and publishes the data into it. Try it on your own and see the result.

Now, you can place this entire setup inside a room or in a constrained space where monitoring is required and then open the *pir* feed in your Adafruit IO Account as shown in Figure 9.13 to remotely monitor any movement within the detection range of the PIR sensor.

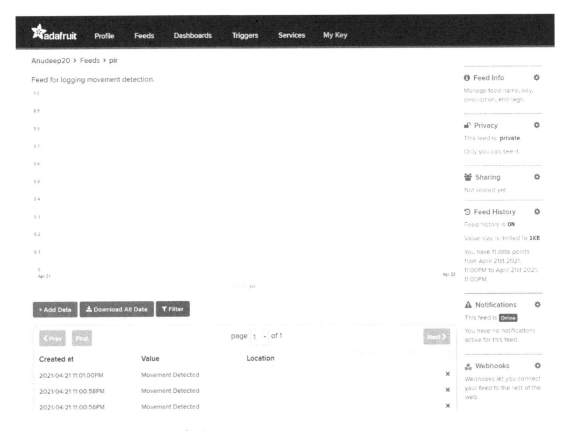

FIGURE 9.13   Data update in *pir* feed

You can find all the resources of this project at this Github link (https://github.com/anudeep-20/30IoTProjects/tree/main/Project%209).

In this project, you have learnt to build a motion detector using a PIR sensor and Adafruit cloud that can be used to remotely monitor movement inside a room or within a constrained space. The same concept can be extended in building a system that can control the electronic appliances inside a room depending on the human presence. You have also learnt about the working of the PIR sensor and the different modes of operation in it. In the next project, you'll build a system that can detect the overflow of liquid in a storage tank and monitor the level of liquid in the storage tank at any point of time using Adafruit Dashboard.

Let's build the next one!

Overflow Detection Using Ultrasonic Sensor

10

The readers are now getting exposed to more and more practical applications. This project is another practical application of preventing the overflow of liquid in a storage tank. Overflow of water from an overhead tank is one of the major problems in the households of developing countries. It is also one of the major reasons for water wastage in households. We could certainly save thousands of litres of water per year if we could alert the user before the tank overflows. With a motive to save water or any other liquid from getting overflown, this project is built.

10.1 MOTIVE OF THE PROJECT

In this project, you'll use an ultrasonic sensor to detect the overflow of liquid in a storage tank and alert the user before an overflow occurs. You'll also remotely monitor the level of liquid in the storage tank using Adafruit feed and Adafruit Dashboard. This project is mainly useful for the storage tanks where the level of liquid will vary and won't be stagnant. Not only in storage tanks but the same concept can be applied in the places where the level of "something" needs to be tracked.

10.2 ULTRASONIC SOUND WAVES

Sound is a vibration that propagates as an acoustic wave through a transmission medium (solid, liquid or gas). Depending on the frequency of transmission, the sound waves are classified into three types. Frequency is measured in Hertz (Hz) which is defined as the number of cycles per second (/s).

- Infrasonic Sound – Sound waves with a frequency of less than 20 Hz.
- Audible Sound – Sound waves with a frequency between 20 Hz and 20,000 Hz.
- Ultrasonic Sound – Sound waves with a frequency greater than 20,000 Hz.

Humans can only hear sounds with frequency in the audible sound range (between 20 Hz and 20,000 Hz). So, the infrasonic and ultrasonic sound waves are inaudible to the human ear but can be heard by some animals as shown in Figure 10.1.

DOI: 10.1201/9781003147169-10

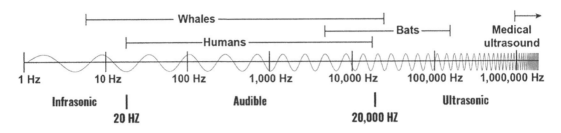

FIGURE 10.1 Classification of sound waves

Ultrasonic sound waves are not different from audible sound waves in their physical properties. They are used in many different fields like non-destructive testing in the manufacturing industry, ultrasound imaging in medicine, ultrasonic fingerprint sensor in electronics and many more. It is also used by the mammal species like bats to locate obstacles and prey.

A more common use of ultrasonic sound waves is in underwater range finding also known as *Sonar*. An ultrasonic pulse is generated by a transmitter in a particular direction, if there is any obstacle in the path of the pulse then it reflects back to the transmitter which will be detected by a detector. The distance of the obstacle from the transmitter is calculated by measuring the time difference between the transmission and receiving the pulse. Sonar is used in many places like finding the depth of the sea bed from the surface, detecting any obstacles in the path of submarines and many more.

In this project, we'll use the same phenomenon used by the sonar in finding the level of liquid in a tank. Sonar generates and receives the ultrasonic waves inside the water. But, in our project, we'll generate the ultrasonic waves above the liquid which will get reflected on the surface of the liquid. The reflected waves are detected at the transmitter to find the level of empty space in the tank as shown in Figure 10.2.

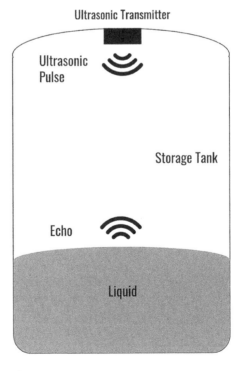

FIGURE 10.2 Liquid level detection

10.3 ULTRASONIC SENSOR

There are many types of ultrasonic sensors available in the market for different use cases and with varied accuracies. In this project, we'll use a low-cost, low-power consuming ultrasonic sensor HC-SR04 as shown in Figure 10.3, which is commonly used in many IoT projects.

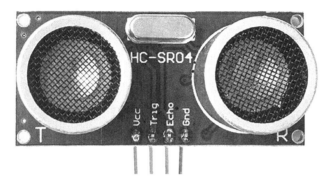

FIGURE 10.3   Ultrasonic sensor (HC-SR04)

HC-SR04 ultrasonic sensor consists of two ultrasonic transducers of which one acts as a transmitter of ultrasonic sound pulses and the other acts as a receiver of these pulses. Similar to any other sensor, the HC-SR04 also has a VCC and a GND pin for powering it. In addition to these two pins, it also has a TRIG and an ECHO pin. The TRIG pin is used for triggering the ultrasonic sound pulses. The ECHO pin outputs a pulse when it receives the reflected ultrasonic sound pulses. The length of this pulse is directly proportional to the time taken to receive the reflected ultrasonic sound pulses. This pulse from the ECHO pin is read by the microcontroller to calculate the distance between the ultrasonic sensor and the reflecting obstacle. For using the HC-SR04 ultrasonic sensor, it is important to know its specifications which are tabulated in Table 10.1.

TABLE 10.1   Specifications of HC-SR04 ultrasonic sensor

WORKING VOLTAGE	DC 5 V
Working current	15 mA
Working frequency	40 Hz
Min range	2 cm
Max range	400 cm
Measuring angle	15 degree
Trigger input signal	10 µs (microseconds) TTL pulse
Echo output signal	Input TTL lever signal and the range in proportion

10.4 WORKING OF ULTRASONIC SENSOR

When the NodeMCU sends a 10 µs HIGH pulse to the TRIG pin of HC-SR04, it generates an 8 cycle ultrasonic burst at 40 kHz. There is a reason for this 8 cycle pattern as it makes the device unique and allows the receiver to differentiate from the ambient ultrasonic noise.

The 8 cycle ultrasonic pulse continues to travel through the air until blocked by any obstacle as shown in Figure 10.4. Meanwhile, the ECHO pin of the HC-SR04 goes HIGH to start forming the beginning of the ECHO-back signal. If the 8 cycle ultrasonic pulse is not reflected, then the ECHO signal will timeout after 38 ms (milliseconds) and goes LOW as shown in Figure 10.5. So, a 38 ms HIGH pulse from the ECHO pin indicates there is no obstacle within the range of the HC-SR04 ultrasonic sensor.

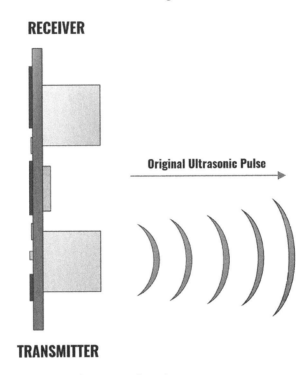

FIGURE 10.4   Ultrasonic pulse when there is no obstacle

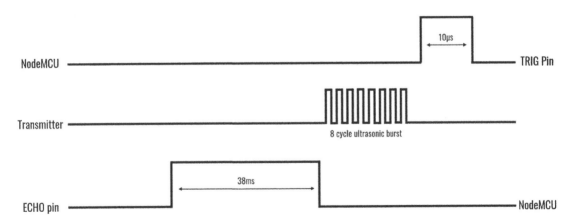

FIGURE 10.5   Pulses generated from NodeMCU and ultrasonic sensor when there is no obstacle

If the transmitted 8 cycle ultrasonic pulse is blocked by an obstacle within the range of the ultrasonic sensor, then this pulse will get reflected. The reflected pulse will be received by the ultrasonic sensor receiver as shown in Figure 10.6. The ECHO pin goes low as soon as the ultrasonic sensor receives the reflected pulse. This will produce a HIGH pulse as shown in Figure 10.7 whose width varies from 150 μs to 25 ms depending on the time it took for the 8 cycle ultrasonic pulse to be received by the ultrasonic sensor.

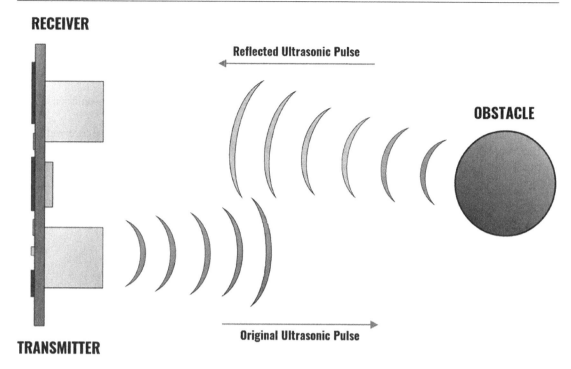

FIGURE 10.6   Ultrasonic pulse where there is an obstacle

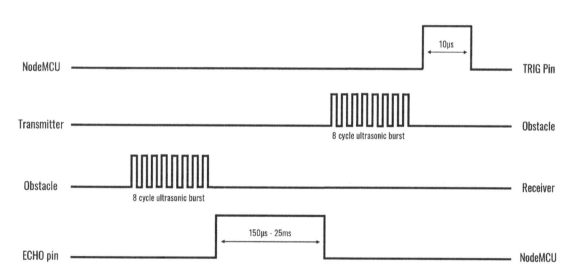

FIGURE 10.7   Pulses generated from NodeMCU and ultrasonic sensor when there is an obstacle

The width of the HIGH pulse generated by the ECHO pin is directly proportional to the time taken for the round trip of 8 cycle ultrasonic pulse. So, to calculate the distance between the ultrasonic sensor and the obstacle, we'll make use of a simple physics formula learnt in high school.

$$Speed = \frac{Distance}{Time} \tag{10.1}$$

Here,

>   *Speed* – It is the speed of the ultrasonic waves in the air which is equal to the speed of the sound in the air, i.e., 34,300 cm/s.
>   *Distance* – It is the unknown term we need to find.
>   *Time* – It is half of the width of the HIGH pulse received from the ECHO pin of the ultrasonic sensor.

Note: We are talking half the width of the HIGH pulse because the width of the HIGH pulse received from the ECHO pin is proportional to the time taken for the round trip of the ultrasonic waves, but we need only the time taken for one way to find the distance between the ultrasonic sensor and the obstacle.

$$34,300 \; \text{cm/s} = \frac{Distance}{Time_{round\text{-}trip} \, / \, 2} \tag{10.2}$$

$$Distance \, (\text{in cm}) = 17,150 * Time_{round\text{-}trip} \, (\text{in second}) \tag{10.3}$$

$$Distance \, (\text{in meter}) = 171.5 * Time_{round\text{-}trip} \, (\text{in second}) \tag{10.4}$$

Either one among the Equations 10.3 or 10.4 can be used to find the distance between the ultrasonic sensor and the obstacle.

10.5 BUZZER

Buzzer is an audio signalling device that is typically used in timers, alarms, computers and many more electronic products. It is usually powered by a DC voltage. Similar to an LED which is used to give visual alerts, Buzzer can be used to give audio alerts in our projects. There are many types of buzzers available in the market, but the most common ones are the piezo buzzer and electromagnetic buzzer.

10.5.1 Piezo Buzzer

Piezo buzzers as shown in Figure 10.8 operate on the principle of the piezoelectric effect. The main component of these buzzers is a piezoelectric element which is composed of a piezoelectric ceramic disc and a metal plate. The piezoelectric ceramic disc has electrodes connected to it which expands and contracts diametrically when an alternating current is passed through it. This produces vibrations in the piezoelectric element and generates sound of a particular frequency. Some of the piezo buzzers have a hollow cavity to produce an audible sound.

The range of operating voltage of piezo buzzers is very large and varies from one manufacturer to another. So, it is recommended to read the operating voltage of the piezo buzzer in the datasheet before usage. Usually, its operating voltage can be between 3 V and 220 V, and it is recommended to use a voltage above 9 V for a louder sound. The piezo buzzers are usually used for haptic feedback when a button is pressed or for a ring or for a beep. It is usually not preferred for continuous sound.

10.5.2 Electromagnetic Buzzer

An electromagnetic buzzer as indicated in Figure 10.9 is composed of an oscillator, vibration diaphragm, magnet, solenoid coil, etc. When power is supplied to the buzzer, the current generated by the oscillator

FIGURE 10.8   Piezo buzzers

FIGURE 10.9   Electromagnetic buzzer

is passed through the solenoid coil which generates a magnetic field. The vibration diaphragm under the interaction of the solenoid coil and magnet vibrates and sounds periodically.

The electromagnetic buzzers are further divided into active and passive buzzers. Both active and passive buzzers look almost the same. Usually, the active buzzers are taller compared to the passive buzzers, this is one of the ways to differentiate them other than by looking at the datasheet. Also, when both the buzzers are facing up, the buzzer with a green circuit board is a passive buzzer and the sealed one with vinyl glue is an active buzzer.

Internally, the main difference between the active and passive buzzers is an oscillator. The active buzzer has an oscillator inside and it can be powered with a DC source as it can convert direct current to pulse signal, whereas the passive buzzer has no oscillator inside and cannot be powered with a DC source.

The range of operating voltage of the electromagnetic buzzers is narrow compared to the range of operating voltage of the piezo buzzers. Usually, the operating voltage of electromagnetic buzzers will be between 1.5 V and 24 V. These buzzers are usually used for delivering continuous sounds like in audio, music and other equipment.

10.6 THE 5V ISSUE OF NodeMCU

The sensors and modules interfaced with NodeMCU in the previous projects will be working as expected at 3.3 V, but the ultrasonic sensor used in this project (HC-SR04) won't work properly at 3.3 V. It works as expected when it is connected to a voltage source equivalent to its operating voltage (i.e., 5 V). Unfortunately, NodeMCU doesn't own a dedicated 5 V output, then what to do?

The first method is to use an external 5 V voltage source for powering the HC-SR04 ultrasonic sensor. If a 5 V voltage source is not available then use a higher external voltage source (like a 9 V Battery) and a 5 V voltage regulator (like LM7805 IC) for powering the HC-SR04 ultrasonic sensor.

Presently in the market, there are many versions of NodeMCU available but the most popular and widely used ones are NodeMCU Amica and NodeMCU v3 (also known as NodeMCU Lolin). You can know your NodeMCU version by seeing the back of the NodeMCU board. Figure 10.10 shows the front and backside of NodeMCU Amica, and Figure 10.11 shows the front and backside of NodeMCU v3 or NodeMCU Lolin.

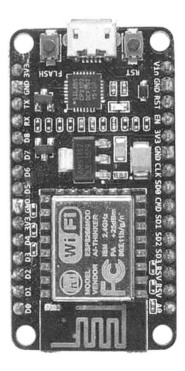

FIGURE 10.10 NodeMCU Amica

FIGURE 10.11   NodeMCU v3 (or NodeMCU Lolin)

There is another way of differentiating the two boards, when the power socket is kept towards you then the third pin from the top-left corner is a *VU pin* for NodeMCU v3 and *RSV pin* (reserved pin) for NodeMCU Amica.

Even though NodeMCU doesn't provide a dedicated 5 V pin, some of the NodeMCU pins are directly tied to the input 5 V power socket. So, these pins will provide the required 5 V output. On the NodeMCU Amica, the *VIN pin* is directly tied to the input power socket and provides 5 V output. On the NodeMCU v3 or NodeMCU Lolin, the *VU pin* provides the required 5 V output. The *VIN pin* on the NodeMCU v3 cannot be used for 5 V output as it fluctuates around 2.2 V output. It is important to remember that the output from the *VIN* or *VU* pin is not as reliable as the 5 V pin of Arduino Uno. The output depends on the power source to which the NodeMCU is connected. In this project, we'll be using the second method i.e, VIN or VU pin of NodeMCU to provide 5V input to the ultrasonic sensor.

10.7 HARDWARE REQUIRED

TABLE 10.2   Hardware required for the project

HARDWARE	QUANTITY
NodeMCU	1
HC-SR04 ultrasonic sensor	1
Buzzer	1
Jumper wires	As required

10.8 CONNECTIONS

You can refer to the list of components required from Table 10.2, followed by Table 10.3, which presents the pin-wise connections. Also, the connection diagram is presented in Figure 10.12, which shall help get a visualization of the connections.

TABLE 10.3   Connection details

NODEMCU	HC-SR04 ULTRASONIC SENSOR	BUZZER
VU or Vin	VCC	
D5	TRIG	
D6	ECHO	
D7		+ve pin
GND	GND	−ve pin

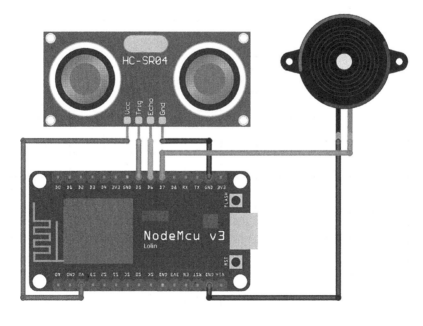

FIGURE 10.12   Connection layout

10.9 ADAFRUIT FEED CREATION

After successfully connecting the components, create a new feed *distance* in your Adafruit account as explained in the previous project. While creating the new feed in Adafruit, fill in the name and description details as shown in Figure 10.13. After successfully creating the *location* feed, it will be added to the list of feeds in your account as shown in Figure 10.14.

FIGURE 10.13   New feed creation

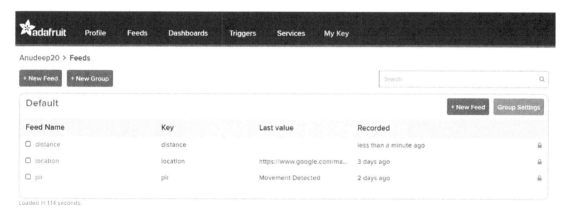

FIGURE 10.14   *distance* feed created

10.10 ADAFRUIT DASHBOARD CREATION

Other than Adafruit feeds, Adafruit IO also provides other features like Dashboard, Trigger and some additional services. Adafruit Dashboard is a GUI tool that can be used to visualize the data in a feed using GUI blocks.

To create a new dashboard, first sign in to your Adafruit IO Account. As soon as you sign in, the first section you'll see is the *Dashboard* section. In the Dashboard section, you'll see all the previously created dashboards in your account. By default, you should have no dashboards created in your account as shown in Figure 10.15.

FIGURE 10.15   Adafruit dashboards

For creating a new dashboard, click on the+*New Dashboard* button and fill in the *Name* and *Description* details of the dashboard and then click the *Create* button as shown in Figure 10.16. Filling *Description* details in the creation of a dashboard is optional. Adafruit allows using all the alphabets and almost all the special characters in naming a Dashboard. But, after you create the dashboard, a *Key* will be generated corresponding to the Dashboard *Name* using only the alphabets in the Dashboard *Name* and the – special character. After a dashboard is successfully created, it'll be added to the list of Dashboards in your account as shown in Figure 10.17.

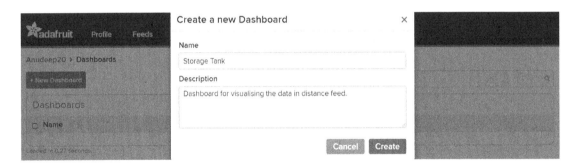

FIGURE 10.16   New dashboard creation

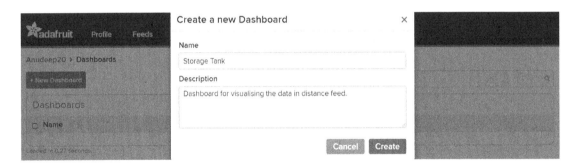

FIGURE 10.17   *Storage Tank* Dashboard created

Click on the newly created *Storage Tank* Dashboard which will open an empty layout in dark mode as shown in Figure 10.18. To change the settings of the dashboard or to add GUI blocks on the dashboard, click on the drop-down beside the gear symbol at the top right corner. In the drop-down, you can see a lot of settings like Edit Layout, Create New Block, Delete Dashboard and many more as shown in Figure 10.19.

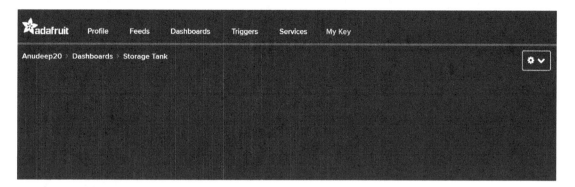

FIGURE 10.18 *Storage Tank* Dashboard

FIGURE 10.19 Dashboard settings

To add a GUI block for visualizing the data of the distance feed, click on *Create New Block* which will open a new window as shown in Figure 10.20. This window contains all the types of GUI blocks that can be added to the layout. For this project, we need a *Gauge* GUI block which is the 4th block in the first row.

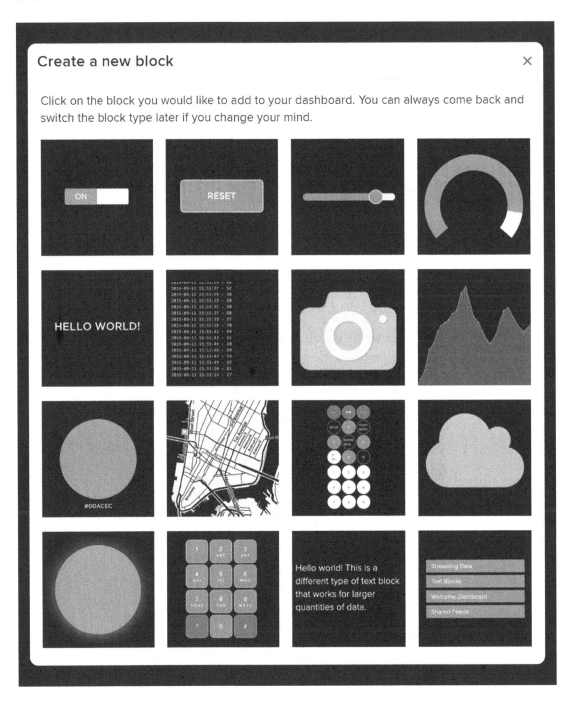

FIGURE 10.20   GUI blocks

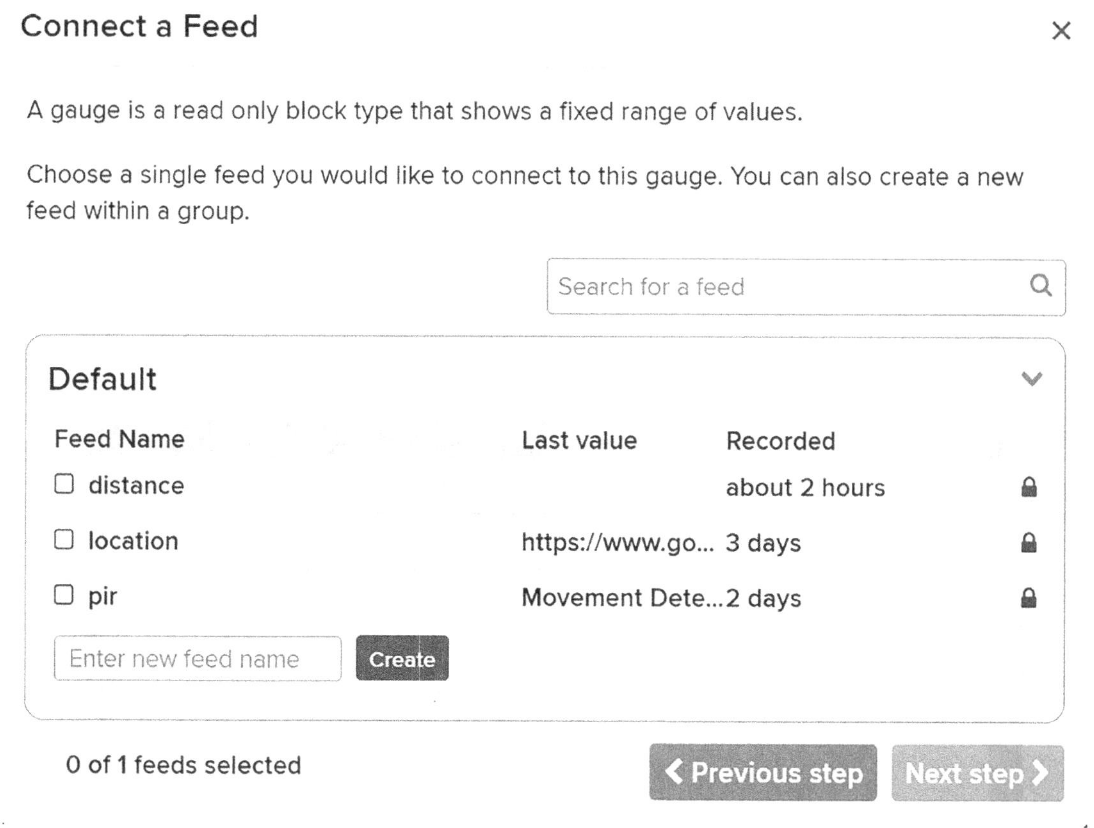

Connect a Feed   ✕

A gauge is a read only block type that shows a fixed range of values.

Choose a single feed you would like to connect to this gauge. You can also create a new feed within a group.

| Search for a feed | Q |

Default   ⌄

Feed Name	Last value	Recorded	
☐ distance		about 2 hours	🔒
☐ location	https://www.go...	3 days	🔒
☐ pir	Movement Dete...	2 days	🔒

| Enter new feed name | **Create** |

0 of 1 feeds selected   **❮ Previous step**   **Next step ❯**

FIGURE 10.21   Connect a feed

By clicking on the *Gauge* GUI block, it'll open a *Connect a Feed* window as shown in Figure 10.21 which will help in connecting the GUI block to a feed. In this project, we need to visualize the data of the *distance* feed. So, check the *distance* feed and click *Next Step* which will open a *Block Settings* window. Fill in all the details as shown in Figure 10.22 and click *Create Block* for creating the *Gauge* GUI block. The Block Title name is optional; the Gauge Min Value and the Gauge Max Value are the minimum and maximum range of the ultrasonic sensor, respectively. Gauge Label is the unit of the distance value that we are uploading to the feed. Low Warning Value and High Warning Value are optional. They are only used to change the colour of the Gauge from green to red if values below the Low Warning Value or values above the High Warning Value are uploaded to the feed.

After the block gets created successfully, it will be added to the layout as shown in Figure 10.23. To view the Layout in full screen click on the Dashboard settings which will show the layout as shown in Figure 10.24. To move the GUI block in the layout, click Edit Layout in the Dashboard settings.

The green/red level and the value at the centre of the Gauge GUI block update in real time as soon as a data point is uploaded to the *distance* feed. Remember that, the green/red level in the gauge is the distance of the ultrasonic sensor from the surface of the liquid. The higher the green level in the gauge, the safer is the tank from an overflow. If the green level in the gauge becomes less than 15, then there is a chance of overflow.

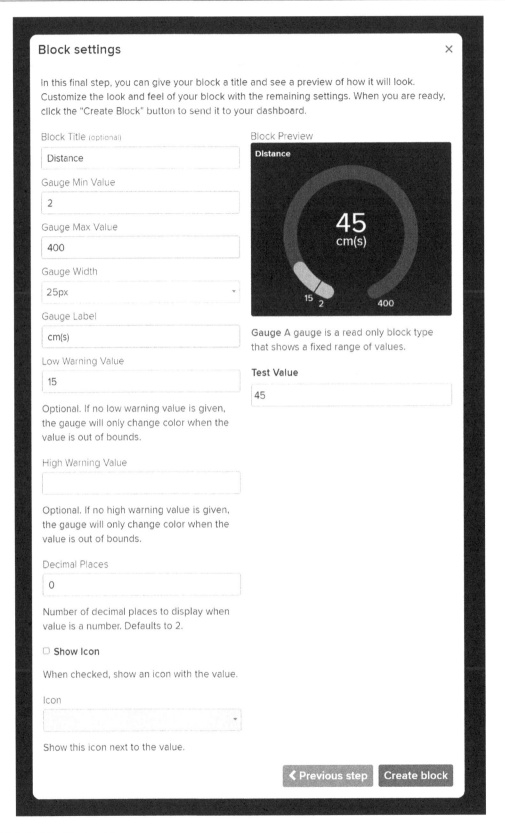

FIGURE 10.22   Block settings

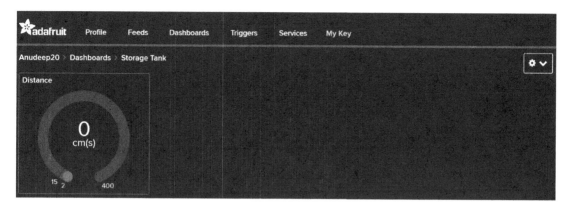

FIGURE 10.23 Gauge GUI block

FIGURE 10.24 *Storage Tank* Dashboard in full screen

10.11 ARDUINO IDE CODE

The code used for connecting the NodeMCU to WiFi and Adafruit is the same code used in the previous projects. The only change is that in the previous project we have uploaded the data to the *pir* feed but in this project, we'll upload the data to the *distance* feed. So, change it accordingly in the second argument of Adafruit_MQTT_Publish(). In the previous code, IO_USERNAME"/feeds/pir" is used which has to be changed to IO_USERNAME"/feeds/distance".

```
#include <ESP8266WiFi.h> //Library for WiFi functionalities
#include <Adafruit_MQTT_Client.h> //Library for Adafruit MQTT Client

//Adafruit Details
```

```
#define server "io.adafruit.com"
#define port 1883
#define IO_USERNAME   "AIO_USERNAME"
#define IO_KEY        "AIO_KEY"

String ssid = "YOUR WIFI NAME"; //SSID of your WiFi
String password = "YOUR WIFI PASSWORD"; //Password of your WiFi

WiFiClient esp; //Create a WiFi Client

//Creation of Adafruit MQTT Client
Adafruit_MQTT_Client mqtt(&esp, server, port, IO_USERNAME, IO_KEY);
//Feed variable for accessing data in Adafruit Feed
Adafruit_MQTT_Publish feed = Adafruit_MQTT_Publish(&mqtt, IO_USERNAME"/feeds/
distance");
```

trigPin and echoPin are used to store the digital pin to which the TRIG and ECHO pin of the ultrasonic sensor is connected to the NodeMCU, respectively. Similarly, buzzer is used for storing the digital pin to which the positive pin of the buzzer is connected.

```
//NodeMCU pin to which Trig pin of Ultrasonic Sensor is connected
const int trigPin = D5;
//NodeMCU pin to which Echo pin of Ultrasonic Sensor is connected
const int echoPin = D6;
const int buzzer = D7; //NodeMCU pin to which +ve pin of Buzzer is connected
```

duration is a long variable used to store the round trip time of the 8 cycle ultrasonic pulse in microseconds. It is preferred to use long variable type for the duration because the time in microseconds for the round trip of the pulse might overflow the int variable. distance stores the distance between the ultrasonic sensor and the surface of the liquid in the storage tank in centimetres. As the range of the ultrasonic sensor is 400 cm, int variable type can be used for distance.

```
long duration; //Stores the round trip time taken by the ultrasonic waves
//Stores the distance of any obstacle from the ultrasonic sensor
int distance;
```

NodeMCU's serial is initialized at 115200 baud rate using Serial.begin(). For triggering an 8 cycle ultrasonic pulse, the NodeMCU must send a signal to the TRIG pin. So, the direction of the trigPin is set as OUPUT. For obtaining the round trip time of the ultrasonic pulse, NodeMCU will read the signal from the ECHO pin. So, the direction of the echoPin is set as INPUT. Similarly, as we are writing voltage on the buzzer, its direction is set as OUTPUT.

```
void setup() {
Serial.begin(115200); //Initialisation of NodeMCU Serial
pinMode(trigPin, OUTPUT); //Defining the direction of TRIG pin
pinMode(echoPin, INPUT); //Defining the direction of ECHO pin
pinMode(buzzer, OUTPUT); //Defining the direction of buzzer

  //Connecting NodeMCU to WiFi
  Serial.println("");
  Serial.print("Connecting to ");
  Serial.println(ssid);
  WiFi.begin(ssid, password);
  while (WiFi.status() != WL_CONNECTED)
```

```
  {
    delay(500);
    Serial.print(".");
  }
Serial.println("");
Serial.println("WiFi connected");
Serial.println(WiFi.localIP());  //Print the local IP address of NodeMCU

//Connecting to MQTT
    Serial.print("Connecting to MQTT");
    while (mqtt.connect()) {
        Serial.print(".");
    }
}
```

For generating an 8 cycle ultrasonic pulse from the HC-SR04 ultrasonic sensor, we need to send a signal (a 10 µs HIGH pulse) to the `trigPin`. At any given point of time, we don't know the state of the `trig-Pin` (whether it's HIGH or LOW). So, to clear the state of the `trigPin`, we'll make it LOW for 2 µs before sending a 10 µs HIGH pulse.

```
void loop() {
  //Clear the trigPin condition
  digitalWrite(trigPin, LOW);
  delayMicroseconds(2);

  //Sets the trigPin HIGH for 10 microseconds for generating Ultrasonic burst
  digitalWrite(trigPin, HIGH);
  delayMicroseconds(10);
  digitalWrite(trigPin, LOW);
```

As soon as a 10 µs HIGH pulse is sent to the `trigPin`, the ultrasonic sensor sends an 8 cycle ultrasonic pulse. After the pulse is released by the transmitter of the ultrasonic sensor, the `echoPin` goes HIGH until the reflected waves are received by the ultrasonic sensor or will timeout after 38 ms.

   `pulseIn()` is an inbuilt function of Arduino IDE which is used to read a pulse (either HIGH or LOW) on a pin. For example, if the second argument of the `pulseIn()` function is given as HIGH, then the function waits for the pin (given in the first argument) to go from LOW to HIGH then starts timing and then waits for the pin to go LOW to stop the timing.

   The `pulseIn()` function will return the length of the pulse in microseconds or returns 0 if no complete pulse is received within the timeout (can be given as a third argument). `pulseIn()` function works perfectly on pulses from 10 microseconds to 3 minutes in length. So, the `pulseIn()` function is used for determining the time duration for which the `echoPin` is HIGH.

   After obtaining the round trip time of the ultrasonic pulse, calculate the distance between the ultrasonic sensor and the surface of the liquid in the storage tank using Equation 10.3 obtained at the end of Section 10.4.

```
  //Reads the echoPin to obtain the ultrasonic wave travel time in
microseconds
  duration = pulseIn(echoPin, HIGH);
  //Calculating the distance from the round trip time
  distance = duration * 0.034 / 2;
```

Any distance which is not in the range of the ultrasonic sensor (2 cm–400 cm) is considered as an invalid distance and is prevented from publishing to the Adafruit feed. If the distance between the ultrasonic sensor and the surface of liquid in the storage tank is less than 15 cm, then it is considered as the liquid is about to overflow and an alert is signalled using the buzzer. Finally, the distance value (in cms) is published to the Adafruit feed.

```
//Invalid distance
if(distance < 2 || distance > 400)
  return;

//About to overflow
if(distance <= 15)
  digitalWrite(buzzer, HIGH);
//Print the distance on the Serial Monitor
Serial.print("Distance: ");
Serial.print(distance);
Serial.println(" cm(s)");

//Publish data to the Adafruit feed
if(feed.publish(distance)){
  Serial.println("Successfully data is Uploaded to the Adafruit.\n");
  delay(2000);
} else
  Serial.println("Sorry, data cannot be Uploaded.\n");

digitalWrite(buzzer, LOW);
}
```

Before uploading the code, don't forget to change your Adafruit username, Adafruit IO key in the place of AIO_USERNAME and AIO_KEY, respectively. Also, change the WIFI_SSID and WIFI_PASSWORD with your WiFi name and password, respectively.

Finally, upload the code to NodeMCU after selecting the appropriate board and COM port. After successfully uploading the code, open the serial monitor and change the baud rate to 115200 to view the data printed on it in a readable format.

Now, you can place this entire setup at the top of the storage tank and open the *Storage Tank* Dashboard in your Adafruit IO Account as shown in Figure 10.25 to remotely monitor the level of the liquid in the storage tank at any point in time.

The Gauge GUI block in the *Storage Tank* Dashboard shows only the latest published value in the *distance* feed. To see the previously published values, you can open the *distance* feed as shown in Figure 10.26.

FIGURE 10.25 *Storage Tank* Dashboard

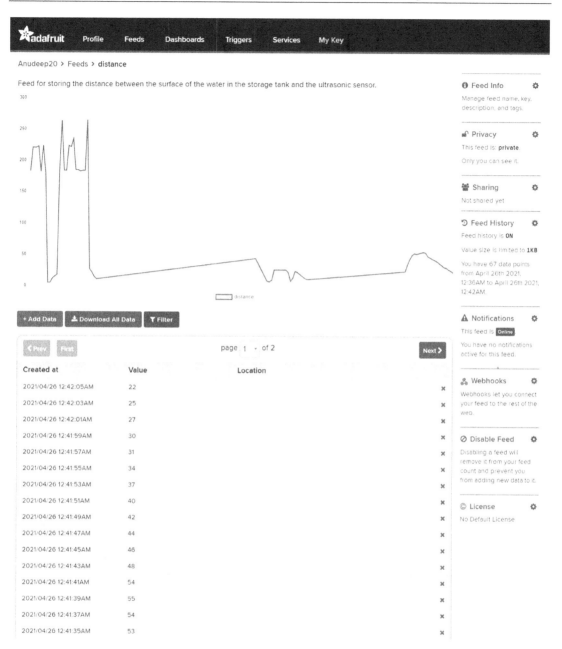

FIGURE 10.26 *distance* feed

You can find all the resources of this project at this Github link (https://github.com/anudeep-20/30IoTProjects/tree/main/Project%2010).

In this project, you have learnt to build a system that can detect the level of liquid in a storage tank using an ultrasonic sensor and alert the user before an overflow. You have also learnt to use Adafruit Dashboard for remotely monitoring the data in a feed. Ultrasonic sensors are not only used in IoT, and they also have a wide range of applications in the field of robotics. The usage of ultrasonic sensor learnt in this project can be used to build obstacle avoidance robots and many more. In the next project, you'll configure and pair HC-05 Bluetooth modules to communicate with each other.

Let's build the next project!

Pairing of Bluetooth Modules

11

The applications of Bluetooth are enormous, and they are used not only in consumer products but also in industrial applications. Bluetooth is used to transfer data by eliminating wired connections. If you remember, in one of the previous projects we have paired a Bluetooth module to the smartphone for transferring information wirelessly. As the smartphone has an intuitive GUI, you could pair the smartphone's Bluetooth to the HC-05 Bluetooth module easily. But how to pair two HC-05 Bluetooth modules to each other? The pairing of two Bluetooth modules has a lot of use cases. In our day-to-day life, we use a lot of Bluetooth devices like TWS (Truly Wireless) earphones. If you are using an older one, then the earphones pair with each other using the concept explained in this project.

11.1 MOTIVE OF THE PROJECT

In this project, you'll configure and pair two HC-05 Bluetooth modules to transfer data between each other. The same concept can be extended to the pairing of multiple Bluetooth modules. Pairing of Bluetooth modules is used in many IoT projects. Even, some of the projects in this book use this concept. If multiple microcontrollers or evaluation boards are in Bluetooth range (10 m range) and need to communicate with each other, then this concept can be employed.

11.2 MODES OF OPERATION OF HC-05 BLUETOOTH MODULE

HC-05 Bluetooth module is an easy-to-use Bluetooth Serial Port Protocol (SPP) module designed for transparent wireless serial connection. It has two different modes of operation.

- Data mode
- Command mode

In all the previous projects, the HC-05 Bluetooth module is used in its default mode (Data mode). But in this project, we'll explore the other mode (Command mode) of the HC-05 Bluetooth module for pairing it with another HC-05 Bluetooth module.

DOI: 10.1201/9781003147169-11

11.2.1 Data mode

Data mode is sometimes referred to as the trans-receiving mode. It is the default operating mode of the HC-05 Bluetooth module. When you buy a new HC-05 Bluetooth module from the market, you'll use it in Data mode.

This mode cannot be used for configuring or changing the settings of the HC-05 Bluetooth module. It can only be used to communicate or transfer data to other Bluetooth devices. The HC-05 Bluetooth module has an onboard LED as shown in Figure 11.1, which is used for knowing the status and operating mode of the Bluetooth.

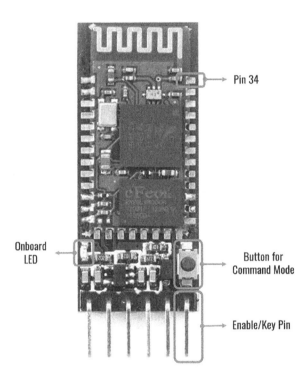

FIGURE 11.1   Components on HC-05 Bluetooth module

If the onboard LED is repeatedly blinking (approximately five times a second), then the HC-05 Bluetooth module is in Data mode and is waiting for another Bluetooth device to pair with it. If the onboard LED is blinking twice in 1 second, then the HC-05 Bluetooth module is in Data mode and is successfully connected to another Bluetooth device.

11.2.2 Command mode

Command mode is also known as AT Command mode or AT mode. AT is the short form of Attention. So, Command mode is sometimes referred to as Attention mode. This operating mode is used for configuring various settings and parameters of the HC-05 Bluetooth module using AT commands. If the onboard LED on the HC-05 Bluetooth module shown in Figure 11.1 blinks once in 2 seconds, then the module is in AT mode or Command mode.

There are mainly three ways for changing the operating mode of the HC-05 Bluetooth module from Data mode to Command mode:

- Using the button on the HC-05 Bluetooth module.
- Using the Enable/Key pin of the HC-05 Bluetooth module.
- Using the 34th pin of IC on the HC-05 Bluetooth module.

11.2.2.1 Using onboard button

Using the onboard button shown in Figure 11.1 is one of the methods for changing the operating mode of the HC-05 Bluetooth module from the Data mode to the Command mode. To change the operating mode from Data mode to Command mode:

- Press and hold the onboard button
- Power the HC-05 Bluetooth module
- Wait for 5 seconds until you see a slower blink rate (one blink in 2 seconds)
- Release the onboard button

In short, power the HC-05 Bluetooth module by pressing the onboard button and don't release the button until you see a slower blink rate (one blink in 2 seconds), which indicates that the module has changed its operating mode to Command mode.

11.2.2.2 Using Enable/Key pin

If you don't want to use the onboard button or if your HC-05 Bluetooth module doesn't have an onboard button, then you can use the Enable/Key pin on the HC-05 Bluetooth module as shown in Figure 11.1. The Enable/Key pin will only be present on the 6-pin HC-05 Bluetooth module. Some HC-05 Bluetooth modules mention the Enable/Key pin as EN as shown in Figure 11.2.

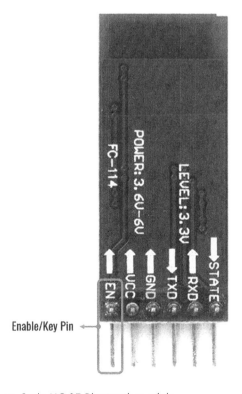

FIGURE 11.2   Enable/Key pin on 6-pin HC-05 Bluetooth module

To change the operating mode of the HC-05 Bluetooth module from Data mode to Command mode, connect the Enable/Key pin to 5 V or 3.3 V before powering the HC-05 Bluetooth module. Wait until the onboard LED blink at the rate of once in 2 seconds indicating the HC-05 Bluetooth module changed its operating mode to Command mode. Now, you can disconnect the Enable/Key pin from the 5 V or 3.3 V supply.

11.2.2.3 Using the 34th pin of IC

You can also change the operating mode of the HC-05 Bluetooth module from Data mode to Command mode using the 34th pin of the IC on the HC-05 Bluetooth module as shown in Figure 11.1. Use this method only if you are unable to change the operating mode of the HC-05 Bluetooth module by using the previously discussed two methods (Sections 11.2.2.1 and 11.2.2.2). This method has to be your last attempt in changing the operating mode of the HC-05 Bluetooth module. The previously discussed two methods can fail in one of the below situations:

- If the HC-05 Bluetooth module is not having an onboard button and Enable/Key pin.
- If the Enable/Key pin present on the HC-05 Bluetooth module is not wired properly to the actual Enable/Key pin on the IC.

So, to change the operating mode of the HC-05 Bluetooth module from Data mode to Command mode, connect the 34th pin on the IC to a 3.3 V supply. Wait until the onboard LED blink at a rate of once in 2 seconds indicating the HC-05 Bluetooth module changed its operating mode to Command mode.

until now, we have discussed the methods to change the operating mode of the HC-05 Bluetooth module from Data mode to Command mode. But vice versa is very easy and simple. To change the operating mode of the HC-05 Bluetooth module from Command mode to Data mode, just disconnect and reconnect the power source. This can also be done by using an AT command so that the power source of the HC-05 Bluetooth module need not be disconnected.

11.3 AT COMMANDS OF HC-05

The Command mode of the HC-05 Bluetooth module is similar to the settings application in your mobile phone, and the AT commands are similar to the various options in the settings application. As a mobile phone is built with a powerful processor and a dedicated display, a GUI application is developed by the mobile phone manufacturer to configure various settings of the mobile phone. But small hardware modules like the HC-05 Bluetooth module do not have any inbuilt powerful processor or a dedicated display. So, the manufacturers of these modules depend on CLI (Command Line Interface) to change the default configurations or settings of these modules.

In a GUI-based settings application, we can see all the available settings and can easily navigate to a particular setting by using the touch screen or physical buttons. But, in the case of CLI, it is not that simple to know or access all the available settings in that module. The manufacturer of these modules provides a datasheet that will help in knowing all the available settings and the relevant commands to access or modify a setting.

The manufacturer of the HC-05 Bluetooth module also provides AT (or Attention) commands to access and modify various default settings of the module. Some of the important AT commands and their use is tabulated in Table 11.1.

Most of the AT commands have dual functionality and can be used for both *set* and *check*. If the AT command is followed by an equal to (=) and a value then it is *set*. But, if the AT command is followed by a question mark (?), then it is *check*. For example,

AT+NAME=My-HC05 sets the display name of your HC-05 Bluetooth module to My-HC05.

AT+NAME? checks the name of the HC-05 Bluetooth module and prints it on the CLI.

TABLE 11.1 AT Commands of HC-05 Bluetooth Module

AT COMMAND	*FUNCTION OF THE AT COMMAND*
AT	Testing command
AT+NAME	Modifying the name of the HC-05 Bluetooth module
AT+ADDR	Knowing the address of the HC-05 Bluetooth module
AT+RESET	Changing the operating mode of the HC-05 Bluetooth module from Command mode to Data mode
AT+ORGL	Erases all the present configurations and restores the default configurations of the HC-05 Bluetooth module

Some of the AT commands have only *set* functionality, some AT commands possess only *check* functionality and some AT commands are execution commands which don't have both *set* and *check* functionality. For example,

AT+PIO has *set* functionality alone.

AT+ADDR has *check* functionality alone.

AT+RESET doesn't have both *set* and *check* functionality and can only be used for execution.

Note: It is not required to remember all the AT commands and their syntax. You can refer to the datasheet whenever required. However, it is suggested to have an overview of all the available AT commands and their functionality to use them appropriately.

You can find all the AT commands of the HC-05 Bluetooth module at this link (https://www.itead. cc/wiki/Serial_Port_Bluetooth_Module_(Master/Slave)_:_HC-05), and the datasheet of the HC-05 Bluetooth module can be downloaded at this link (https://www.estudioelectronica.com/wp-content/uploads/2018/09/istd016A.pdf).

In a GUI settings application, you can only change a particular setting to one of the available options and there will be no chance of getting an error unless there is a bug in that application. In the case of CLI, you are not restricted and free to enter any command. But, all the commands you enter need not be correct. If a wrong command is entered, then an error is printed on the CLI. Similarly, if a wrong command is entered in the Command mode of the HC-05 Bluetooth module, an error code is printed on the CLI depending on the mistake. Table 11.2 shows the explanation relating to all the error codes which will help in understanding the error.

TABLE 11.2 Error Codes of HC-05 Bluetooth Module

ERROR CODE	*EXPLANATION*
0	Command error/Invalid command
1	Results in default value
2	PSKEY write error
3	Device name is too long (>32 characters)
4	No device name specified (0 length)
5	Bluetooth address NAP is too long
6	Bluetooth address UAP is too long
7	Bluetooth address LAP is too long
8	PIO map not specified (0 length)
9	Invalid PIO port number entered
A	Device class not specified (0 length)
B	Device class too long
C	Inquire access code not a specified (0 length)
D	Inquire access code too long
E	Invalid inquire access code entered

(Continued)

TABLE 11.2 (*Continued*)   Error Codes of HC-05 Bluetooth Module

ERROR CODE	EXPLANATION
F	Pairing password not specified (0 length)
10	Pairing password too long (>16 characters)
11	Invalid role entered
12	Invalid baud rate entered
13	Invalid stop bit entered
14	Invalid parity bit entered
15	No device in the pairing list
16	SPP not initialized
17	SPP already initialized
18	Invalid inquiry mode
19	Inquiry timeout occurred
1A	Invalid/Zero length address entered
1B	Invalid security mode entered
1C	Invalid encryption mode entered

For sending AT commands to an HC-05 Bluetooth module, it has to be connected to any computing device with a display (like PC). There are many ways to connect HC-05 Bluetooth module to a PC; among them, the two most common methods make use of USB to TTL serial converter or any UART supported evaluation board (like Arduino Uno, NodeMCU).

11.4 CONFIGURATION USING USB TO TTL SERIAL CONVERTER

For pairing two Bluetooth modules, one Bluetooth module has to be configured as a master and the other Bluetooth module has to be configured as a slave. USB to TTL serial converter and Tera Term can be used to configure two HC-05 Bluetooth modules as Master and Slave using AT commands. This Master-Slave configuration will help in pairing the two HC-05 Bluetooth modules. Some books and articles refer to the pairing of the Bluetooth modules as Master-Slave configuration. So, it is important to remember this point and not to get confused that Master-Slave configuration is a different concept.

11.4.1 Hardware required

TABLE 11.3   Hardware required for the project

HARDWARE	QUANTITY
USB to TTL serial converter	1
HC-05 Bluetooth module	2
Jumper wires	As required

11.4.2 Connections

You can refer to the list of required components from Table 11.3, followed by Table 11.4, for the pin-wise connections. Also, the connection diagram is presented in Figure 11.3, which shall help get a visualization of the connections.

TABLE 11.4   Connection details

USB TO TTL SERIAL CONVERTER	HC-05 BLUETOOTH MODULE
5V	VCC
TXD	RXD
RXD	TXD
GND	GND

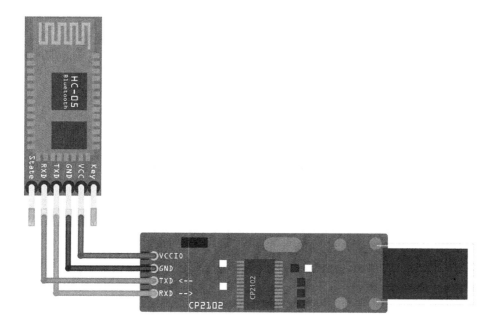

FIGURE 11.3   Connection layout

11.4.3 Tera term setup

Open the *Device Manager* after connecting the setup to your PC as shown in Figure 11.3. Expand the *Ports* section and note down the COM port assigned to your USB to TTL serial converter. Here, COM6 is assigned to our CP2102 USB to TTL serial converter as shown in Figure 11.4. Now, open Tera Term. If you haven't installed Tera Term previously, then refer to the detailed explanation of Tera Term installation in Project 8 and install it.

After opening the Tera Term, select *Serial* among the two radio buttons at the left and then from the drop-down, select the COM port assigned to your USB to TTL serial converter which you have noted before. Then click *OK* as shown in Figure 11.5.

Now, you'll be seeing a CLI as shown in Figure 11.6. To send AT commands to the HC-05 Bluetooth module using this CLI, you need to change some settings. First, go to *Setup > Terminal* as shown in Figure 11.7, which will open a new window. Change *Transmit* to *CR+LF* from the drop-down, check the box beside *Local echo* and then click *OK* as shown in Figure 11.8.

Next, go to *Setup > Serial port* as shown in Figure 11.9 which will open a new window. Change the *Speed* to *38400* from the drop-down and click *New setting* as shown in Figure 11.10. This will change the baud rate of the Tera Term CLI to 38400 bits per second. Now, the Tera Term CLI is all set for sending the AT commands to the HC-05 Bluetooth module in Command mode.

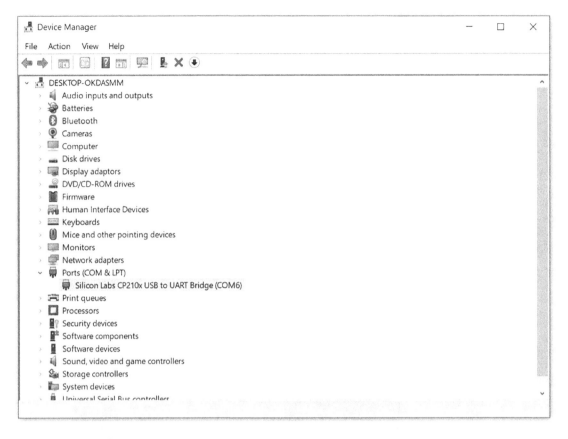

FIGURE 11.4   Device manager

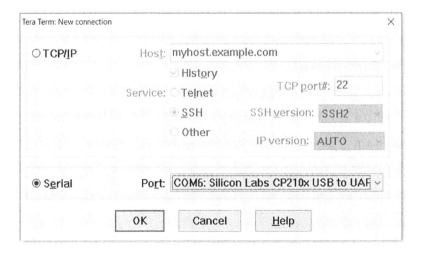

FIGURE 11.5   Selection of COM port

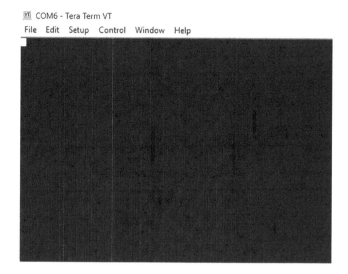

FIGURE 11.6 Tera term CLI

FIGURE 11.7 Terminal

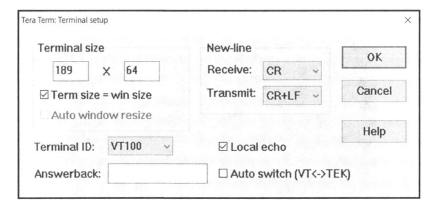

FIGURE 11.8   Terminal setup

FIGURE 11.9   Serial port

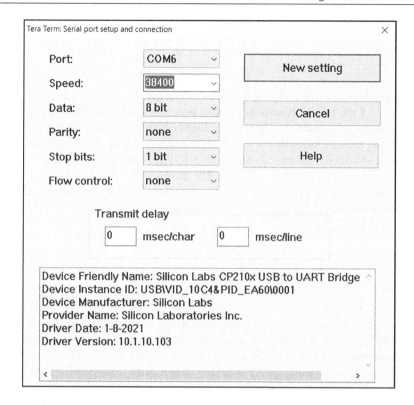

FIGURE 11.10 Serial port setup and connection

Remember: The HC-05 Bluetooth module works at different baud rates in different operating modes. In Data mode, the HC-05 Bluetooth module can operate at all the valid baud rates like 9600, 38400, 115200 and many more. But in the Command mode, the HC-05 Bluetooth module cannot operate at all the valid baud rates and can only operate at 38400 baud rate (bits/second). So, to send AT commands in Command mode to the HC-05 Bluetooth module, the CLI has to be configured at 38400 baud rate.

11.4.4 AT Commands for slave

For configuring the HC-05 Bluetooth module as Slave, the following AT commands are used. The first command, AT is used to check whether

- The HC-05 Bluetooth module is in Command mode or not.
- The CLI settings are properly configured or not.
- The hardware is properly working or not.
- The HC-05 Bluetooth module and the USB to TTL serial converter are properly connected or not.

If all the above-mentioned conditions are satisfied, then the HC-05 Bluetooth module will return OK or else nothing will be returned. The AT+RMAAD command is used to release the module from any previously paired devices. If the release is successful, then it'll return OK or else an error code will be returned.

```
AT
OK
AT+RMAAD
OK
```

The AT+NAME? command is used to know the name of the HC-05 Bluetooth module. This command will return the name of the HC-05 Bluetooth module along with OK. The existing name of our HC-05 Bluetooth module is H-C-2010-06-01. To identify the slave HC-05 Bluetooth module easily, its name has to be changed. So, the AT+NAME=HC-05_SLAVE command is used. This command will change the name of the slave HC-05 Bluetooth module to HC-05_SLAVE. If changing the name is successful, then it'll return OK or else an error code will be returned. The third and fourth commands are optional, but it is recommended to follow them as it'll be easier to differentiate the slave HC-05 Bluetooth module and the master HC-05 Bluetooth module just by seeing their names.

```
AT+NAME?
+NAME:H-C-2010-06-01
OK
AT+NAME=HC-05_SLAVE
OK
```

The AT+PSWD? command is used for seeing the password of the HC-05 Bluetooth module. This command will return the password of the HC-05 Bluetooth module along with OK. The existing password of our HC-05 Bluetooth module is 1234. If you want to change the default password of the HC-05 Bluetooth module, it can be easily changed using the AT+PSWD="<YOUR NEW PASSWORD>" command where the new password should be written in the place of <YOUR NEW PASSWORD>. This command will return OK if the password is successfully changed or else an error code is returned.

```
AT+PSWD?
+PIN:"1234"
OK
```

The AT+UART? command is used to know the baud rate, stop bit and parity of the HC-05 Bluetooth module. This command will return the baud rate, stop bit and parity of the HC-05 Bluetooth module along with OK. The existing baud rate, stop bit and parity of our HC-05 Bluetooth module is 9600,0 and0, respectively. If you want to change the default baud rate, stop bit and parity of the HC-05 Bluetooth module, it can be easily changed using the AT+UART="<BAUD RATE>,<STOP BIT>,<PARITY>" command where the new baud rate, stop bit and parity should be written in the place of <BAUD RATE>, <STOP BIT> and <PARITY>, respectively. This command will return OK if the parameters are successfully changed or else an error code is returned.

```
AT+UART?
+UART:9600,0,0
OK
```

The AT+ROLE? command is used to know the role of the HC-05 Bluetooth module. This command will return one of the following numbers, 0, 1 or 2 along with OK. By using the returned number as listed in Table 11.5, the role of the HC-05 Bluetooth module can be easily determined. The role number of our HC-05 Bluetooth module is 0, which implies that its role is a slave.

TABLE 11.5 Number to role relation

NUMBER	ROLE
0	Slave
1	Master
2	Slave-Loop

To change the role of the HC-05 Bluetooth module, the AT+ROLE=<ROLE-NUMBER> command can be used where the <ROLE-NUMBER> has to be replaced with the new role number. This command will return OK if the role is successfully changed or else an error code is returned.

```
AT+ROLE?
+ROLE:0
OK
```

Figure 11.11 shows the general format of the Bluetooth MAC address. It is divided into three parts:

- Non-significant Address Part (NAP) – It contains the first 2 bytes (16 bits) of the MAC address and is used in frequency hopping synchronization frames.
- Upper Address Part (UAP) – It contains the next 1 byte (8 bits) of the MAC address and is used for seeding in various Bluetooth specification algorithms.
- Lower Address Part (LAP) – It contains the remaining 3 bytes (24 bits) of the MAC address. It is used to uniquely identify the Bluetooth device in every transmitted frame.

The UAP and LAP together make the significant address part (SAP) of the Bluetooth MAC Address. The NAP and UAP are provided by OUI (Organizationally Unique Identifier), and the LAP is assigned by the vendor.

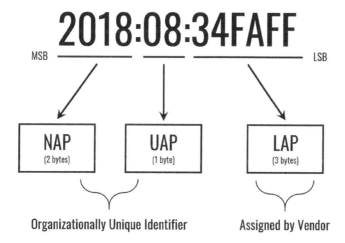

FIGURE 11.11   Format of Bluetooth MAC address

The AT+ADDR? command is used for knowing the MAC Address of the HC-05 Bluetooth module. This command will return the MAC address of the HC-05 Bluetooth module by eliminating the leading zeros along with OK. The MAC address of our HC-05 Bluetooth module is 2018:08:34FAFF, but 2018:8:34FAFF is returned by eliminating the leading zero before 8. Note this returned MAC address by replacing colon (:) with a comma (,), and we need to note this MAC address as 2018,08,34FAFF.

```
AT+ADDR?
+ADDR:2018:8:34FAFF
OK
```

After finishing the configuration, send the AT+RESET command which is used to change the operating mode of the HC-05 Bluetooth module from Command mode to Data mode without disconnecting the power. If the change of operating mode is successful, then OK is returned or else an error code is returned.

```
AT+RESET
OK
```

Note: For pairing two Bluetooth modules using Master-Slave configuration, some parameters of Master and Slave Bluetooth modules must be the same. The password and baud rate of both Master and Slave Bluetooth modules must be the same for pairing and communication between them.

11.4.5 AT Commands for master

For configuring the HC-05 Bluetooth module as Master, the following AT commands are used. Most of the AT commands used for configuring the master HC-05 Bluetooth module are used while configuring the slave HC-05 Bluetooth module. So, only the new AT commands will be explained in this section.

```
AT
OK
AT+RMAAD
OK
AT+NAME?
+NAME:H-C-2010-06-01
OK
AT+NAME=HC-05_MASTER
OK
```

The default password and baud rate of the master HC-05 Bluetooth module are the same as the slave HC-05 Bluetooth module. So, these parameters are not changed.

```
AT+PSWD?
+PIN:"1234"
OK
AT+UART?
+UART:9600,0,0
OK
```

The command AT+ROLE=1 is used to change the role of the HC-05 Bluetooth module to master. After the role of the HC-05 Bluetooth module is successfully changed, its operating mode will automatically change from Data mode to Command mode. So, again the operating mode of the HC-05 Bluetooth module must be changed from Data mode to Command mode to enter the next set of AT commands.

```
AT+ROLE?
+ROLE:0
OK
AT+ROLE=1
OK
```

The AT+CMODE? command is used to know the connecting mode of the HC-05 Bluetooth module. This command will return one of the following numbers, 0, 1 or 2 along with OK. By using the returned number as in Table 11.6, the connecting mode of the HC-05 Bluetooth module can be easily determined. The cmode number of our HC-05 Bluetooth module is 1, which implies that it can connect to any address.

TABLE 11.6 Number to CMode relation

NUMBER	CMODE
0	Connect Fixed Address
1	Connect Any Address
2	Slave-Loop

To pair the master HC-05 Bluetooth module to the slave HC-05 Bluetooth module alone, the cmode number of the master HC-05 Bluetooth module has to be changed such that it'll connect only to a fixed address. So, the AT+CMODE=0 command is used. If changing the cmode is successful, then OK is returned or else an error code is returned. The pairing of master and slave HC-05 Bluetooth modules will work even if the cmode is 1. But, at the time of their pairing, no other Bluetooth module must be in the range of the master HC-05 Bluetooth module or else it might pair with the other Bluetooth module.

```
AT+CMODE?
+CMODE:1
OK
AT+CMODE=0
OK
```

The AT+BIND? command is used to know the address to which the HC-05 Bluetooth module is bound. This is the Bluetooth address to which the HC-05 Bluetooth module tries to pair as soon as it is powered. By default, any HC-05 Bluetooth module is bound to 0000:00:000000 fixed address. For the master HC-05 Bluetooth module to pair with the slave HC-05 Bluetooth module as soon as it is powered, its binding address must be changed to the MAC address of the slave HC-05 Bluetooth module. So, the AT+BIND=2018,08,34FAFF command is used. If the change of the binding address is successful, then OK is returned or else an error code is returned.

```
AT+BIND?
+BIND:0:0:0
OK
AT+BIND=2018,08,34FAFF
OK
AT+RESET
OK
```

After finishing the configuration, power the master and slave HC-05 Bluetooth modules. After waiting for some time, both the HC-05 Bluetooth modules get paired. Pairing can be easily confirmed by seeing the onboard LED on both the HC-05 Bluetooth modules. They both will blink at the same time and at the same blink rate. Both the HC-05 Bluetooth modules will flash once every two seconds.

11.5 CONFIGURATION USING ARDUINO UNO

If you don't have a USB to TTL serial converter or if you don't want to use a USB to TTL serial converter, you can use any UART supported microcontrollers or evaluation boards (like Arduino Uno, NodeMCU, etc.) to configure the HC-05 Bluetooth modules as Master and Slave. In the previous section, we have used USB to TTL serial converter and Tera Term for configuring the HC-05 Bluetooth modules as Master and Slave. But in this section, we'll use an Arduino Uno board and Arduino IDE for configuring the HC-05 Bluetooth modules as Master and Slave.

11.5.1 Hardware required

TABLE 11.7 Hardware required for the project

HARDWARE	QUANTITY
Arduino Uno	1
HC-05 Bluetooth module	2
Jumper wires	As required

11.5.2 **Connections**

You can refer to the list of required components from Table 11.7, followed by Table 11.8, for the pin-wise connections. Also, the connection diagram is presented in Figure 11.12, which shall help get a visualization of the connections.

TABLE 11.8   Connection details

ARDUINO UNO	HC-05 BLUETOOTH MODULE
5 V	VCC
TX	TXD
RX	RXD
GND	GND

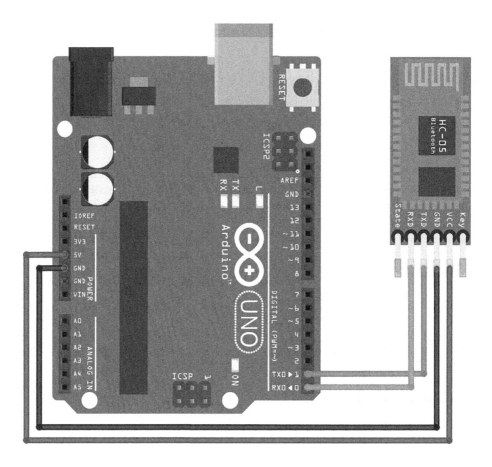

FIGURE 11.12   Connection layout

11.5.3 Arduino IDE code and setup

For sending AT commands to an HC-05 Bluetooth module, an empty Arduino code must be uploaded to the Arduino Uno board. Before uploading the code, don't forget to disconnect the Rx and Tx connections.

```
void setup() {
  // put your setup code here, to run once:
}

void loop() {
  // put your main code here, to run repeatedly:
}
```

After successfully uploading the code, reconnect the Rx and Tx pins to the Arduino Uno board. Also, don't forget to change the operating mode of the HC-05 Bluetooth module to Command mode by using one of the methods explained in Section 11.2.2. Next, open the Serial monitor and select *Both NL & CR* and *38400 baud* from the drop-down at the bottom right corner as shown in Figure 11.13. Now, the Arduino IDE serial monitor is all set for sending the AT commands to the HC-05 Bluetooth module in Command mode.

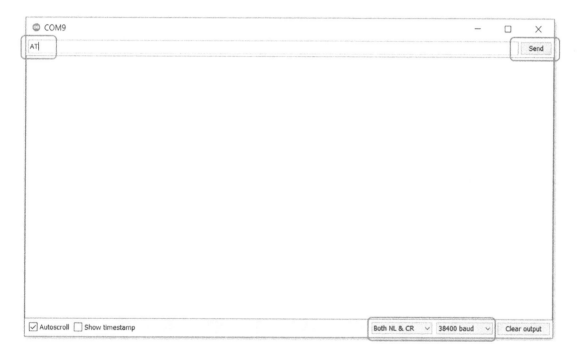

FIGURE 11.13   Arduino IDE serial monitor configurations

You can type the AT commands in the text box which is at the top of the serial monitor and click *Send* button as shown in Figure 11.13. One of the disadvantages of using the Arduino IDE serial monitor is that you can only see the returned output from the HC-05 Bluetooth module but the entered command cannot be seen.

11.5.4 AT Commands for slave

The same AT commands used in Section 11.4.4 are used to configure an HC-05 Bluetooth module as slave using Arduino Uno. As the serial monitor of Arduino IDE doesn't show the entered AT commands, you can only see the returned messages as shown in Figure 11.14. These messages will confirm whether the execution of AT command is successful or not.

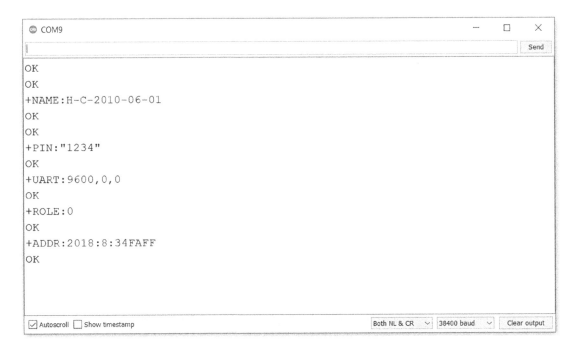

FIGURE 11.14 Returned messages of slave AT commands

11.5.5 AT Commands for master

The same AT commands used in Section 11.4.5 are used to configure an HC-05 Bluetooth module as master using Arduino Uno. As the serial monitor of Arduino IDE doesn't show the entered AT commands, you can only see the returned messages as shown in Figure 11.15. These messages will confirm whether the execution of AT command is successful or not.

After finishing the configuration, power the master and slave HC-05 Bluetooth modules. After waiting for some time, both the HC-05 Bluetooth modules get paired. Now, you can send data from one HC-05 Bluetooth module to another HC-05 Bluetooth module. The pairing of two HC-05 Bluetooth modules will work only if both of them are in 10 meters range (or Bluetooth range).

By using AT commands, you can also configure multiple slaves to connect to a single master. This has very little application in IoT projects, and we would suggest you to explore this concept on your own.

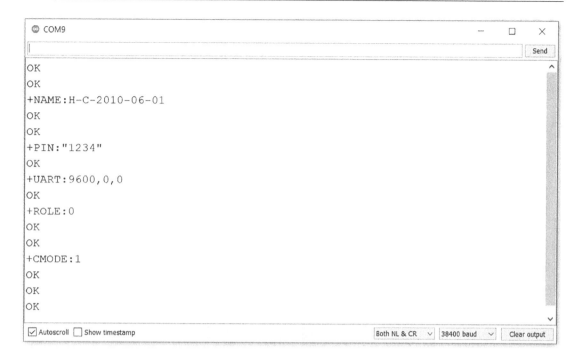

FIGURE 11.15   Returned messages of master AT commands

11.6 TRY IT

This is a small exercise to test your understanding. In this exercise, you'll pair two HC-05 Bluetooth modules and control the circuitry at one HC-05 Bluetooth module with another HC-05 Bluetooth module. At one HC-05 Bluetooth module, you'll have an LED connected to an Arduino board, and at the other end, you'll have another HC-05 Bluetooth module using which you'll send commands to control the LED. To understand the architecture better, refer to Figure 11.16.

Just for the sake of exercise, an LED is controlled but you can control any sensor using the same concept. At the second end (right side in Figure 11.16), we have used a USB to TTL serial converter, but you can also use another Arduino board to send commands. Try to complete the exercise on your own. If you are unable to solve it, don't worry! The below hardware table, connection diagram and code is the solution for it. Explanation of the code is not required as it is easily understandable.

11.6.1 Hardware required

TABLE 11.9   Hardware required for the project

HARDWARE	QUANTITY
Arduino Uno	1
USB to TTL serial converter	1
HC-05 Bluetooth module	2
LED	1
Resistor (220 Ω)	1
Jumper wires	As required

11.6.2 Connections

You can refer to the list of required components from Table 11.9. The connections details for the left-side circuitry in Figure 11.16 are tabulated in Table 11.10, and the connections for the right-side circuitry in Figure 11.16 are tabulated in Table 11.4.

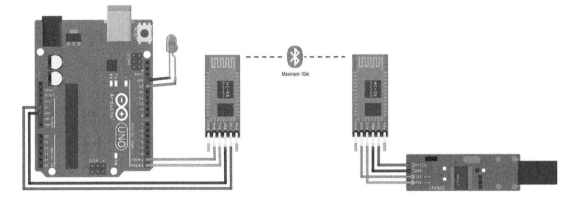

FIGURE 11.16   Connection layout

TABLE 11.10   Connection details for left-side circuitry

ARDUINO UNO	HC-05 BLUETOOTH MODULE	LED
5V	VCC	
RXD	TXD	
TXD	RXD	
13		+ve pin (long pin)
GND	GND	−ve pin (short pin)

11.6.3 Arduino IDE code for left-side circuitry

```
int LED = 13; //Arduino Uno pin to which the LED's +ve pin is connected
String recvCommand = ""; //Stores the received command

void setup() {
  Serial.begin(9600); //Begin the Serial communication
  pinMode(LED, OUTPUT); //Set the direction of LED as OUTPUT
}

void loop() {
  //Checks whether any information is available in Serial or not
  if(Serial.available() > 0){
    recvCommand = Serial.readString(); //Reads the string
    recvCommand.trim(); //Removes all the extra spaces at the beginning and
at the end of the String.
  }

  //For switching ON the LED
  if(recvCommand == "ON" || recvCommand == "on"){
    digitalWrite(LED, HIGH);
    Serial.println("LED is successfully switched ON.");
    recvCommand = "";
  }
```

```
//For switching OFF the LED
if(recvCommand == "OFF" || recvCommand == "off"){
  digitalWrite(LED, LOW);
  Serial.println("LED is successfully switched OFF.");
  recvCommand = "";
  }
}
```

Upload the code to Arduino Uno after selecting the appropriate board and COM port. After successfully uploading the code, open the serial monitor and select the appropriate baud rate to see the data printed on the serial monitor.

11.6.4 Tera term CLI setup for right-side circuitry

Setting up the Tera Term CLI is very simple and easy. The setup process is the same as explained before in Section 11.4.3 with few changes. First, open Tera Term and select the COM port of your USB to TTL serial converter as shown in Figure 11.5 and click *OK*. This will open a new Tera Term CLI as shown in Figure 11.6.

Next, go to *Setup > Terminal* as shown in Figure 11.7 and change *Transmit* to *CR+LF* from the drop-down, check the box beside *Local echo* and then click *OK* as shown in Figure 11.8. Finally, go to *Setup > Serial port* as shown in Figure 11.9 and change the *Speed* to *9600* from the drop-down and click *New setting* as shown in Figure 11.17. This will change the baud rate of the Tera Term CLI to 9600 bits per second. Now, the Tera Term CLI is all set for sending commands to control the LED.

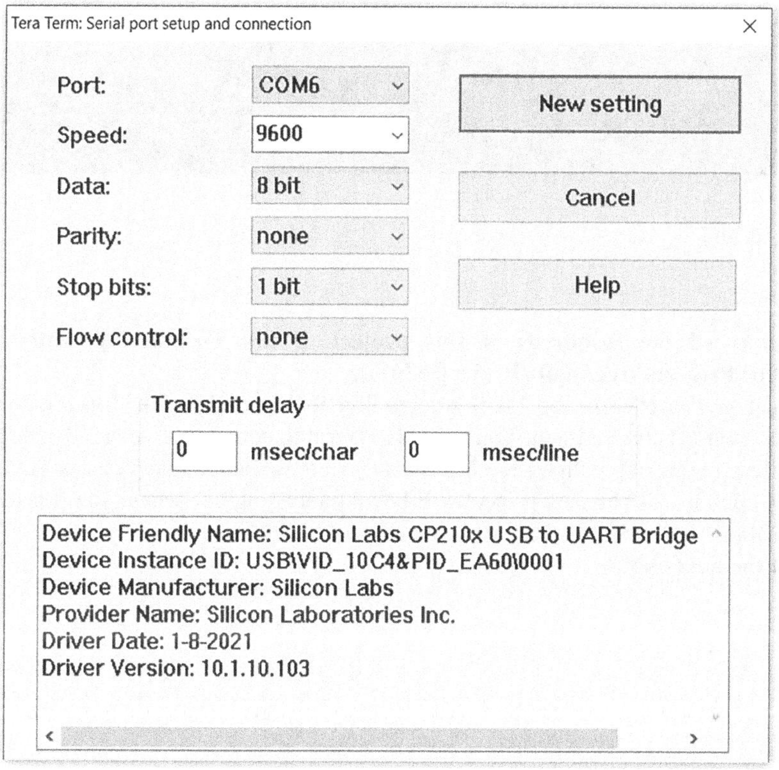

FIGURE 11.17   Serial port setup and connection

11.6.5 Result

After uploading the code to Arduino Uno and finishing the Tera Term setup process, power the Arduino Uno board and USB to TTL serial converter. Wait for some time until both the HC-05 Bluetooth modules get paired.

After successfully pairing the two HC-05 Bluetooth modules, send an ON command using the Tera Term CLI to the Arduino Uno. This should switch ON the LED, and LED is successfully switched ON. message should be printed on the Tera Term CLI as shown in Figure 11.18. Next, send an OFF command using the Tera Term CLI to the Arduino Uno. This should switch OFF the LED, and LED is successfully switched OFF. message should again be printed on the Tera Term CLI as shown in Figure 11.18.

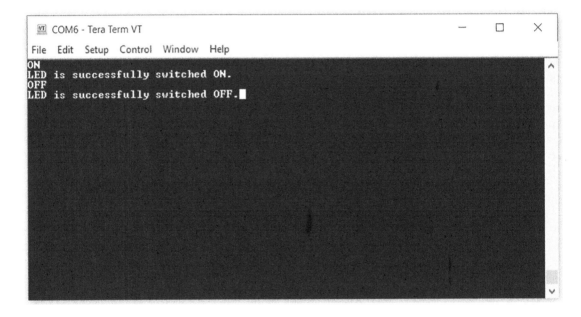

FIGURE 11.18   Output

You can find all the resources of this project at this Github link (https://github.com/anudeep-20/30IoTProjects/tree/main/Project%2011).

In this project, you have learnt about various operating modes in an HC-05 Bluetooth module and different methods to switch between them. You have also learnt to configure two HC-05 Bluetooth modules as Master and Slave for pairing with each other. This concept is used at many places to transfer data and remotely control appliances. The next project will be an interesting one where you'll learn to upload the code to an evaluation board without a cable (wirelessly).

Let's build the next one!

Upload Your Code Over The Air (OTA)

12

Until the previous project, you have uploaded code to an evaluation board (Arduino Uno or NodeMCU) using a cable. But, in this project, you'll upload code to Arduino Uno and NodeMCU without using a cable (wirelessly). The process of uploading code without using a cable is known as Over The Air (OTA) code upload. The code can be uploaded using Bluetooth protocol or WiFi protocol or any other wireless protocol. In the case of NodeMCU, it has an onboard WiFi chip. So, the code can be uploaded OTA using WiFi. But, in the case of Arduino Uno, it doesn't have any onboard WiFi or Bluetooth chip. So, external modules (like HC-05 Bluetooth module, ESP8266 WiFi chip) need to be connected for uploading the code. This project is divided into two parts depending on the wireless protocol used for uploading the code:

- OTA code upload using Bluetooth
- OTA code upload using WiFi

OTA code upload is generally done in the case of small microcontrollers due to difficulty in uploading the code using a cable/wires or if the microcontroller is fixed on a PCB (Printed Circuit Board) or for faster and easier code updates. This is an important concept in IoT and electronics as a whole.

12.1 OTA CODE UPLOAD USING BLUETOOTH

12.1.1 Motive of the project

In this part of the project, you'll upload code to an Arduino Uno board OTA. You'll use an HC-05 Bluetooth module to receive the code wirelessly from the Arduino IDE to the Arduino Uno board.

12.1.2 Bootloader

If you take a bare microcontroller that is just manufactured from a factory and power it up, nothing will happen except the internal clock keeps ticking. There will be no program to run after getting power.

A bootloader is nothing but a program that gets loaded when a microcontroller is booted (powered). Without a bootloader, there'll be no program to run after the bare microcontroller is powered. To burn a new program into the flash memory, you need to burn it through a special programming interface on the bare microcontroller using specialized equipment. This equipment will be costly, cannot be carried everywhere, and everyone cannot afford it. This will eventually reduce the usability of a microcontroller.

To avoid this, most of the microcontroller boards will be pre-loaded with a program known as bootloader. This will allow us to load a new program (or code) in more convenient ways like using a USB cable. When the microcontroller is powered or reset, the bootloader will check if there is an upload request

DOI: 10.1201/9781003147169-12

or not. If there is an upload request, it will burn the new program into the flash memory or else it will run the last program that is present in the flash memory.

So, for running a new code on a microcontroller or an evaluation board, its bootloader must burn the new code into its flash memory. The baud rate at which the bootloader burns the code into the flash memory is different for different boards. Some of the Arduino boards and their respective baud rates are listed in Table 12.1.

TABLE 12.1   Arduino boards and their bootloader baud rates

ARDUINO BOARD	BAUD RATE
Arduino Pro or Pro Mini (5 V, 16 MHz) w/ATmega328P	57600
Arduino Pro or Pro Mini (3.3 V, 8 MHz) w/ATmega328P	57600
Arduino Pro or Pro Mini (5 V, 16 MHz) w/ ATmega168	19200
Arduino Pro or Pro Mini (3.3 V, 8 MHz) w/ ATmega168	19200
Arduino Yún	57600
Arduino/Genuino Uno	115200
Arduino Duemilanove or Diecimila w/ ATmega328P	57600
Arduino Duemilanove or Diecimila w/ATmega168	19200
Arduino Nano w/ATmega328P	57600
Arduino Nano w/ATmega168	19200
Arduino/Genuino Mega w/ATmega2560	115200
Arduino Mega w/ATmega1280	57600
Arduino Leonardo	57600
Arduino Leonardo ETH	57600
Arduino/Genuino Micro	57600
Arduino Esplora	57600
Arduino Mini	115200
Arduino Mini w/ATmega168	19200
Arduino Ethernet	115200
Arduino Fio	57600
Arduino BT	19200
LilyPad Arduino	57600
LilyPad Arduino w/ATmega168	19200
Arduino NG or older	19200
Arduino Robot Control	57600
Arduino Robot Motor	57600
Arduino Yún Mini	57600
Arduino Industrial 101	57600
Linino One	57600
Arduino Uno WiFi	115200

When we upload code to an evaluation board using a cable and Arduino IDE, the Arduino IDE software automatically takes care of the baud rate (using the board information selected at *Tools > Board* before uploading) at which the code needs to be sent to the bootloader so that the code gets burnt into the flash memory.

So, if we need to send code to the Arduino Uno over Bluetooth, the Bluetooth module connected to the Arduino Uno must receive and send the code to the bootloader at a 115200 baud rate (which is the baud rate of Arduino Uno's bootloader).

12.1.3 Setup of HC-05 Bluetooth module

The code can be transferred to the Arduino Uno from the Arduino IDE using HC-05 Bluetooth module in two ways:

- If you are only having a single HC-05 Bluetooth module, it can be directly paired with the PC for sending the code.
- If your PC doesn't support pairing with an HC-05 Bluetooth module and if you have two HC-05 Bluetooth modules, then they can be configured as Master and Slave at 115200 baud rate for sending the code.

As the second method is already explained in the previous project, we'll use the first method for sending the code OTA to Arduino Uno. Before pairing the HC-05 Bluetooth module to your PC, it has to be configured with the below AT commands.

```
AT+NAME=HC-05_OTA
AT+UART=115200,0,0
AT+POLAR=1,0
```

AT+NAME=HC-05_OTA command is only used for identifying the HC-05 Bluetooth module configured for OTA upload. It is an optional AT command. As explained in the previous project, AT+UART=115200,0,0 sets the baud rate to 115200, stop bit to 0 and parity to 0. AT+POLAR=1,0 is used to control the PIO8 and PIO9 pins on the chip of HC-05 Bluetooth module. Figure 12.1 shows the pins on the chip of HC-05 Bluetooth module and highlights the PIO8 and PIO9 pins.

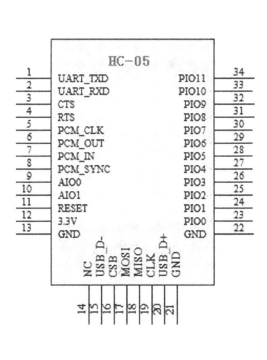 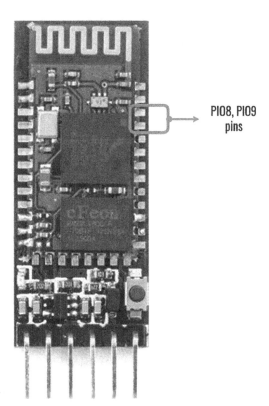

FIGURE 12.1   Pins on the HC-05 Bluetooth module chip

12.1.4 Pairing HC-05 Bluetooth module with PC

Most PCs can pair with an HC-05 Bluetooth module. Pairing an HC-05 Bluetooth module to a PC is similar to pairing with a mobile phone. Follow the below steps to pair an HC-05 Bluetooth module to a PC.

- Open Bluetooth Settings and switch on the Bluetooth pairing and visibility on your PC as shown in Figure 12.2.

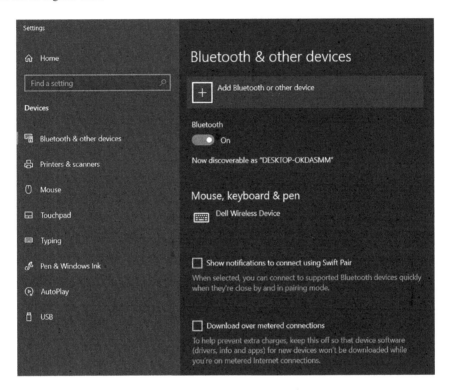

FIGURE 12.2   Switch ON Bluetooth pairing and visibility

- Search for the available devices by clicking the *Add Bluetooth or other device* button. Among the available devices (shown in Figure 12.3), pair with the HC-05 Bluetooth module which has been configured for OTA code upload (HC-05_OTA). Enter the password and click *Connect* as shown in Figure 12.4. After the HC-05 Bluetooth module successfully gets connected to the PC, it'll be listed in the *Other devices* column as shown in Figure 12.5.

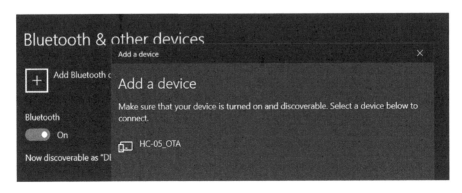

FIGURE 12.3   Available Bluetooth devices

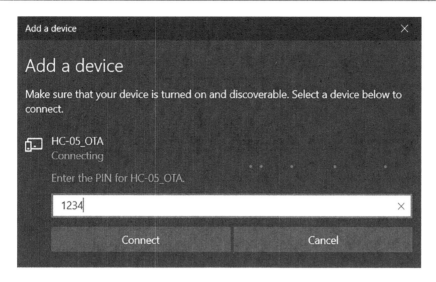

FIGURE 12.4   Connecting to the Bluetooth module

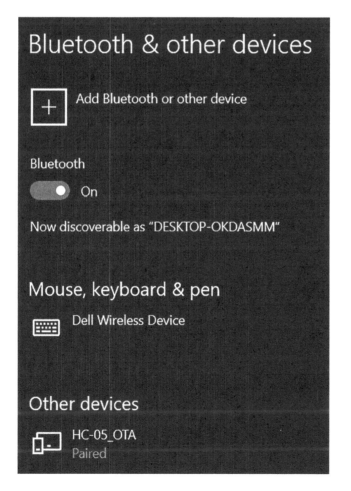

FIGURE 12.5   Listed in other devices

- If you go to *More Bluetooth options* on your PC and navigate to the *COM ports* tab, you can see that two COM ports are assigned to the HC-05_OTA Bluetooth module as shown in Figure 12.6. The same information can also be seen using your Device Manager as shown in Figure 12.7. Among the two assigned COM ports, one is used for incoming data and the other is used for outgoing data.

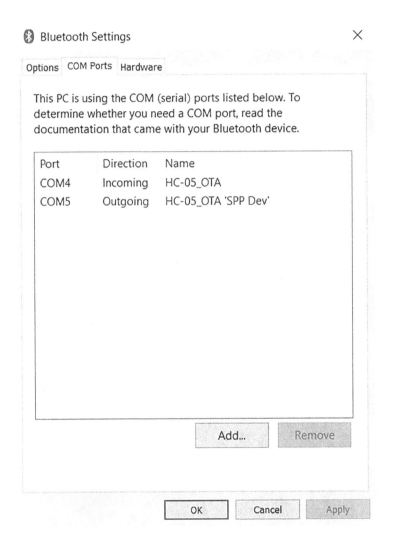

FIGURE 12.6   Bluetooth settings

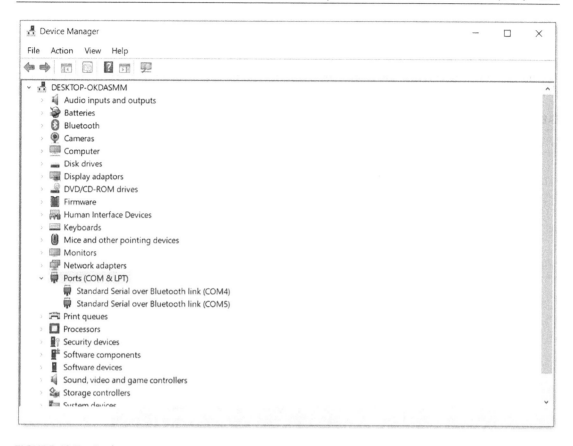

FIGURE 12.7   Device manager

12.1.5 Hardware required

In this project, we'll upload LED Blink code OTA to Arduino Uno. But you can upload any code OTA to an Arduino Uno board using Bluetooth.

TABLE 12.2   Hardware required for OTA code upload using Bluetooth

HARDWARE	QUANTITY
Arduino Uno	1
HC-05 Bluetooth module	1
LED	1
10 µF electrolytic capacitor	1
Jumper wires	As required

If a 10 µF electrolytic capacitor is not available, you can also use a 1 µF electrolytic capacitor. If both of them are not available, you can try using any other available electrolytic capacitor values.

12.1.6 Connections

You can refer to the list of components required from Table 12.2, followed by Table 12.3, which presents the pin-wise connections. Also, the connection diagram is presented in Figure 12.8, which shall help get a visualization of the connections.

TABLE 12.3   Connection details for OTA code upload using Bluetooth

ARDUINO UNO	HC-05 BLUETOOTH MODULE	10 μF ELECTROLYTIC CAPACITOR	LED
5 V	VCC		
RXD	TXD		
TXD	RXD		
	State	+ve pin	
RESET		−ve pin	
9			+ve pin (long pin)
GND	GND		−ve pin (short pin)

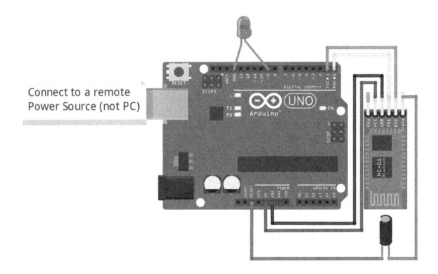

Connect to a remote Power Source (not PC)

FIGURE 12.8   Connection layout for OTA code upload using Bluetooth

12.1.7 Arduino IDE code

This is the same code with minor changes that is used in the 1st Project for blinking an LED using Arduino Uno. Hence, explanation of the code is not required.

```
int LEDPin = 9; //Arduino Digital Pin 9
void setup()
{
  Serial.begin(115200);      //Initialize the Arduino's Serial
  pinMode(LEDPin,OUTPUT);    //Set the direction of LEDpin
  digitalWrite(LEDPin,LOW); //Initially, keep the LED OFF
}
void loop()
{
  digitalWrite(LEDPin,HIGH);  // Write HIGH to ON the LED
```

```
  Serial.println("LED is switched ON");
  delay(1000);                      //Wait for a second

  digitalWrite(LEDPin,LOW);    // Write LOW to OFF the LED
  Serial.println("LED is switched OFF");
  delay(1000);                      // Wait for a second
}
```

Before uploading the code, select the appropriate Board and select a COM port among the two COM ports (COM4 and COM5) assigned to the paired HC-05_OTA Bluetooth module as shown in Figure 12.9. Generally, the COM port assigned in the outgoing direction is used for sending the data (code). In our case, COM5 is assigned as the outgoing direction COM port. If the code doesn't get uploaded OTA using the outgoing direction COM port (COM5), then try uploading the code by using the incoming direction COM port (COM4).

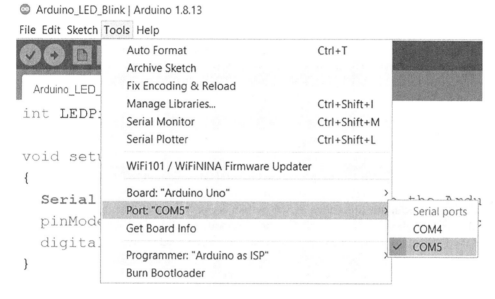

FIGURE 12.9   COM port selection

After successfully uploading the code, you can see the blinking of LED connected to the Digital Pin 9 of Arduino Uno. Also, open the serial monitor and select 115200 baud rate to see the data printed on it.

12.1.8 General errors in OTA upload

ser _ open(): can't open device error shown in Figure 12.10 is one of the most common errors you will be seeing while uploading code to Arduino OTA. It can be due to one of the following reasons:

- When the COM port of the paired Bluetooth module is used by some other applications for viewing the data (E.g., Tera Term) or used by serial monitor/serial plotter of Arduino IDE.
- The Bluetooth module might be sending or receiving data from some other device. (E.g., The HC-05 is connected to a mobile App and data being visualized over there.)
- The Bluetooth module is not properly connected or not powered ON.
- If none of the above reasons is the cause, then wait for some time and upload the code or switch off and switch on the Bluetooth module.

FIGURE 12.10   Error uploading the code OTA

12.2 OTA CODE UPLOAD USING WiFi

12.2.1 Motive of the project

In this part of the project, you'll upload code to a NodeMCU board OTA. You'll use the onboard WiFi module (ESP8266) on the NodeMCU to receive the code wirelessly from the Arduino IDE.

The OTA code upload using WiFi is much more advanced than the OTA code upload using Bluetooth. It provides more customization and advanced security features. Unlike the OTA code upload using Bluetooth, it can upload code OTA to any module in the WiFi network and is not restricted to 10 m range. Another important feature of OTA code upload using WiFi is that it can upload code OTA to multiple modules in the same network from a central location.

12.2.2 Installation of python

To upload code OTA using WiFi, python needs to be installed on your PC. If python is already installed on your PC, then you can skip this section. For downloading Python 2 or Python 3 software, go to the official python website at this link (https://www.python.org/downloads/). The top section of the Python Downloads webpage shows only the latest Python 3 software release. If you want to download an older release of the Python 3 software or Python 2 software, then you need to go to the releases section at the bottom of the same page.

In this project, we'll download and install the latest release of the Python 3 software. The latest release of the Python 3 software may not satisfy the required dependencies of the OTA code upload using WiFi for everyone. If it doesn't work for you, try using an older release of Python 3 software or Python 2 software.

Run the Python 3 executable file after downloading it. This will open a new window as shown in Figure 12.11. Check the box beside *Add Python 3.9 to PATH* and select *Custom installation*. This will redirect to another window as shown in Figure 12.12. All the listed features in this window are optional, check the box beside the optional features you need and click *Next*. Now, you'll be redirected to an Advanced Options selection window as shown in Figure 12.13. Select all the Advanced Options you need and choose a convenient installation location. Don't forget to select the *Add Python to environment variables* option before clicking *Install*. Wait for some time (about 5 minutes) for the installation of python to complete. After successful installation, you'll see a *Setup was successful* message as shown in Figure 12.14.

In the latest macOS (formerly, Mac OS X or OS X), Python 2 is pre-installed. Even if it is not installed, it can be easily installed using the homebrew missing software package manager. If homebrew is not installed on your Mac OS, it is highly recommended to install it. It'll help in installing many software packages easily. You can install the homebrew missing software manager by following the steps in this link (https://brew.sh/#install).

Python 3 software can be easily installed using homebrew by running the following command in the macOS Terminal.

FIGURE 12.11   Custom installation of Python

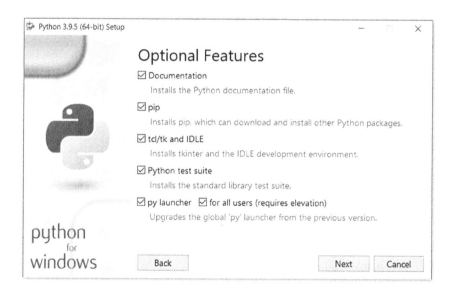

FIGURE 12.12   Choosing the optional features

```
brew install python3
```

If you don't want to use homebrew, Python 2 and Python 3 software for macOS can be downloaded from this link (https://www.python.org/downloads/mac-osx/). The procedure for python installation in macOS is similar to the installation in windows OS.

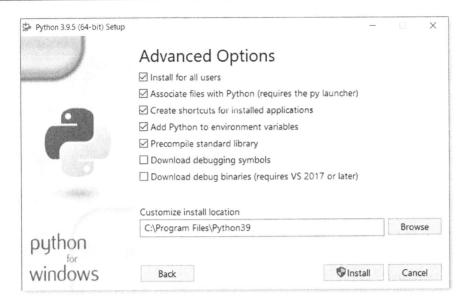

FIGURE 12.13 Customizing the Advanced Options

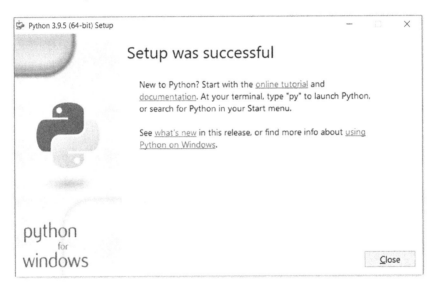

FIGURE 12.14 Completion of Python installation

12.2.3 Hardware required

In this project, we'll upload LED Blink code OTA to NodeMCU. But you can upload any code OTA to a NodeMCU board using WiFi.

TABLE 12.4 Hardware required for OTA code upload using WiFi

HARDWARE	QUANTITY
NodeMCU	1
LED	1
Jumper wires	As required

12.2.4 Connections

You can refer to the list of components required from Table 12.4, followed by Table 12.5, which presents the pin-wise connections. Also, the connection diagram is presented in Figure 12.15, which shall help get a visualization of the connections.

TABLE 12.5   Connection details for OTA code upload using WiFi

NODEMCU	LED
D4	+ve pin (long pin)
GND	−ve pin (short pin)

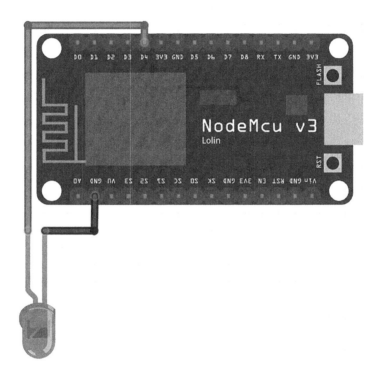

FIGURE 12.15   Connection layout for OTA code upload using WiFi

12.2.5 Updating NodeMCU firmware for OTA capability

The NodeMCU module (Amica or Lolin) by default doesn't have an OTA code upload capability. So, the OTA firmware code must be uploaded to the NodeMCU serially for enabling the OTA code upload capability. It is mandatory to initially upload the OTA firmware code serially for uploading the next set of codes OTA to NodeMCU.

Install the ArduinoOTA library or download the library from this link (https://github.com/esp8266/Arduino/tree/master/libraries/ArduinoOTA) and paste it in the *Documents* > *Arduino* > *libraries* on your PC. Now, open your Arduino IDE and open the BasicOTA code from the *Examples* as shown in Figure 12.16. If your Arduino IDE detects the architecture of the ArduinoOTA library, then it'll be available under *File > Examples > ArduinoOTA > BasicOTA* or else it'll be available under *File > Exampl es > INCOMPATIBLE > ArduinoOTA > BasicOTA.*

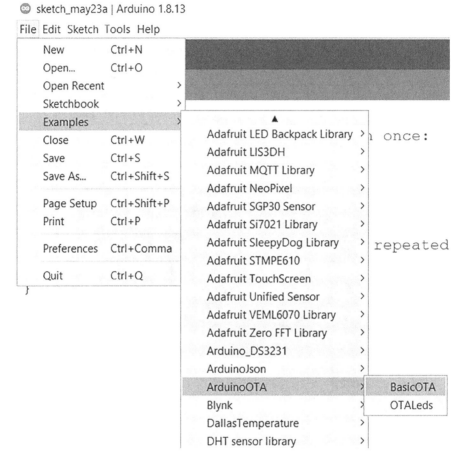

FIGURE 12.16   BasicOTA code

After opening the BasicOTA code, change the `your-ssid` and `your-password` fields with your network name and password respectively, and upload the code serially (using a cable).

Alternatively, without downloading the library, you can directly upload the BasicOTA code provided in the resources (https://github.com/anudeep-20/30IoTProjects/tree/main/Project%2012) by changing the `your-ssid` and `your-password` fields with your network details.

After successfully uploading the code, open your serial monitor and change the baud rate to 115200 to see the local IP address assigned to your NodeMCU as shown in Figure 12.17. If you don't see an IP address printed on the serial monitor, then press the RST button on NodeMCU and wait for some time for the NodeMCU to print its local IP address.

Now, if you go to *Tools > Port* in your Arduino IDE, you can see a new Port listed under *Network Ports* as shown in Figure 12.18. This will be of the format `esp8266-<ChipID> at <NodeMCU IP Address>`. If you don't find it, restart your Arduino IDE.

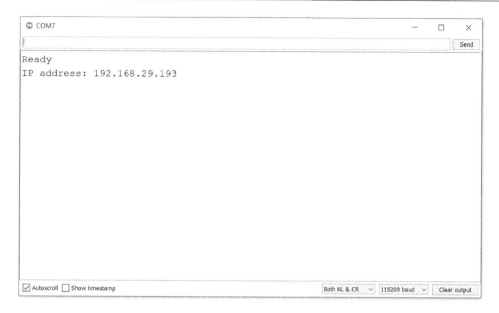

FIGURE 12.17 IP address of NodeMCU

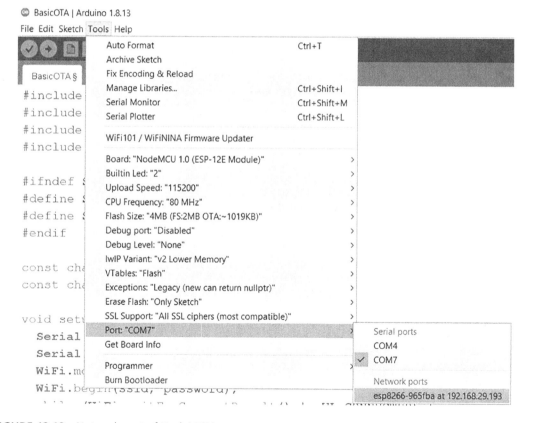

FIGURE 12.18 Network port of NodeMCU

12.2.6 Uploading code OTA to NodeMCU

Whenever you upload a new code to NodeMCU, the BasicOTA code must be added to that code. Otherwise, your NodeMCU will lose the OTA code upload capability and you cannot upload new codes OTA. So, it is recommended to modify the new code before uploading.

Suppose, if you have forgotten to upload the modified code and have uploaded the code without the BasicOTA code, then you need to again serially upload the BasicOTA code to regain the OTA code upload capability to your NodeMCU.

This is the modified LED Blink code that needs to be uploaded to NodeMCU for having a continued OTA code upload capability. Except the two highlighted code snippets, the entire code is the BasicOTA code. Hence, explanation of the code is not required.

```
#include <ESP8266WiFi.h>
#include <ESP8266mDNS.h>
#include <WiFiUdp.h>
#include <ArduinoOTA.h>

const char* ssid = "YOUR WIFI NAME"; //SSID of your WiFi
const char* password = "YOUR WIFI PASSWORD"; //Password of your WiFi

int LED = D4; //NodeMCU Pin to which LED is connected
void setup() {
  pinMode(LED, OUTPUT); //Set the direction of LED as OUTPUT
  digitalWrite(LED, LOW); //Initailly the LED is set to OFF

  Serial.begin(115200);
  Serial.println("Booting");
  WiFi.mode(WIFI_STA);
  WiFi.begin(ssid, password);
  while (WiFi.waitForConnectResult() != WL_CONNECTED) {
    Serial.println("Connection Failed! Rebooting...");
    delay(5000);
    ESP.restart();
  }
// Port defaults to 8266
// ArduinoOTA.setPort(8266);

// Hostname defaults to esp8266-[ChipID]
// ArduinoOTA.setHostname("myesp8266");

// No authentication by default
// ArduinoOTA.setPassword("admin");

// Password can be set with it's md5 value as well
// MD5(admin) = 21232f297a57a5a743894a0e4a801fc3
//ArduinoOTA.setPasswordHash("21232f297a57a5a743894a0e4a801fc3");

  ArduinoOTA.onStart([]() {
    String type;
    if (ArduinoOTA.getCommand() == U_FLASH)
      type = "sketch";
    else // U_SPIFFS
      type = "filesystem";

    // NOTE: if updating SPIFFS this would be the place to unmount SPIFFS
using SPIFFS.end()
    Serial.println("Start updating " + type);
```

```
  });
  ArduinoOTA.onEnd([]() {
    Serial.println("\nEnd");
  });
  ArduinoOTA.onProgress([](unsigned int progress, unsigned int total) {
    Serial.printf("Progress: %u%%\r", (progress / (total / 100)));
  });
  ArduinoOTA.onError([](ota_error_t error) {
    Serial.printf("Error[%u]: ", error);
    if (error == OTA_AUTH_ERROR) Serial.println("Auth Failed");
    else if (error == OTA_BEGIN_ERROR) Serial.println("Begin Failed");
    else if (error == OTA_CONNECT_ERROR) Serial.println("Connect Failed");
    else if (error == OTA_RECEIVE_ERROR) Serial.println("Receive Failed");
    else if (error == OTA_END_ERROR) Serial.println("End Failed");
  });
  ArduinoOTA.begin();
  Serial.println("Ready");
  Serial.print("IP address: ");
  Serial.println(WiFi.localIP());
}
void loop() {
  ArduinoOTA.handle();

  //Blinking of LED
  digitalWrite(LED, HIGH);

  delay(1000);
  digitalWrite(LED, LOW);
  delay(1000);
}
```

Note: The delay() function used in the code pauses the execution of the program in the NodeMCU. If an OTA code upload request is sent when the execution of the program is paused, there is a chance that the request might be missed by your NodeMCU. This is a rare case, but mostly the code gets uploaded. So, we have not avoided the delay() function in the code. If the delay() function is causing trouble in uploading the subsequent codes OTA to NodeMCU, use the below alternate code.

The difference between the alternate code and the previous code is only at two places. The below two code snippets must be replaced with the two highlighted code snippets in the previous code respectively. Only use the alternate code if you are facing an issue with the delay() function in uploading the subsequent codes OTA to NodeMCU.

LED is used for storing the digital pin to which the LED is connected. prevMillis is an unsigned long variable used to store the time (in milliseconds) at which the LEDState has been changed last time. currMillis stores the current execution time. delayMillis is the time (in milliseconds) between the blinks of an LED. LEDState stores the current state of the LED.

```
int LED = D4; //NodeMCU Pin to which LED is connected
unsigned long prevMillis = 0; //Previous time at which LED state is updated
unsigned long currMillis = 0; //Stores the current execution time
int delayMillis = 1000; //Delay between Blink
bool LEDState = LOW; //State of LED
```

If the difference between the current execution time (currMillis) and the time at which the LEDState changed last time (prevMillis) is greater than 1,000 milliseconds (1 second), then the state of the LED is reversed.

```
currMillis = millis(); //Record the current execution time
if(currMillis - prevMillis >= delayMillis){
    LEDState = not(LEDState); //Change the state of LED
    digitalWrite(LED, LEDState); //Write the new state of LED
    prevMillis = currMillis; //Record the previous LED state changed time
}
```

Finally, upload the code to NodeMCU OTA after selecting the appropriate board and network port. When you are uploading a code OTA using a network port for the first time, you'll be seeing a Security Alert as shown in Figure 12.19. If you have your NodeMCU and PC on a Private network, then select only the *Private networks* and click *Allow access*. If you have your NodeMCU and PC on a Public network, then read the risks of allowing an app through a Firewall. If you have understood the risks and ready to accept them, then select both *Private networks* and *Public networks* and click *Allow Access*. Always remember that using a Public network will expose your devices and there will be a greater chance of malware attack.

If you don't give the access, the code doesn't get uploaded OTA and throws an error as shown in Figure 12.20.

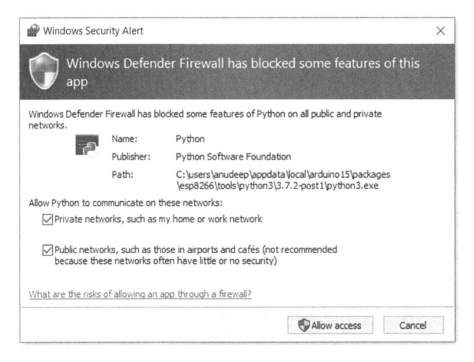

FIGURE 12.19   Security alert

FIGURE 12.20   No response error

After the code gets uploaded successfully, you can see the blinking of LED connected to the D4 pin of NodeMCU. Remember! If you are on a Network port, you cannot open the Serial monitor to see the serial data.

In the resources (https://github.com/anudeep-20/30IoTProjects/tree/main/Project%2012), we have provided a template for uploading new codes OTA to NodeMCU. It is recommended to write your new codes in that template and upload it OTA for having continued OTA code upload capability for your NodeMCU.

12.2.7 Customization and security features

The BasicOTA code provides a lot of customization and security features for uploading the code to NodeMCU OTA. These features are disabled and can be enabled depending on our requirement. The customizable features provided by the BasicOTA code are

- Port number
- Hostname
- Password in plain text
- Password in MD5 Hash

```
// Port defaults to 8266
// ArduinoOTA.setPort(8266);

// Hostname defaults to esp8266-[ChipID]
// ArduinoOTA.setHostname("myesp8266");

// No authentication by default
// ArduinoOTA.setPassword("admin");

// Password can be set with it's md5 value as well
// MD5(admin) = 21232f297a57a5a743894a0e4a801fc3
// ArduinoOTA.setPasswordHash("21232f297a57a5a743894a0e4a801fc3");
```

To change the port number from which the code is uploaded OTA to NodeMCU, the `ArduinoOTA.setPort()` function is used. By default, 8266 port number is used. It is recommended to not change the default port number unless it is occupied by some other application.

To change the name of the Network port, the `ArduinoOTA.setHostname()` function is used. By default, `myesp8266` is used as the hostname for the network port. Even though you change the hostname, the entire network port name won't be changed. For understanding the relation between the hostname and the network port name refer to Table 12.6.

TABLE 12.6   Relation between Hostname and Network Port name

HOSTNAME	NETWORK PORT NAME
myesp8266	myesp8266-<CHIPID> at <Local IP Address>
MyNewHostName	MyNewHostName-<CHIPID> at <Local IP Address>

We have set the new hostname as `MyNewHostName` using `ArduinoOTA.setHostname("MyNewHostName")` function and uploaded the code. This also changed the network port name in the *Tools > Port* as shown in Figure 12.21.

If a password is not set for the OTA code upload to NodeMCU, then anyone in the same network of your NodeMCU can upload any code OTA. If you are on a Private network, then this may not be a huge problem. But, if you are operating your NodeMCU on a Public network, then anyone can hack into your NodeMCU and can steal the data or damage your board. To avoid this, it is preferred to set a password.

You can set a password in plain text using the `ArduinoOTA.setPassword()` function. By default, there will be no authentication to your OTA upload code. We have set a new password as `mynewpassword` using the `ArduinoOTA.setPassword("mynewpassword")` function and uploaded the code. While you upload the code this time, the Arduino IDE doesn't ask for a password. But, when you upload the code next time, a new window pops up in the Arduino IDE to enter a password as shown in Figure 12.22.

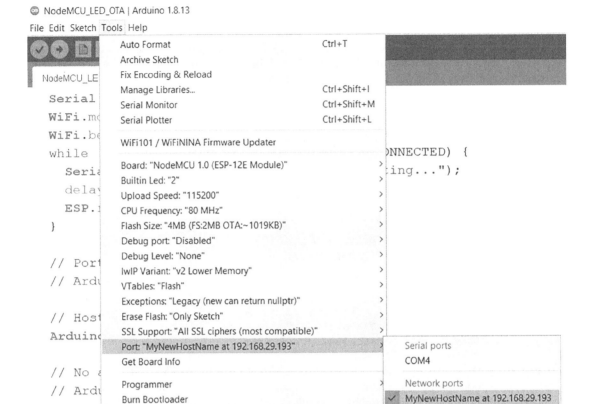

FIGURE 12.21 New network port name

FIGURE 12.22 Enter password

Enter the password and click *Upload*. If the entered password is correct, then the Arduino IDE prints Authenticating…OK on the bottom terminal and the code gets uploaded successfully as shown in Figure 12.23. If the entered password is wrong, then the Arduino IDE prints Authentication Failed on the bottom terminal and prompts to enter the password again. If the user cancels the password window, then an error is printed as shown in Figure 12.24.

FIGURE 12.23   Successful code upload

FIGURE 12.24   Error while uploading code

If you want to share your OTA upload code with others, then keeping the password in plain text may not be a good idea. The plain text password in the `ArduinoOTA.setPassword()` function can be easily read by others, and there won't be any use in setting a password for OTA code upload. To avoid this, the BasicOTA code has provided a very useful function `ArduinoOTA.setPasswordHash()` that can set the password in MD5 Hash.

It is theoretically impossible to retrieve the original string from an MD5 Hash. So, even if you mention the MD5 Hash password in the OTA upload code, nobody can find out the original password unless you tell them.

A readable string can be easily converted to an MD5 Hash online using this website (*https://www. md5hashgenerator.com/*). By using this website, `mynewpassword` is converted to `8e70383c69f7a-3b7ea3f71b02f3e9731` MD5 Hash. Now, `ArduinoOTA.setPasswordHash("8e70383c69f7a3 b7ea3f71b02f3e9731")` can be used to set `mynewpassword` as the password for OTA upload code.

When you upload a code with an MD5 Hash password for the first time, you won't be asked to authenticate unless you have set a password before. But, when you try to upload a code next time, a window will pop up as shown in Figure 12.22. Here, you can enter the password in plain text (Not in MD5 Hash) for authentication. Suppose, if you forgot the password. Don't worry! Again upload the BasicOTA code serially to the NodeMCU. This will clear the previous code and will reset all the settings to default.

Each of the explained four functions can be used independently for enabling their respective features. These functions must be placed inside the `void setup()` for enabling their respective features.

You can find all the resources of this project at this Github link (https://github.com/ anudeep-20/30IoTProjects/tree/main/Project%2012).

In this project, you have learnt to upload code OTA (without any cable) to an evaluation board using Bluetooth and WiFi protocol. This is an interesting technology and can be used in many IoT and electronic applications. In the next project, you'll learn to log sensor data continuously to Google Sheets for better analysis.

Let's build the next project!

Log Sensor Data to Google Spreadsheets 13

Data analytics is one domain that is very useful in the enhancement of IoT systems. It is the process of examining data to draw insights about the information they contain. In IoT, data analytics is used to analyze sensor behaviour and improve its quality. The microcontrollers or evaluation boards used in IoT have very little storage capacity and computing power. So, the large amount of sensor data cannot be stored in the evaluation boards and need to be stored on a local PC or on the cloud. As always connecting the evaluation board to the PC for recording the sensor data might not be possible, so the best method is to store the sensor data on a cloud. This data can be extracted and analyzed for improving the quality of the sensor.

13.1 MOTIVE OF THE PROJECT

In this project, you'll continuously log temperature sensor data along with timestamp to a Google Spreadsheet. This project can be modified to continuously log any sensor data to Google Spreadsheets. There are many advantages of logging sensor data to Google Spreadsheets over Adafruit cloud storage.

- Unlike Adafruit cloud storage, there is no minimum delay required between the uploading of two consecutive data points. So, more data points can be uploaded to Google Spreadsheets in the same time period. It should be understood that there will be an inevitable small delay between the upload of any two data points. This delay might be caused due to various reasons like the Internet speed, faulty hardware or problem in the Google servers.
- Unlike Adafruit cloud storage, multiple parameters of a sensor can be stored in a single sheet.
- Any number of sensors' data can be stored in a single spreadsheet. Data of different sensors can be separately stored in different sheets of a spreadsheet.

This concept can be used in many IoT projects to record and analyze the sensor's data. It is also used in many industries to continuously track, record and analyze the data from an industrial-grade sensor which will help in its quality assessment.

13.2 TEMPERATURE SENSOR

A temperature sensor is an electronic device that is used to measure the degree of hotness or coldness of an environment or an object. Temperature sensors are mainly used for measuring the environmental temperature, temperature of liquids and temperature inside containers. They are used in all systems where temperature can affect their functionality. For example, if a smartphone is heavily used, its processor gets heated up eventually reducing its performance and can also damage the internal circuitry because of overheating. To prevent this, all smartphones have an internal temperature sensor that will monitor the

DOI: 10.1201/9781003147169-13

temperature of the smartphone. If the smartphone is getting overheated, either a warning message will be shown or the phone automatically gets switched off. This will prevent the smartphone from damage.

The temperature sensors are further divided into two types depending on the mode of operation.

- Contact-based temperature sensors – These sensors need to be placed in contact with the object for measuring the temperature.
- Non-contact-based temperature sensors – These sensors measure the temperature of the object without having contact with them. They remotely detect the IR energy emitted by an object and send it to electronic circuitry for temperature detection.

There are many temperature sensors available in the market which can be used for different IoT applications. All these temperature sensors can be easily interfaced with the commonly used evaluation boards (like Arduino Uno, NodeMCU).

- DHT series (DHT11, DHT22) – These digital temperature sensors can measure both temperature and relative humidity.
- LM series (LM35, LM335, LM34) – These are linear temperature sensors that come directly calibrated in one of the temperature units. LM35 comes directly calibrated in Celcius (°C), LM34 comes directly calibrated in Fahrenheit (°F) and LM335 comes directly calibrated in Kelvin (K). The analog output from these sensors is directly proportional to the temperature. These sensors are very easy to use and have a large temperature detection range. LM75 is a special temperature sensor in the LM series. It can be interfaced using I²C (SDA and SCL lines) as well.
- BM series (BMP180, BME280) – These are barometric sensors that also measure temperature. These sensors are mostly used when pressure, humidity and temperature are to be measured together.
- DS18B20 temperature sensor – It is a one-wire digital temperature sensor that offers good accuracy. It is also available in a waterproof version which is ideal for measuring the temperature of liquids.
- MLX series (MLX90614, MLX90615) – These are contactless infrared digital temperature sensors that can be used to measure the temperature of objects without any physical contact.

In this project, we'll use an LM35 temperature sensor (shown in Figure 13.1). It is a contact-based temperature sensor that is widely used in IoT applications for measuring environmental temperature.

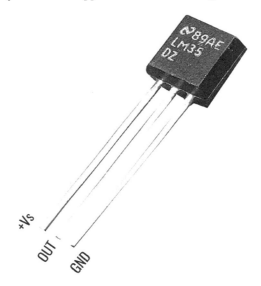

FIGURE 13.1   LM35 temperature sensor

LM35 or LM35DZ is a precision temperature sensor manufactured by Texas Instruments. Some of the specifications of LM35 are tabulated in Table 13.1. Refer to the datasheet (https://www.ti.com/lit/ds/symlink/lm35.pdf) of LM35 to know its complete properties, features and specifications.

TABLE 13.1   Specifications of LM35 Temperature Sensor

Input voltage	−0.2 V to 35 V
Output voltage	−1 V to 6 V
Output current	10 mA
Temperature detection range	−55°C to 150°C
Accuracy	±0.5°C (at 25°C)

13.3 HARDWARE REQUIRED

TABLE 13.2   Hardware required for the project

HARDWARE	QUANTITY
NodeMCU	1
LM35 temperature sensor	1
Jumper wires	As required

13.4 CONNECTIONS

You can refer to the list of required components from Table 13.2, followed by Table 13.3, for the pin-wise connections. Also, the connection diagram is presented in Figure 13.2, which shall help you to get a visualization of the connections.

TABLE 13.3   Connection details

NODEMCU	LM35 TEMPERATURE SENSOR
3.3 V	+Vs
A0	Out
GND	GND

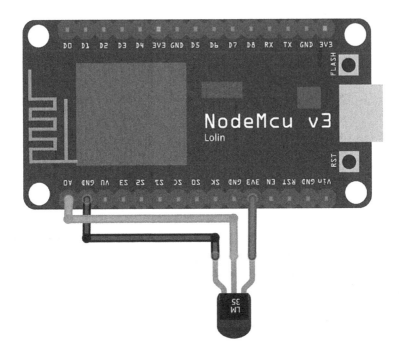

FIGURE 13.2   Connection layout

13.5 GOOGLE SPREADSHEET SETUP

The obvious thing that is needed to upload data to a Google Spreadsheet is a Google user account. If you don't have one, you can easily sign up at this link (https://accounts.google.com/SignUp?hl=en) by providing your details as shown in Figure 13.3.

After successfully creating the Google Account or if you previously have a Google Account, sign in to it and open the Google Spreadsheets (https://docs.google.com/spreadsheets/u/0/) website. Click on the+symbol above *Blank* as shown in Figure 13.4 to create a new blank spreadsheet.

After opening the new spreadsheet, change the spreadsheet name (top-left) from *Untitled SpreadSheet* to *SensorData* and the Sheet name (bottom-left) from *Sheet1* to *LM35Data*. The spreadsheet and 1st sheet can be renamed to any name. Changing the name of the spreadsheet and 1st sheet is not mandatory. It is only for your convenience.

In the first column of the *LM35Data* sheet, write *Date, Time, Temperature (°C)* and *Condition* in four consecutive columns (1A,˙ 1B, 1C, 1D cells) as shown in Figure 13.5. If you want to make the spreadsheet accessible to other people, then click *Share* at the top-right corner and change the settings.

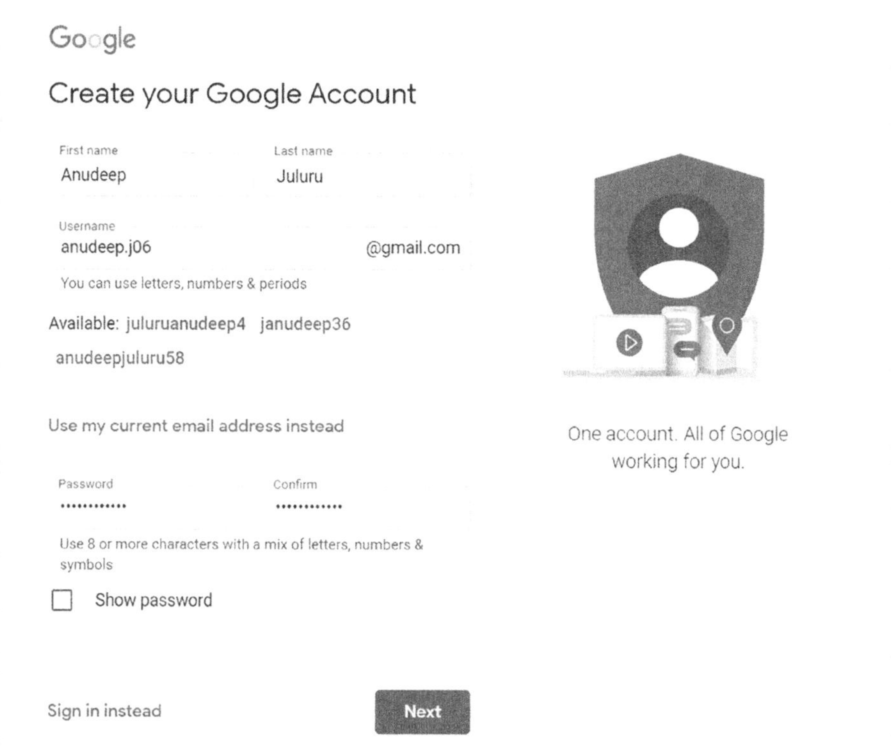
Google
Create your Google Account
First name
Anudeep
Last name
Juluru
Username
anudeep.j06
@gmail.com
You can use letters, numbers & periods
Available: juluruanudeep4 janudeep36
anudeepjuluru58
Use my current email address instead
Password
Confirm
Use 8 or more characters with a mix of letters, numbers & symbols
Show password
Sign in instead
Next
One account. All of Google working for you.

FIGURE 13.3 Creating a Google Account

Sheets
Search
Start a new spreadsheet
Template gallery
Blank
To-do list
Annual budget
Monthly budget
2021 Calendar
2020 Calendar
Recent spreadsheets
Owned by anyone
Last modified
No spreadsheets yet
Click + to create a new spreadsheet

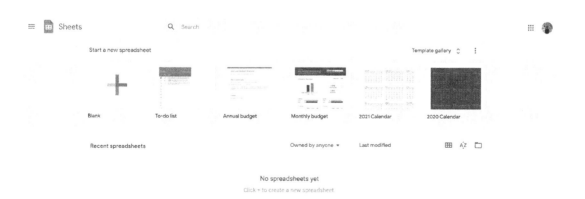

FIGURE 13.4 Creating a new blank spreadsheet

FIGURE 13.5   *SensorData* spreadsheet

13.6 GOOGLE APPS SCRIPT SETUP

Apps Script is a cloud scripting platform developed by Google for lightweight application development for Google Workspace (G Suite). It is a rapid application development platform that makes it fast and easy to create business applications that integrate with Google Workspace using JavaScript cloud scripting language.

For uploading data to the spreadsheet, we need to write a JavaScript code in Google Apps script. This code will receive the data from the NodeMCU and will write it on the *LM35Data* sheet. For opening *Script editor* from the Google Spreadsheets, go to *Tools > Script editor* as shown in Figure 13.6.

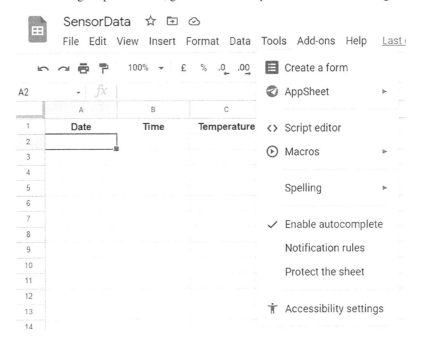

FIGURE 13.6   Script editor

If you have signed in to multiple Google Accounts in your browser, the *Script editor* doesn't open and throws an error as shown in Figure 13.7. So, sign out from all the other Google Accounts and then try opening it.

Google Drive

Sorry, unable to open the file at present.

Please check the address and try again.

Get stuff done with Google Drive

Apps in Google Drive make it easy to create, store and share online documents, spreadsheets, presentations and more.

Learn more at drive.google.com/start/apps.

FIGURE 13.7   Error opening Google Apps script

After opening the Google Apps script editor, change the script name (top-left) from *Untitled project* to *SensorData Script*. The JavaScript code used for uploading data to the spreadsheet needs to be written in the text editor shown in Figure 13.8. If you don't know JavaScript, don't worry! It is very easy to code and understand JavaScript.

FIGURE 13.8   Google Apps script

13.7 GOOGLE APPS SCRIPT CODE

The code written in Google Apps script can be published as a web app only if it satisfies the following conditions:

- It should contain a doGet(e) or doPost(e) function.
- Each of the function must return a HTML service HTMLOutput object or a Content service TextOutput object.

If the user sends an HTTP GET request to the web app, then doGet(e) function is executed or if the user sends an HTTP POST request to the web app then doPost(e) function is executed. The e argument in these functions represents the event parameter which contains the data of any request parameters.

In this project, we are sending temperature data from the NodeMCU to the web app. So, we will use a doGet(e) function to receive the data. The e argument contains the parameters sent by the NodeMCU in the form of a JSON. If you are unaware of JSON (JavaScript Object Notation) format, then it is recommended to read about it at this link (https://developer.mozilla.org/en-US/docs/Web/JavaScript/Reference/Global_Objects/JSON) before proceeding further.

Logger.log() is used to print data on the Execution log. It is similar to Serial.print() printing data on serial monitor. As Logger.log() cannot print a JSON object, it needs to be converted to a string using JSON.stringify().

In JavaScript, there is no need for declaring a variable by the type of data stored in it. var can be used to declare any kind of data type (int, float, string, etc.). result is used to store the string which needs to be returned by the doGet() function. Initially, it is assumed that parameters are not received by storing Failed receiving Parameters in the result variable.

```
function doGet(e) {
  Logger.log(JSON.stringify(e));
  var result = "Failed receiving Parameters";
```

If there is no data in the e argument or if there is no parameter data in the e argument, then No Parameters is stored in the result. If the e argument contains parameter data, then this data is written on to the *SensorData* spreadsheet.

GSheetID stores the ID of the *SensorData* Google Spreadsheet. You can get the ID of your Google Spreadsheet from its URL as shown in Figure 13.9. Copy this ID and paste it in the place of YOUR GOOGLE SPREADSHEET ID.

LM35DataSheet points to the *LM35Data* sheet in the *SensorData* spreadsheet. If you have renamed the 1st sheet other than *LM35Data*, then copy that name in the place of LM35Data in .getSheetByName("LM35Data"). newRow points to the row after the last filled row in the *LM35Data* sheet. rowData stores the data to be written in the newRow.

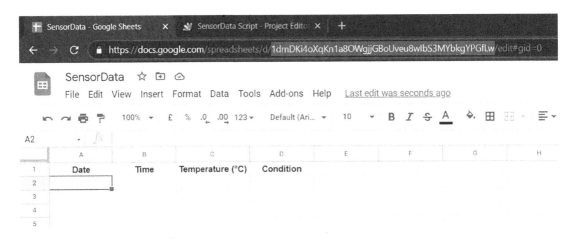

FIGURE 13.9　Google Spreadsheets ID

```
if (e == null || e.parameter == null) {
  result = "No Parameters";
}
else {
  var GSheetID = "YOUR GOOGLE SPREADSHEET ID";
```

```
    var LM35DataSheet = SpreadsheetApp.openById(GSheetID).
getSheetByName("LM35Data");
    var newRow = LM35DataSheet.getLastRow() + 1;
    var rowData = [];
```

currDate stores a Date() object which contains the present date and time information. You can know more about the Date() object and its related functions at this link (https://developer.mozilla.org/en-US/docs/Web/JavaScript/Reference/Global_Objects/Date).

.toLocaleDateString()function is used to obtain the date information from the currDate. .toLocaleTimeString()function is used to obtain the time information from the currDate. This information is stored in the rowData array.

```
    var currDate = new Date();
    rowData[0] = currDate.toLocaleDateString();
    rowData[1] = currDate.toLocaleTimeString();
```

for loop is used to iterate through all the parameters in the e argument. All the parameters will be available at e.parameter in the form of string data type. .replace() is used to replace a substring in the string with another substring. For example, "How Is You?".replace("Is", "Are") replaces the Is substring with Are substring.

The first argument in the replace, /["']/g is a regular expression (also known as RegEx or RegExr). Regular expressions are patterns used to match character combinations in a string. Go to these websites (https://developer.mozilla.org/en-US/docs/Web/JavaScript/Guide/Regular_Expressions, https://regexr.com/) to know more about regular expressions, for creating a new regular expression or for the explanation of a known regular expression. The regular expression /["']/g used in the code is explained below:

/ – Start of the regular expression
[– Start of the character set
" – Matches " character
' – Matches ' character
] – End of character set
/ – End of the regular expression
g – Global Expression flag. It is used to find all the matches (not just the first one).

In short, / / are used at the beginning and end of the regular expression. [] contains the character set that needs to be matched inside them. g is used for finding all the matches of the character set.

```
for (var param in e.parameter) {
    var value = (e.parameter[param]).replace(/["']/g, "");
```

Nested if-else statements are used to write the parameters in the correct order to rowData. You can also use a switch statement for the same purpose. If the parameter is temperature, then it is written to the 2nd index of rowData, or if the parameter is condition, then it is written to the 3rd index of rowData. If the parameter is not among these two, then nothing is stored in the rowData.

getRange() is used to point a particular cell or a range of cells. It can be used in four different ways:

- getRange(row, column) – returns a particular cell with the row number as row and column number as column.
- getRange(row, column, numRows) – returns a range of cells in a single column. The cells in the row number row to numRows of column number column are returned.
- getRange(row, column, numRows, numColumns) – returns a range of cells in multiple rows and multiple columns. The cells in the row number row to numRows of column number column to numColumns are returned.

- `getRange(a1Notation)` – returns the range of cells (single or multiple rows or columns) mentioned in the `a1Notation`. If the `a1Notation` is C2, then data in the 3rd column and 2nd row cell is returned. If the `a1Notation` is B1:D5, then data in all the cells from 2nd column 1st row to 4th column 5th row are returned.

The numbers in the `.getRange()` function represent the actual row numbers and columns numbers in the spreadsheet. Unlike the array indices, they don't start at 0. `.setBackground()` is used to change the background of a cell or a range of cells.

```
if(param == "temperature") {
    rowData[2] = value;
    result = "Data written on column C Successfully.";
} else if(param == "condition") {
    rowData[3] = value;

    if (value == "COLD") {
      LM35DataSheet.getRange(newRow, 4).setBackground("#66CDAA");
    } else if (value == "NORMAL") {
      LM35DataSheet.getRange(newRow, 4).setBackground("#7CFC00");
    } else if (value == "HOT") {
      LM35DataSheet.getRange(newRow, 4).setBackground("#DC143C");
    }
    result += "\nData written on column D Successfully.";
} else {
  result += "\nUnsupported Parameter Detected.";
}
}
```

`.setValues()` is used for writing data in a cell or range of cells. Finally, the `doGet()` function returns the `result` using a `ContentService.createTextOutput()` object as explained at the beginning of this section.

```
    LM35DataSheet.getRange(newRow,1,1,rowData.length).setValues([rowData]);
}
    return ContentService.createTextOutput(result);
}
```

Save the code using *Ctrl+S* (on Windows) or *⌘-S* (on macOS) or use the save symbol at the top of the text editor. Click Deploy (top-right) and select *New Deployment* as shown in Figure 13.10. This will open a new window as shown in Figure 13.11. Click the gear symbol beside the *Select type* and select *Web app* among the options in the dropdown.

Fill in the *Description* of your Web app, select *Execute as Me(<YOUR EMAIL ID>)*, *Who can Access* as *Anyone* and then Click *Deploy* as shown in Figure 13.12. Next, click *Authorize access* as shown in

FIGURE 13.10 New deployment

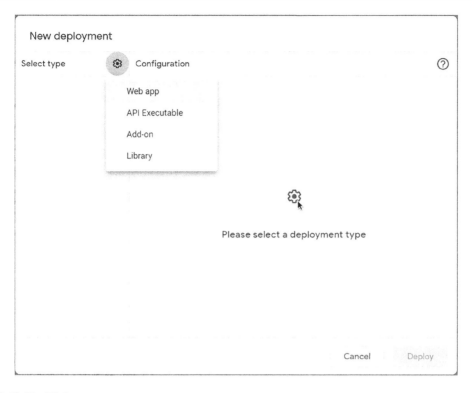

FIGURE 13.11   Web app

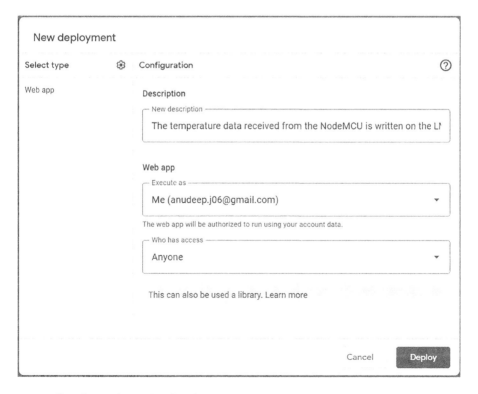

FIGURE 13.12   Fill in the configuration details

FIGURE 13.13   Authorize access

Figure 13.13. This will open a new browser window as shown in Figure 13.14. Among the listed Google Accounts, select the Google Account from which you have deployed this web app.

Now, you'll see a warning message *Google hasn't verified this app* as shown in Figure 13.15. Don't worry on seeing this warning message. The web app you deployed is not dangerous and doesn't steal any of your or others' data. This warning message is shown to all the web apps which haven't gone through the Google verification process. This process is recently launched by Google to protect user's data.

Click *Advanced* and then click *Go to SensorData Script (unsafe)* as shown in Figure 13.16. This will redirect you to a new page as shown in Figure 13.17. Click *Allow* to allow the *SensorData Script* web app to access your spreadsheets in Google Drive. Now, you'll see *Deployment successful* message as shown in Figure 13.18. Copy the *Deployment ID* and click *Done*.

Remember: Whenever you make changes to the code, save it and then deploy it as a new version of web app for the changes to be applied.

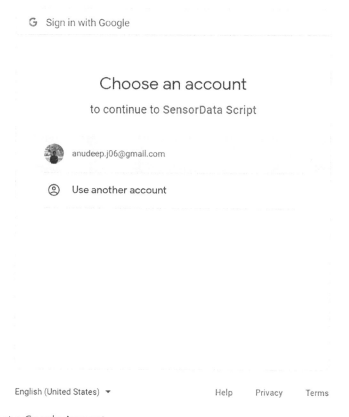

FIGURE 13.14   Select a Google Account

For deploying the new version of the web app, click *Deploy* and select *Manage deployments* from the dropdown as shown in Figure 13.19. Select the version as *New version*, fill in the *Description* and click *Deploy* as shown in Figure 13.20. Now, you'll be able to see *Deployment successfully updated* message as shown in Figure 13.21. Copy the *Deployment ID* and click *Done*.

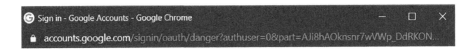

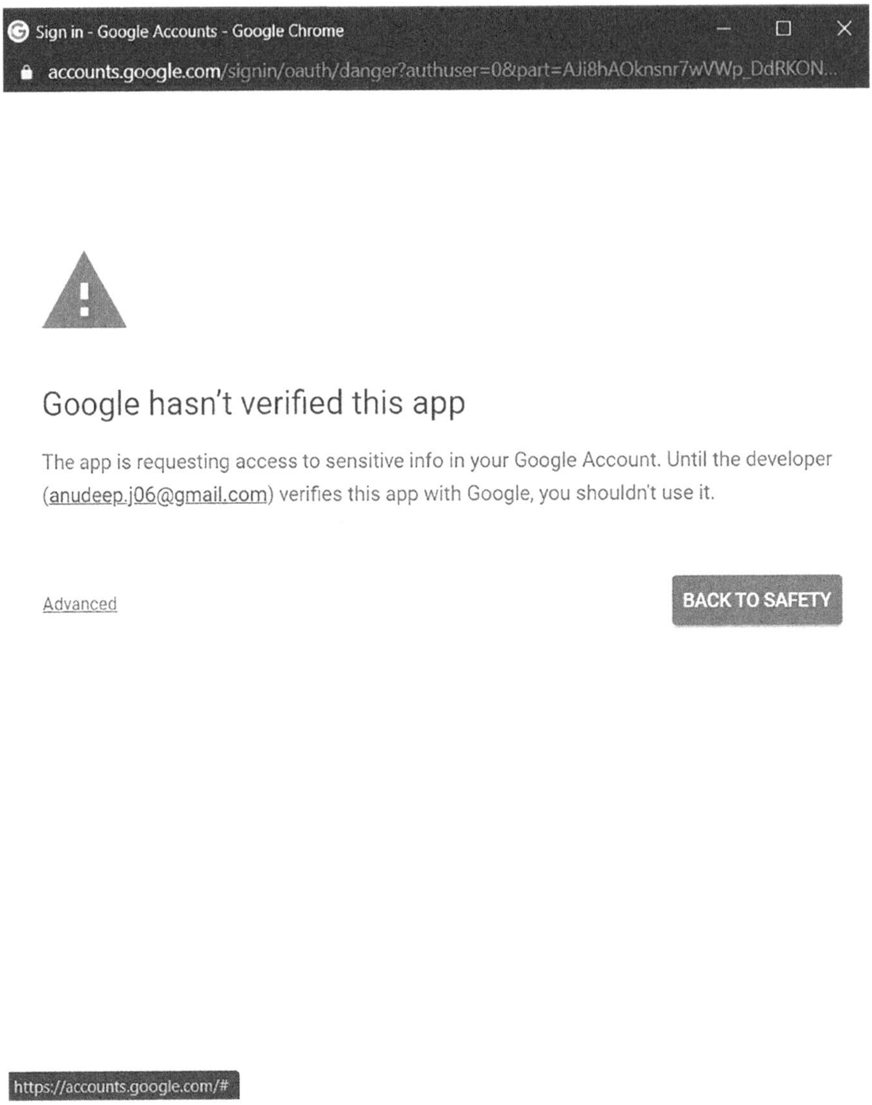
Sign in - Google Accounts - Google Chrome
accounts.google.com/signin/oauth/danger?authuser=0&part=AJi8hAOknsnr7wVWp_DdRKON...
Google hasn't verified this app
The app is requesting access to sensitive info in your Google Account. Until the developer (anudeep.j06@gmail.com) verifies this app with Google, you shouldn't use it.
Advanced
BACK TO SAFETY
https://accounts.google.com/#

FIGURE 13.15 Warning message

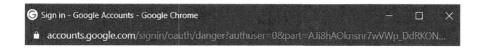

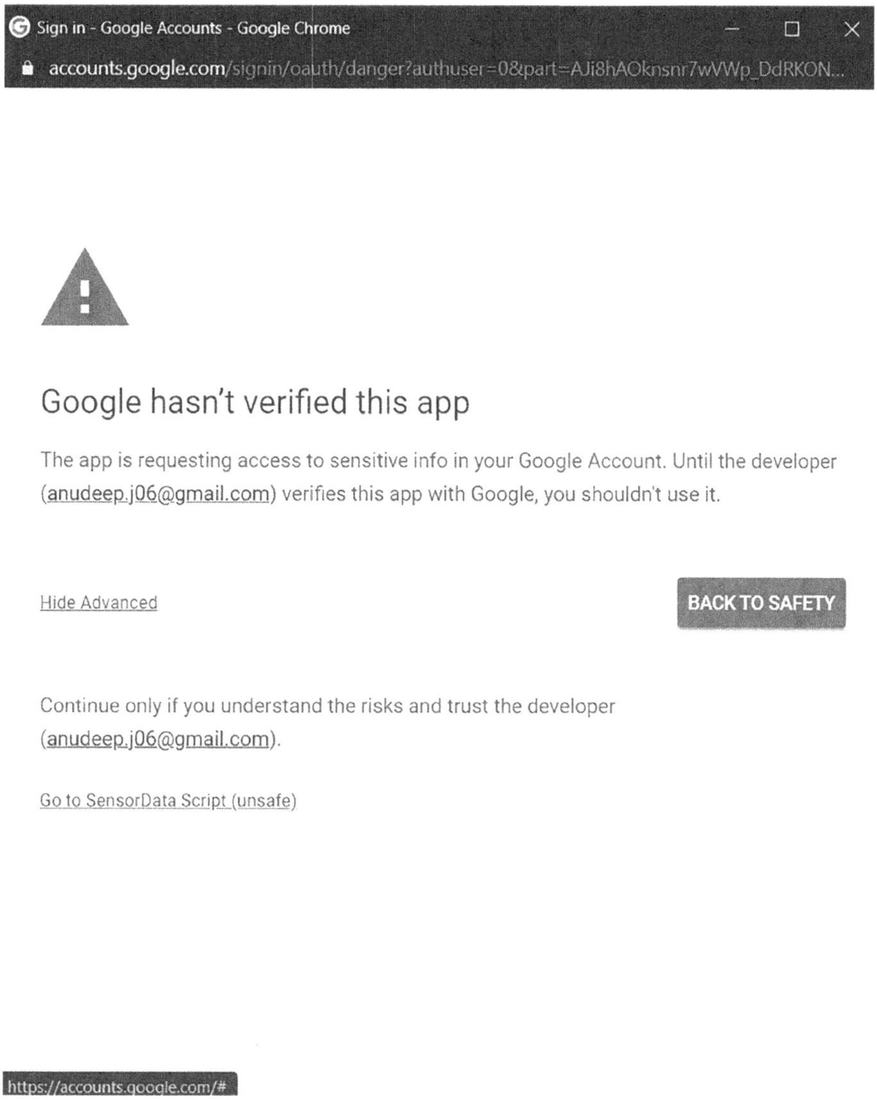
Sign in - Google Accounts - Google Chrome
accounts.google.com/signin/oauth/danger?authuser=0&part=AJi8hAOknsnr7wVWp_DdRKON...
Google hasn't verified this app
The app is requesting access to sensitive info in your Google Account. Until the developer (anudeep.j06@gmail.com) verifies this app with Google, you shouldn't use it.
Hide Advanced
BACK TO SAFETY
Continue only if you understand the risks and trust the developer (anudeep.j06@gmail.com).
Go to SensorData Script (unsafe)
https://accounts.google.com/#

FIGURE 13.16 Advanced section of warning message

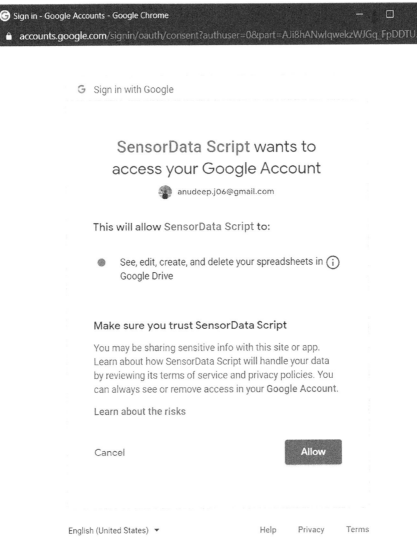

FIGURE 13.17   Allow access

FIGURE 13.18   Deployment successful

FIGURE 13.19   Manage deployments

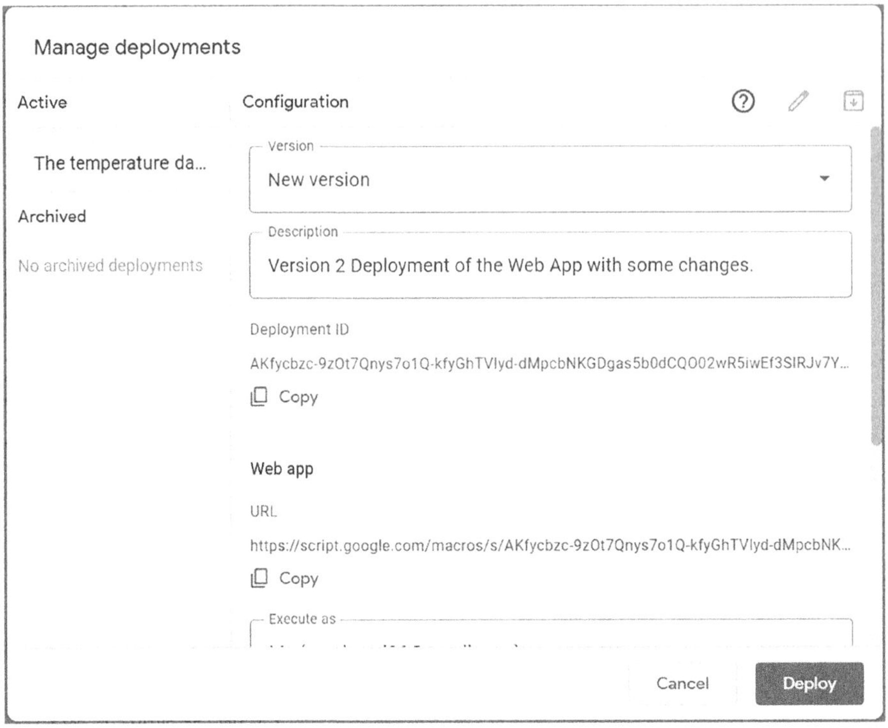

Manage deployments

Active   Configuration   ⓘ   ✎   ⊕

    The temperature da...   ┌─ Version ──────────────────────────────────┐
                               │ New version   ▾   │
Archived   └──┘

No archived deployments   ┌─ Description ──────────────────────────────┐
                               │ Version 2 Deployment of the Web App with some changes. │
                               └──┘

                               Deployment ID

                               AKfycbzc-9zOt7Qnys7o1Q-kfyGhTVlyd-dMpcbNKGDgas5b0dCQO02wR5iwEf3SIRJv7Y...
                               ⧉ Copy

                               Web app

                               URL

                               https://script.google.com/macros/s/AKfycbzc-9zOt7Qnys7o1Q-kfyGhTVlyd-dMpcbNK...
                               ⧉ Copy

                               ┌─ Execute as ──────────────────────────────

                                                          Cancel   **Deploy**

FIGURE 13.20   Configuration of the new version

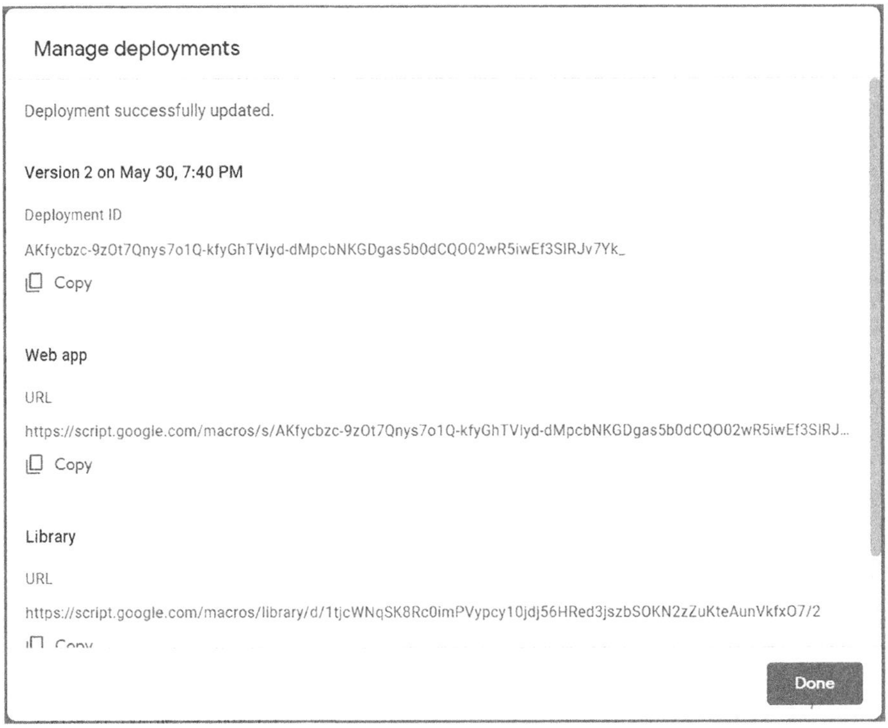

Manage deployments

Deployment successfully updated.

Version 2 on May 30, 7:40 PM

Deployment ID

AKfycbzc-9zOt7Qnys7o1Q-kfyGhTVlyd-dMpcbNKGDgas5b0dCQO02wR5iwEf3SIRJv7Yk_
⧉ Copy

Web app

URL

https://script.google.com/macros/s/AKfycbzc-9zOt7Qnys7o1Q-kfyGhTVlyd-dMpcbNKGDgas5b0dCQO02wR5iwEf3SIRJ...
⧉ Copy

Library

URL

https://script.google.com/macros/library/d/1tjcWNqSK8Rc0imPVypcy10jdj56HRed3jszbSOKN2zZuKteAunVkfxO7/2
⧉ Copy

                                                                      Done

FIGURE 13.21   New version deployment successful

13.8 TESTING OF WEB APP

After deploying the web app, we need to test it to know whether it's working properly or not. To test the web app without NodeMCU, you need the web app URL shown in Figure 13.18. If you didn't take a note of it in the previous step, click *Deploy* and select *Test Deployments* as shown in Figure 13.22, and there you can find the web app URL.

FIGURE 13.22   Test deployments

Open a new tab or browser window and paste the URL. At the end of URL add *?temperature=25&condition=NORMAL* and click Enter. If there is no error in the code and the web app is working as expected, then you'll see *Data written successfully* message as shown in Figure 13.23. Now, go to the *SensorData* spreadsheet and see the data written on the 2nd row of *LM35Data* sheet as shown in Figure 13.24. If there is any error in the code or in the deployment of the web app, then you'll see an exception or error as shown in Figure 13.25.

FIGURE 13.23   Successful working of web app

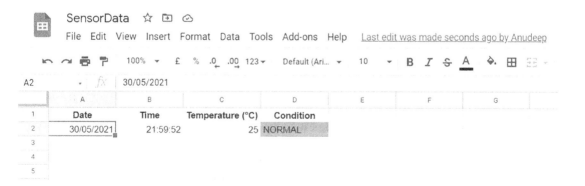

FIGURE 13.24   Data written on *LM35Data* sheet

Exception: Unexpected error while getting the method or property openById on object SpreadsheetApp. (line 10, file "Code")

FIGURE 13.25 Error in web app

13.9 ARDUINO IDE CODE

ESP8266WiFi.h library is used for connecting the NodeMCU to the WiFi. NodeMCU (or ESP8266) is designed to easily send and receive HTTP requests over the Internet. But sending HTTPS (HTTP Secure) requests using NodeMCU is not easy. HTTPS requests are nothing but HTTP requests over TLS (formerly SSL) connection. In an HTTPS connection, the data sent and received from the PC is encrypted and protected. To easily send and receive HTTPS requests over the Internet using NodeMCU, WiFiClientSecure.h library is used.

```
#include <ESP8266WiFi.h> //Library for WiFi functionality
#include <WiFiClientSecure.h> //Library for secure connections
```

Similar to Arduino Uno, NodeMCU also has an analog pin (A0). As the LM35 temperature sensor is an analog sensor, it is connected to the A0 analog pin of NodeMCU. ssid and password are used to store your WiFi name and password, respectively.

```
#define LM35Pin A0 //NodeMCU Pin to which LM35 is connected

const char* ssid = "YOUR WIFI NAME"; //SSID of your WiFi
const char* password = "YOUR WIFI PASSWORD"; //Password of your WiFi
```

host is the address to which the HTTPS request is sent. As we are sending an HTTPS request to the Google Apps script web app, the host will be script.google.com. HTTPSPort stores the port through which NodeMCU is connected to the web app. By default, HTTP uses port 80 and HTTPS uses port 443. deploymentID stores the Deployment ID of your web app. Replace Your Deployment ID with the Deployment ID which you have copied in Section 13.7.

```
//Google SpreadSheets Details
const char* host = "script.google.com";
const int HTTPSPort = 443;
String deploymentID = "Your Deployment ID";
```

requestURL stores the URL that needs to be sent to the web app. clientReply stores the reply from the web app. VRef is used to store the voltage to which the LM35 temperature sensor's +V$_s$ pin is connected. temperatureVal is used to store the analog readings read from the LM35 temperature sensor's OUT pin. resolution is used to calculate and store the resolution with which the temperatureVal must be multiplied to obtain the correct temperature in °C. count is used to store the number of requests sent to the web app from the start of the program execution.

```
String requestURL; //Stores the request URL for sending to web app
String clientReply; //Stores the reply from the web app
```

```
int VRef = 3.3; //Voltage to which LM35 is connected
float temperatureVal; //Used to store the read temperatures
float resolution = (VRef*100)/1024.0; //Calculation of resolution
String condition; //Stores the temperature condition

int count = 0; //Counting the number of requests sent to web app
```

espClient is a WiFiClientSecure object used for creating a secure client that can connect to a specific IP address and port. In the void setup(), NodeMCU's serial is initialized at 115200 baud rate using Serial.begin(). As we are reading analog values from the LM35 temperature sensor, the direction of LM35Pin is set as INPUT. The code for connecting the NodeMCU to WiFi is the same as explained in the previous projects.

```
WiFiClientSecure espClient; //Create a secure WiFi client

void setup() {
  Serial.begin(115200); //Initialisation of NodeMCU Serial
  //Defining the direction of LM35 Temperature sensor
  pinMode(LM35Pin,INPUT);

  //Connecting NodeMCU to WiFi
  Serial.println("");
  Serial.print("Connecting to ");
  Serial.println(ssid);
  WiFi.begin(ssid, password);
  while (WiFi.status() != WL_CONNECTED)
  {
    delay(500);
    Serial.print(".");
  }
  Serial.println("");
  Serial.println("WiFi connected");
  //Print the local IP address of NodeMCU
  Serial.println(WiFi.localIP());
```

WiFiClientSecure library can be used to establish a secure connection between web app and NodeMCU in three ways:

- Using a root certificate authority (CA) cert
- Using a root CA cert plus a client cert and key
- Using a pre-shared key (PSK)

There are two issues with using one of these methods. Certificates and fingerprints of web pages to which the NodeMCU needs to be connected must be extracted and stored manually. Another major issue is that these fingerprints change once in a year and the certificates expire. So, whenever there is a change in fingerprint, the code needs to be updated manually. There is no way to overcome these problems in the case of secure connections.

We require a secure connection mostly when we are transferring sensitive data (like personal data, passwords, etc.) over the Internet. Secure connections for transferring less sensitive or less important data is overkill. In our application, we are only sending temperature and room condition data to the web app. There is no big problem even if this data is compromised. So, we can use the .setInsecure() function which will establish a connection between NodeMCU and the web app without validating the certificate. If you want to connect the NodeMCU to the web app through a secure connection, then read the official docs (https://arduino-esp8266.readthedocs.io/en/2.4.0/esp8266wifi/client-secure-examples.html).

```
    espClient.setInsecure(); //Used for HTTPS requests
}
```

`.connected()` is used to know whether the connection between NodeMCU and the web app is established or not. If the connection is broken or not established, then `.connect(host, HTTPSPort)` is used to connect NodeMCU to the web app.

```
void loop() {
    //Connecting to Client
    if(!espClient.connected()){
      Serial.print("Connecting to Client");
      while (!espClient.connect(host, HTTPSPort)){
        Serial.print(".");
      }
      Serial.println("");
    }
```

Similar to `digitalRead()`, `analogRead()` reads the analog values from an analog pin. `temperatureVal` stores the analog values read from the `LM35Pin` analog pin and then multiplies with the `resolution` to obtain the correct temperature in °C.

```
    //Reading Temperature values from LM35 Temperature sensor
    temperatureVal = analogRead(LM35Pin);
    temperatureVal = temperatureVal*resolution;
    Serial.print("Temperature: ");
    Serial.print(temperatureVal);
    Serial.println(" °C");
```

`condition` is assigned based on the temperature. If the temperature is less than or equal to 20°C, then it is considered COLD. If the temperature is between 20°C and 32°C, then it is considered NORMAL. If the temperature is greater than 32°C, then it is considered HOT.

```
    //Assigning condition based on Temperature
    if(temperatureVal <= 20)
      condition = "COLD";
    else if(temperatureVal <= 32)
      condition = "NORMAL";
    else
      condition = "HOT";
```

`requestURL` contains the deployment ID of your web app and the parameters to be sent (temperature and condition). `espClient.print()` is used to send an HTTP request header to the web app. `Connection` can be `close` or `keep-alive`. `close` will end the connection, and the connection needs to be re-established for sending another request. `keep-alive` keeps the connection open, and the requests can be sent without establishing a connection again.

```
    //Request URL for sending to web app
    requestURL = "/macros/s/" + deploymentID + "/exec?temperature=" +
String(temperatureVal) + "&condition=" + condition;

    //HTTPS request sent to web app
    espClient.print("GET " + requestURL + " HTTP/1.1\r\n" +
               "Host: " + host + "\r\n" +
               "User-Agent: BuildFailureDetectorESP8266\r\n" +
               "Connection: Keep-Alive\r\n\r\n");

    Serial.println("Request sent");
```

`espClient.readString()` is used to read incoming data from the web app. The data returned by the web app is very huge and contains a lot of information (HTTP response headers and HTML code). The required information to know whether the request from NodeMCU is successful or not will be present in the first line of HTTP response header. We can easily read the first line of the HTTP response header alone using `espClient.readStringUntil('\n')`, but the remaining data will still be present in the NodeMCU's buffer. So, in the next cycle of `void loop()`, the `espClient.readStringUntil('\n')` will read the first line of the buffer data (second line of the previous HTTP response header). This will continue until the complete buffer data is read. So, to read the next HTTP response header, the NodeMCU's buffer data must be cleared.

Previously (before Arduino IDE 1.0), `.flush()` function is used to clear any incoming buffer data. Now, the same function is used to wait for the transmission of outgoing data. In the present version of Arduino IDE, the buffer data can be cleared only after reading it.

Depending on the success of the request, different HTTP response header and HTML code will be returned by the web app. The HTTP response header starting with HTTP/1.1 302 is considered as request successful, and any other HTTP response header is considered as failure. Some of the usually seen HTTP response headers' first line are

- `HTTP/1.1 302 Moved Temporarily`
- `HTTP/1.1 500 Internal Server Error`
- `HTTP/1.1 200 OK`

To know more about the HTTP request and response headers, read the documentation (https://developer.mozilla.org/en-US/docs/Glossary/HTTP_header). If you want to know the complete incoming data returned by the web app, see the *returnedData.docx* file in the resources of this project (https://github.com/anudeep-20/30IoTProjects/tree/main/Project%2013).

```
//Reading the reply from web app
  clientReply = espClient.readString();
  if(clientReply.startsWith("HTTP/1.1 302")){
    Serial.println("Data Published Successfully");
  } else {
    Serial.println("Data Publish Failed");
  }
```

Reading the complete response using `.readString()` function takes almost 7 seconds. So, if you want to upload data to Google Spreadsheet, then there will be a minimum delay of 7 seconds between two consecutive data points. This time gap might be a problem if you want to record the sensor data at a higher frequency. To upload sensor data with a minimal delay, you can comment out the above highlighted code and uncomment the following code. This will also prevent you from knowing whether the request sent by NodeMCU is successful or not.

There is no restriction of minimum delay that has to be given after each request. But, if there is no delay, then the NodeMCU sends multiple HTTP requests one after the other to the web app and the web app will not have enough time to process all of them. So, some or many requests might be dropped by the web app. To avoid this situation, it is better to give at least a 1 second delay (ideally 2 seconds).

Sending continuous requests without reading, the web app's response is causing a peculiar problem. After 70 requests, the NodeMCU is unable to send any more requests to the web app. To avoid this problem, the NodeMCU is restarted after every 70 requests using `ESP.restart()`.

```
/*
  if(count >= 70)
    ESP.restart(); //Restart the NodeMCU

  delay(2000);
```

```
        count += 1;
    */
}
```

Finally, upload the code to NodeMCU after selecting the appropriate board and COM port. After successfully uploading the code, open the serial monitor and change the baud rate to 115200 to see whether the NodeMCU is connected to the WiFi and web app or not. If the NodeMCU is successfully connected to the WiFi and web app then open your Google Spreadsheet to see the temperature data being uploaded to it as shown in Figure 13.24.

You might have a doubt that what is the need for sending two parameters (temperature and condition) to the web app as we can determine the condition using the temperature values in the Google Apps script code. It is done for two reasons:

- The first and the main reason is to show you how to send multiple parameters (>1) to the web app.
- The second reason is to make the dependency on the Google Apps script code minimal. The Google Apps script code will be on the server, and it is not changed as frequently as NodeMCU's code. If you want to change the thresholds at which the conditions (COLD, NORMAL, HOT) change, then you need to change the code in the Google Apps script and deploy a new version. So, it is better to have the condition part of the code in NodeMCU as you can update the code in the NodeMCU whenever needed.

For analysis purpose and to know whether the system works for continuous data upload or not, we have recorded the temperature of our room continuously for more than 24 hours using the LM35 temperature sensor. If you want to see the data and make some analysis out of it, see the *SensorData.xlsx* file in the resources of this project (https://github.com/anudeep-20/30IoTProjects/tree/main/Project%2013).

13.10 TRY IT

You can use this project in almost all of your IoT projects where sensor data needs to be stored. You can add location details to the sensor data by adding a GPS sensor to your project. This will help in geotagging your sensor data.

You can also store data of multiple physical parameters of a sensor in a single spreadsheet by assigning different sheets for different physical parameters. For example, a BME280 sensor can give temperature, humidity, pressure and altitude data. So, the data of different physical parameters can be stored in different sheets of a single spreadsheet. Try this on your own by modifying the code in Google Apps script and NodeMCU.

You can find all the resources of this project at this Github link (https://github.com/anudeep-20/30IoTProjects/tree/main/Project%2013).

In this project, you have learnt to upload sensor data continuously to Google Spreadsheet without any data limit or delay limit between two consecutive data points. You have also learnt about different types of temperature sensors and some basics of JavaScript coding. The next project will act as a bridge between electronics and electrical projects. In the next project, you'll control an electric light bulb using NodeMCU and Adafruit Dashboard.

Let's build the next one!

Controlling Electrical Appliances Using Relay

14

In our daily life, we use more electrical devices than electronic devices starting from an electric bulb to air conditioner. We'll control them manually by switching them ON or OFF, but at times we need to control them remotely. For example, we might forget to switch OFF the electric bulb/fan and go out or we might want to switch ON the air conditioner before we reach home. There are many more cases where we want to control the electrical appliances remotely. This project will help you in remotely controlling electrical appliances using IoT.

14.1 MOTIVE OF THE PROJECT

In this project, you'll control an electric light bulb using Relay and Adafruit Dashboard. Until the previous project, you have only controlled electronic modules and sensors but did not control any electrical device or appliance. This project acts as a bridge between electronics and electrical projects. It helps you in controlling any electrical device or appliance using an electronic evaluation board or microcontroller.

Warning: In this project, you need to use high AC voltage for powering an electric bulb. If not used properly, high AC voltage is very dangerous and can cause

- Physical damage to the equipment
- Fire or any other dangerous hazards
- Serious injuries or death

So, the entire project must be performed with utmost care, and we take no responsibility for any of your precipitous actions.

14.2 RELAY

Relay is nothing but an electromechanical switch that connects or disconnects a circuit. Switches that you generally see at your home are controlled manually for connecting or disconnecting a circuit. But a relay is controlled using an electrical signal.

For controlling an LED, you can directly connect it to any IO pin of a microcontroller (or an evaluation board) and program it to send signals for turning the LED ON and OFF. But the same procedure cannot be followed for controlling a 10 W electric bulb or any electrical equipment because they cannot be interfaced with any electronic microcontroller. So, a relay is used to bridge the gap between them.

DOI: 10.1201/9781003147169-14

Relay can be used to connect or disconnect a high voltage or high current circuit using a small signal from a microcontroller.

There are many types of relay available in the market which are used for different applications (IoT, industrial, automotive, etc.). Some of them are electromagnetic relays, latching relays, non-latching relays, solid-state relays, reed relays, small signal relays, time delay relays, thermal relays, frequency relays, polarized relays and many more. Each of them has a different type of working mechanism.

Similar to a normal switch, a relay has one or more poles, each of the pole's electrical contact can be thrown at one of the terminals by using small electronic signals. In this project, we'll use an active low SPDT (single-pole double-throw) 5 V DC 1 channel relay module as shown in Figure 14.1 to control an electric bulb using NodeMCU and Adafruit Dashboard.

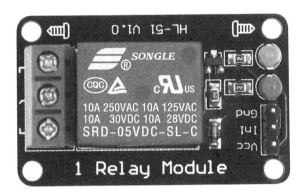

FIGURE 14.1   5 V DC 1 channel relay module

An SPDT relay has a single pole whose electrical contact can be thrown at either of the two terminals as shown in Figure 14.2. The single pole is a common terminal, and the other two terminals are normally open (NO) and normally closed (NC).

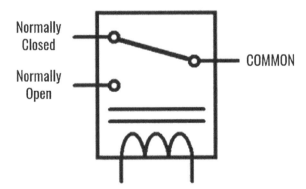

FIGURE 14.2   SPDT (single-pole double-throw) relay

Internally, an electromechanical relay energizes the coil wound on an iron core to move an armature using magnetic flux as shown in Figure 14.3. The movement of the armature is used to move the common between normally open and normally closed. For an active low relay, NO will be connected to common only when the coil is energized or else it will be disconnected. Similarly, NC will be connected to common only when the coil is not energized or else it will be disconnected. For an active high relay, everything is reversed when compared to an active low relay. All of these results are tabulated in Table 14.1 for your easier understanding.

Note: In the whole book, we'll only use active low relay modules wherever necessary and do not use any active high relay module. So, whereever relay module is mentioned in this book, it means it is an active low relay module.

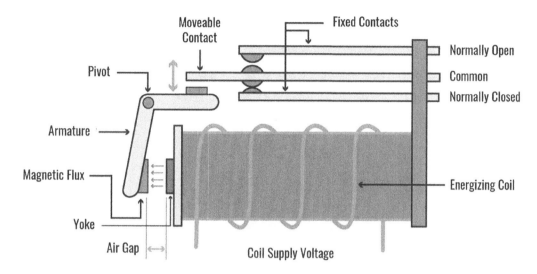

FIGURE 14.3   Components in an electromechanical relay

TABLE 14.1   Connection between Common, NO and NC

	ACTIVE LOW		ACTIVE HIGH	
	ENERGIZEDCOIL	NON-ENERGIZED COIL	ENERGIZED COIL	NON-ENERGIZED COIL
	Common			
Normally Open (NO)	Connected	Disconnected	Disconnected	Connected
Normally Closed (NC)	Disconnected	Connected	Connected	Disconnected

14.3 HOW TO CHECK THE COMMON, NC AND NO TERMINALS?

In most IoT projects, 5 V DC relay modules are used for controlling electrical equipment. These relay modules are mainly manufactured in China, and most of them have names of terminals in the Chinese language. So, for a person who doesn't know the Chinese language (either traditional or simplified), it is difficult to understand the terminal names. So, one of the smart ways is to use Google lens and Google Translate as shown in Figures 14.4–14.6 for identifying the terminals.

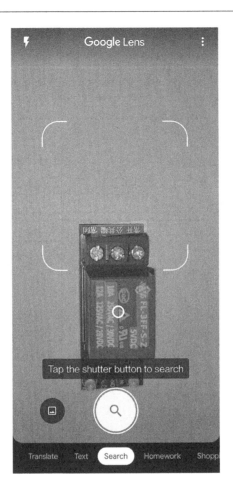

FIGURE 14.4   Google lens

Some of the 5 V DC relay modules don't have any terminal names and uses signs to indicate the terminals as shown in Figure 14.1. It is easy to decode the signs and find the names of the terminals. The middle pin of any 5 V DC relay will be the common terminal. The sign corresponding to the common terminal will be tilted to any one of the two sides. The tilted side of the common terminal sign is the NC terminal, and the other one is NO terminal. In Figure 14.1, the top-left terminal is NO and the bottom left terminal is NC.

Few 5 V DC relay modules do not indicate both sign and names of the terminals at the front. So, check for the sign or names of the terminal at the back of the module. If you don't find any indication for knowing the terminals at both front and back of the module, then you can use a multimeter for knowing the NO and NC terminals of the relay module.

A multimeter is a measuring device that is used to measure various electrical properties like DC voltage, AC voltage, current, resistance and many more. Figure 14.7 shows a commonly used low-cost digital multimeter along with explanation of each section functionality. Always remember to keep the multimeter knob at OFF after usage because keeping the multimeter ON might drain the battery or cause battery leakage which might damage the whole multimeter.

When the relay module is not powered, the common terminal and NC terminal will be connected (or short-circuited). So, using the continuity option (diode or WiFi symbol) and probes in the multimeter, the short-circuited terminals in the relay module can be easily found. Among the short-circuited terminals, the terminal other than the middle one (common terminal) is the NC terminal. If a terminal at one end is NC, then the terminal at the other end will be NO.

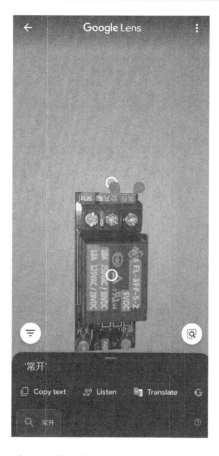

FIGURE 14.5   Selecting the text to be translated

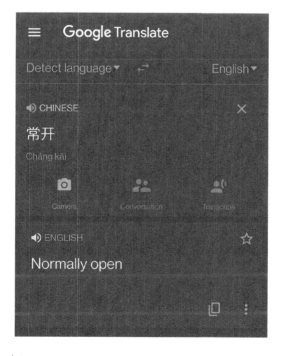

FIGURE 14.6   Google translate

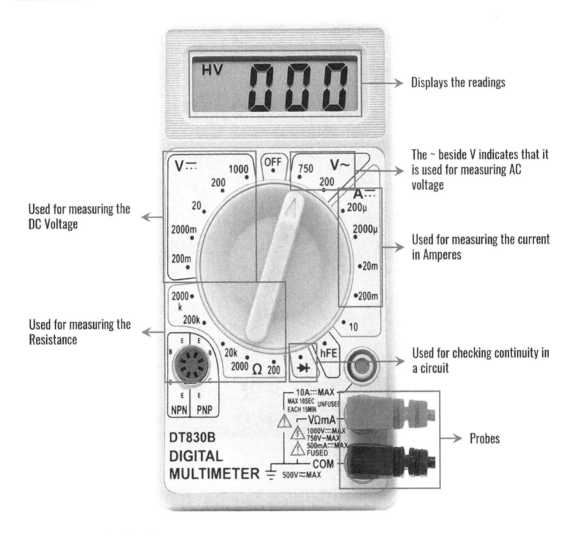

FIGURE 14.7 Digital multimeter

14.4 ELECTRIC LIGHT BULB

An electric light bulb is a light bulb that produces visible light using electricity (electric power). There are many types of electric light bulbs (or electric bulbs) available in the market like fluorescent light bulbs, incandescent light bulbs, halogen incandescent light bulbs, LED light bulbs, compact fluorescent bulbs (CFLs) and many more. Figure 14.8 shows some of the types of electric bulbs. For this project, you can use any type of electric bulb . These electric bulbs can be easily connected to an AC power supply using a holder and two-pin socket as shown in Figure 14.9.

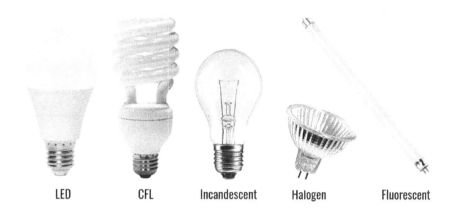

LED CFL Incandescent Halogen Fluorescent

FIGURE 14.8 Types of electric bulbs

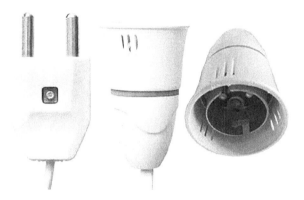

FIGURE 14.9 Two-pin socket and electric bulb holder

14.5 HARDWARE REQUIRED

TABLE 14.2 Hardware required for the project

HARDWARE	QUANTITY
NodeMCU	1
5 V DC relay	1
Electric bulb	1
Bulb holder	1
Two-pin plug	1
Screwdriver	1
Multi-strand wires	As required
Jumper wires	As required

14.6 CONNECTIONS

TABLE 14.3   Connection details

NODEMCU	5 V DC RELAY	ELECTRIC BULB AND BULB HOLDER	TWO-PIN PLUG
3V3	VCC		
D1	IN		
	NC	1st wire	
	Common		1st pin
		2nd wire	2nd pin
GND	GND		

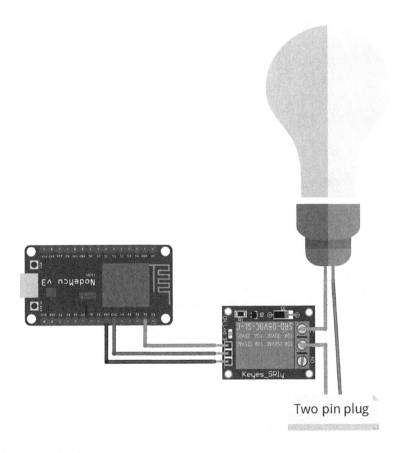

Two pin plug

FIGURE 14.10   Connection layout

The list of required components for the project is tabulated in Table 14.2, followed by Table 14.3, which presents the pin-wise connections. Also, the connection diagram is shown in Figure 14.10, which shall help get a visualization of the connections.

Some bulb holders available in the market are pre-connected with wires and some aren't. If your bulb holder is not connected with wires, then connect two multi-strand wires to its two ends. Connect the other end of these wires to a two-pin plug with the help of a screwdriver. Now, fix an electric bulb in the bulb holder and connect the two-pin plug to an AC power supply. If the bulb doesn't glow, then there might be some problem with your connections or electric bulb.

If the bulb glows, then the connection between the bulb holder and the two-pin plug is perfect. Now, cut one of two multi-strand wires between the bulb holder and the two-pin plug as shown in Figure 14.11. Connect one end of the cut wire to the NC terminal of the relay and the other end to the common terminal as shown in Figure 14.12.

FIGURE 14.11   Cut the wire

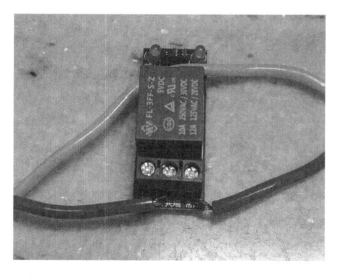

FIGURE 14.12   Connect the wire to relay

14.7 ADAFRUIT FEED AND ADAFRUIT DASHBOARD

Previously you have used the Adafruit Dashboard only for viewing the sensor data. But, in this project, you'll use the Adafruit dashboard for remotely controlling an electric bulb. First, create an Adafruit feed named *Electric Bulb* as shown in Figure 14.13. Next, create an Adafruit dashboard named *Electric Bulb*

Control as shown in Figure 14.14. Creation of an Adafruit feed and Adafruit dashboard is clearly explained in Sections 8.13 and 10.10, respectively. Refer to those sections if you haven't followed them before.

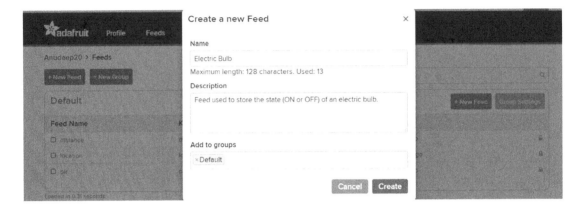

FIGURE 14.13   New Adafruit feed creation

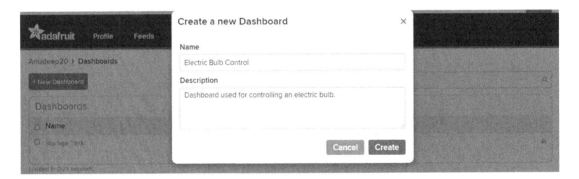

FIGURE 14.14   New Adafruit dashboard creation

Open the newly created *Electric Bulb Control* dashboard and add a toggle GUI block to the layout. For adding a new GUI block, click the Dashboard settings (gear icon at the top-right corner) and select + *Create New Block* from the drop-down. This will open a list of GUI blocks that can be added to the layout. Among them, select the Toggle block as shown in Figure 14.15. On selecting the Toggle block, a list of feeds available in your Adafruit account that can be connected to the Toggle block will open. Among them, select the newly created *Electric Bulb* and click *Next step >* as shown in Figure 14.16 which will open the Block settings window. Fill in the details as shown in Figure 14.17 and click *Create block*. The Block Title name is optional, write the Button On Text as *ON*, Button Off Text as *OFF* and the Test Value is optional, you can enter either ON or OFF. After clicking *Create block*, the Toggle GUI block will be added to the layout as shown in Figure 14.18.

Now, you can click the toggle GUI block to switch between ON and OFF. Presently, the state of the Toggle switch is in OFF state. Clicking the toggle switch once will change its state from OFF to ON and will add ON data to the *Electric Bulb* feed as shown in Figure 14.19. So, whenever you click the toggle switch its state changes and the changed state (either ON or OFF) will be added to the *Electric Bulb* feed. The last added data to the *Electric Bulb* feed can be subscribed by the NodeMCU to control the electric bulb

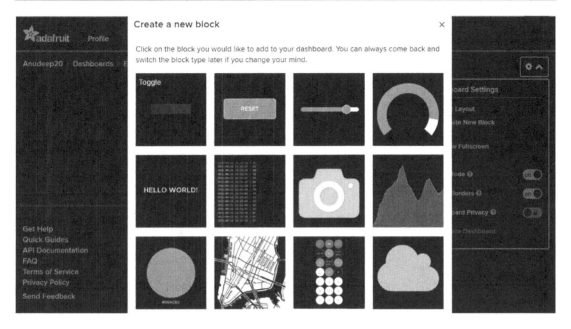

FIGURE 14.15   Selection of Toggle block

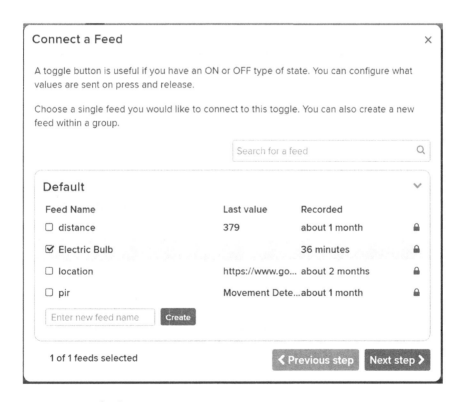

FIGURE 14.16   Connect a feed

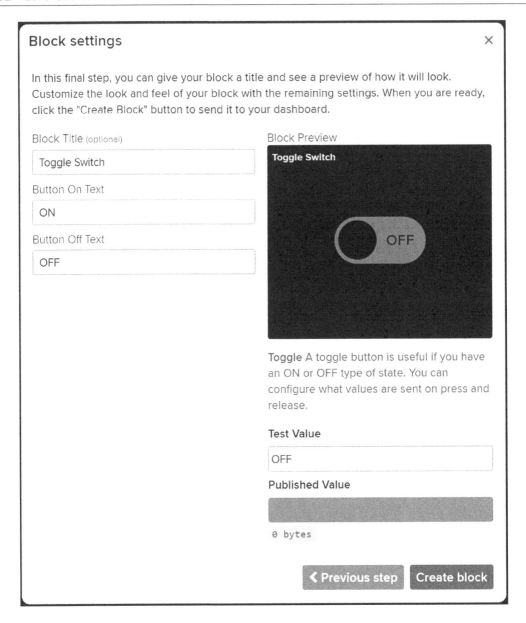

Block settings ×

In this final step, you can give your block a title and see a preview of how it will look. Customize the look and feel of your block with the remaining settings. When you are ready, click the "Create Block" button to send it to your dashboard.

Block Title (optional)

Toggle Switch

Button On Text

ON

Button Off Text

OFF

Block Preview

Toggle Switch

OFF

Toggle A toggle button is useful if you have an ON or OFF type of state. You can configure what values are sent on press and release.

Test Value

OFF

Published Value

0 bytes

‹ Previous step   **Create block**

FIGURE 14.17   Block settings

FIGURE 14.18   Toggle GUI block

Anudeep20 > Feeds > **Electric Bulb**

Feed used to store the state (ON or OFF) of an electric bulb.

Appears on Dashboards

- Electric Bulb Control

FIGURE 14.19 *Electric Bulb* feed

14.8 ARDUINO IDE CODE

The same libraries and code used in the previous project are used for connecting the NodeMCU to WiFi and Adafruit.

```
#include <ESP8266WiFi.h> //Library for WiFi functionalities
#include <Adafruit_MQTT_Client.h> //Library for Adafruit MQTT Client

//Adafruit Details
#define server "io.adafruit.com"
#define port 1883
#define IO_USERNAME  "AIO_USERNAME"
#define IO_KEY       "AIO_KEY"

String ssid = "WIFI_SSID"; //SSID of your WiFi
String password = "WIFI_PASSWORD"; //Password of your WiFi

WiFiClient esp; //Create a WiFi Client
```

Previously you have only published data to an Adafruit feed but never subscribed to an Adafruit feed. In this project, you'll subscribe to the *Electric Bulb* feed for reading the data in it. Similar to `Adafruit_MQTT_Publish`, `Adafruit_MQTT_Client.h` library provides `Adafruit_MQTT_Subscribe` for subscribing to a particular feed. EBFeed is an `Adafruit_MQTT_Subscribe` object used for subscribing to the *Electric Bulb* feed. The second argument of `Adafruit_MQTT_Subscribe()` is a combination of your Adafruit IO username and the key name of the feed. The key name of the *Electric Bulb* feed is `electric-bulb` which can be seen in Figure 14.20.

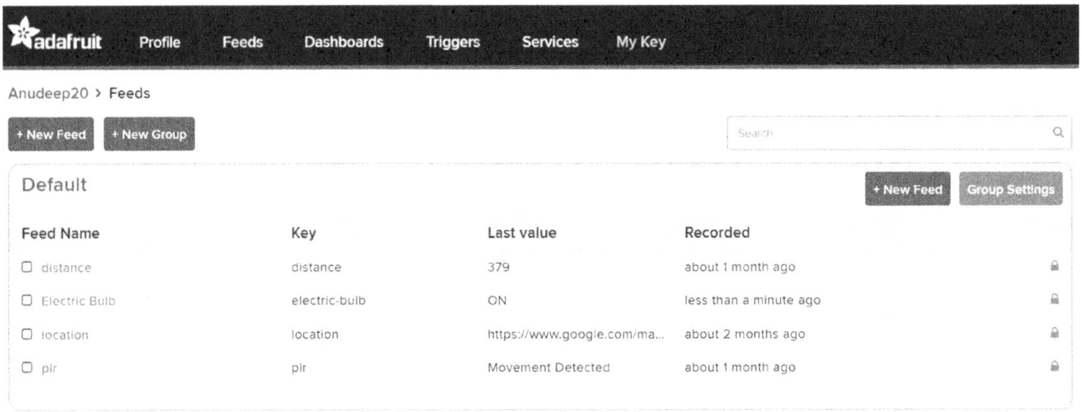

FIGURE 14.20   Key name of *Electric Bulb* feed.

```
//Creation of Adafruit MQTT Client
Adafruit_MQTT_Client mqtt(&esp, server, port, IO_USERNAME, IO_KEY);
//Variable for subscribing data in an Adafruit Feed
Adafruit_MQTT_Subscribe EBFeed = Adafruit_MQTT_Subscribe(&mqtt, IO_USERNAME"/
feeds/electric-bulb");
```

`relayPin` is used to store the digital pin to which IN pin of relay is connected to the NodeMCU. `dataFeed` is used to store the last added data in the subscribed feed. The `Serial` of NodeMCU is initialized at 115200 baud rate using `Serial.begin()`. As we are writing voltages on the `relayPin`, its direction is set as OUTPUT. Initially, the electric bulb is set to OFF state by writing LOW to `relayPin`.

```
int relayPin = D1; //Initialize the Relay pin
String dataFeed; //Stores data read from the feed

void setup()
{
  Serial.begin(115200); //Initialisation of NodeMCU Serial
  pinMode(relayPin, OUTPUT); //Defining the direction of relay
  digitalWrite(relayPin, LOW); //Initially the electric bulb is set to OFF

  //Connecting NodeMCU to WiFi
  Serial.println("");
  Serial.print("Connecting to ");
  Serial.println(ssid);
```

```
WiFi.begin(ssid, password);
while (WiFi.status() != WL_CONNECTED)
{
  delay(500);
  Serial.print(".");
}
Serial.println("");
Serial.println("WiFi connected");
Serial.println(WiFi.localIP()); //Print the local IP address of NodeMCU
```

You can subscribe to as many feeds as you want depending on the memory constraints of the evaluation board or microcontroller you are using. By default, the maximum number of subscriptions allowed is 5, but you can change it by modifying the #define MAXSUBSCRIPTIONS 5 line in the *Adafruit_MQTT.h* file located at *Arduino > libraries > Adafruit_MQTT_Library*.

For every feed you want to subscribe to, add .subscribe(&<FEED OBJECT>) in the void setup(). If you are writing NodeMCU and Adafruit connection code in the void setup(), then remember to add this line before the connection code.

```
//Subscribing to Electric Bulb Feed
mqtt.subscribe(&EBFeed);
}
```

Previously Adafruit connection code is written in the void setup(). This will check the connection between NodeMCU and Adafruit only once. But, if the connection is lost in between, we need to reset the NodeMCU for re-establishing the connection. To avoid this, we'll check for the connection between NodeMCU and Adafruit every time the loop is executed.

mqtt.connected() checks the connection between NodeMCU and Adafruit. If they are not connected, then mqtt.connect() is used for establishing the connection.

```
void loop() {
  //Connecting to MQTT
  if(!mqtt.connected()){
    Serial.print("Connecting to MQTT");
    while (mqtt.connect()){
      Serial.print(".");
    }
    Serial.println("");
  }
```

A pointer to the Adafruit _ MQTT _ Subscribe object is created to determine from which subscription the data is received. In our case, we have only one subscription so we don't need it. But, if you are subscribing to multiple feeds, then it is required.

mqtt.readSubscription(<TIME OUT>) will wait up to TIME OUT milliseconds for a reply from Adafruit. If it doesn't receive, then it'll time out and return 0 which will exit the while loop. But, if a reply is received from Adafruit, then it'll compare with each of the subscribed feed objects. In our case, mqtt.readSubscription() will wait for 2,000 milliseconds for a reply from Adafruit. If it doesn't receive a reply within 2,000 milliseconds, then it'll time out and will exit from the while loop. But, if a reply is received within 2,000 milliseconds, then it'll be compared with subscribed *Electric Bulb* feed object (&EBFeed). If they match, then the last added data to the *Electric Bulb* feed can be read using EBFeed.lastread.

The EBFeed stores the last added data point alone. Also, there is a limit to the size of the data read from the feed. By default, the size is set to 20 bytes. This data size is sufficient for most IoT applications unless you want to read huge chunks of data. However, this can be changed by editing the #define SUBSCRIPTIONDATALEN   20 line in the *Adafruit_MQTT.h* file. If the received data from the subscribed feed is ON, then the electric bulb is switched ON by writing HIGH on relayPin. Similarly, if the received data from the subscribed feed is OFF, then the electric bulb is switched OFF by writing LOW on relayPin.

```
//Read data from the Feed
Adafruit_MQTT_Subscribe *subscription;
while ((subscription = mqtt.readSubscription(2000))) {
  if (subscription == &EBFeed) {
    dataFeed = (char *)EBFeed.lastread;

    Serial.print("Received Data: ");
    Serial.print(dataFeed);

    //Control the electric bulb using relay
    if(dataFeed == "ON")

      digitalWrite(relayPin, HIGH);
    if(dataFeed == "OFF")
      digitalWrite(relayPin, LOW);
  }
 }
}
```

Before uploading the code, don't forget to change your Adafruit username, Adafruit IO key in the place of AIO _ USERNAME and AIO _ KEY, respectively. Also, change the WIFI _ SSID and WIFI _ PASSWORD with your WiFi name and password, respectively.

Finally, upload the code to NodeMCU after selecting the appropriate board and COM port. After successfully uploading the code, open the serial monitor and change the baud rate to 115200 to view the data received from the subscribed feed (*Electric Bulb*).

Now, connect the two-pin plug of the electric bulb to an AC power supply and open the *Electric Bulb Control* dashboard in your Adafruit IO account. Click the Toggle Switch to see the state of the electric bulb toggle between ON and OFF.

14.9 TRY IT

You can also generate the same functionality by connecting the cut wire of the bulb holder between common and NO terminals. But the code uploaded to NodeMCU needs to be changed accordingly. Try this on your own. If you are unable to do it, don't worry! just change digitalWrite(relayPin, HIGH) to digitalWrite(relayPin, LOW) and vice versa in the code uploaded to NodeMCU.

Generally, controlling an electric bulb using Adafruit dashboard might be a little inconvenient compared to controlling the same electric bulb using a mobile application. So, build a similar dashboard using Blynk mobile application (Blynk app) and control the electric bulb. You'll learn about the setup and usage of the Blynk mobile application in the next project. If you are unable to solve it, refer to the below setup and Arduino IDE code for the solution.

14.9.1 Blynk Setup

Click+*New Project* on the main screen of the Blynk app to create a new project. Enter the name of the project as *Blub control*, choose the device as *NodeMCU*, connection type as *WiFi* from the drop-down and click *Create* as shown in Figure 14.21. After successfully creating the project, the Blynk app will send an Auth Token for the *Bulb control* project to your registered Email ID.

Open the *Bulb control* project and click create (+ symbol at the top-right corner) to open *Widget Box* as shown in Figure 14.22. From the *CONTROLLERS* section, add a *Button* for 200 energy. After the *Button* controller gets added to the *Bulb control* project as shown in Figure 14.23, click *BUTTON* for editing its settings. Enter the name of the *Button* controller as *Electric Bulb*, select the *PIN* as *D1* in the *OUTPUT*, select the *MODE* as *SWITCH* and click the back button for the settings to apply as shown in Figure 14.24.

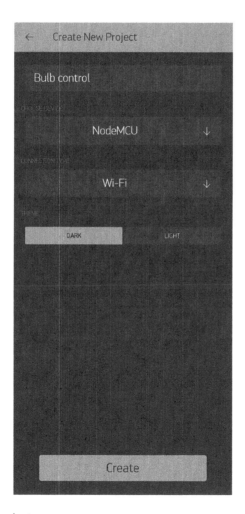

FIGURE 14.21 Create new project

FIGURE 14.22 Widget box

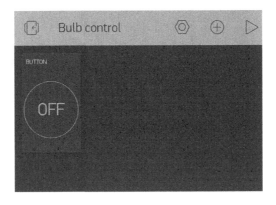

FIGURE 14.23 Bulb control project

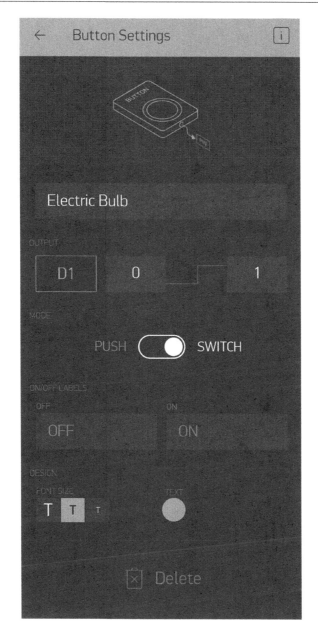

FIGURE 14.24 Button settings

14.9.2 Arduino IDE code

```
#include <ESP8266WiFi.h>
#include <BlynkSimpleEsp8266.h>

char authToken[] = "YOUR AUTH TOKEN";
char ssid[] = "YOUR WIFI NAME";
char password[] = "YOUR WIFI PASSWORD";
```

```
void setup()
{
  Serial.begin(115200);
  Blynk.begin(authToken, ssid, password);
}

void loop()
{
  Blynk.run();
}
```

Replace YOUR AUTH TOKEN with the Blynk auth token for *Bulb control* project received in your email and YOUR WIFI NAME and YOUR WIFI PASSWORD with your WiFi credentials. Upload the code after selecting the appropriate board and COM port. After successfully uploading the code, open the Blynk app and run the *Bulb control* project. Similar to Adafruit Dashboard, you can click the *ELECTRIC BULB* Toggle button shown in Figure 14.25 for changing the state of the electric bulb.

FIGURE 14.25   *Bulb control* project

   You can find all the resources of this project at this Github link (https://github.com/ anudeep-20/30IoTProjects/tree/main/Project%2014).
   In this project, you have learnt to interface an electric bulb to a small evaluation board using a relay. You have also learnt to subscribe to an Adafruit feed for reading the data stored in it. The methodology used in this project can be extended to control any electrical device using small evaluation boards or microcontrollers. In the next project, you'll build a smart irrigation system using a combination of sensors and Blynk mobile application which will help farmers remotely monitor and water their farms.
   Let's GO to the next project!

Smart Irrigation System Using Blynk

15

Agriculture plays an important role in the development and growth of any country. In developing countries, agriculture constitutes the major source of livelihood and income for most of the population. The use of latest technologies in agriculture can improve the yield and income of the farmers. Presently, many companies are working on technologies like image processing of leaves, weather tracking, satellite imagery and many more which helps the farmers in the early detection of diseases, knowing the soil quality and crops that give a high yield. This project helps the readers to build a simple smart irrigation system using IoT that can be used by the farmers to constantly monitor their farms.

15.1 MOTIVE OF THE PROJECT

In this project, you'll build a smart irrigation system for farmers using a soil moisture sensor, DHT11 sensor, rain sensor and Blynk mobile application to remotely monitor and water their farm using a water pump.

The watering of plants inside a farm needs to be done appropriately. Overwatering or under watering of plants inside the farm might damage the entire farm. So, the situation of the farm needs to be assessed appropriately before watering. Unless the farmers visit the farm, they don't know its condition and some farmers cannot assess the farm even after visiting it. So, to help farmers assess the parameters of their farm, soil moisture and DHT11 sensors are used.

Farmers cannot stay at the farm 24×7 to check its condition, so a mobile application can be used to remotely monitor the farm and control the water pump. Blynk mobile application (app) will help in building custom dashboards (projects) that can receive and send data to an evaluation board or microcontroller over the Internet. No mobile app development knowledge is required for building dashboards with Blynk mobile application. It is a drag-and–drop-based application that is easy and straightforward to use. We have also included a rain sensor in the system to keep the farmer updated about the rain through the dashboard in the Blynk mobile application. To make the project simple, we'll use a mini DC submersible water pump in place of an AC water pump. The same system can also be implemented to automatically water indoor gardens or plants depending on their soil parameters.

15.2 SOIL MOISTURE SENSOR

Soil moisture sensors measure the volumetric water content in soil indirectly by using electrical resistance or dielectric constant. Different soil moisture sensors are available in the market for different purposes, and each of them makes use of a different detection mechanism. The sensor shown in Figure 15.1 is the most commonly used soil moisture sensor in IoT. It measures soil moisture based on electrical resistance.

DOI: 10.1201/9781003147169-15

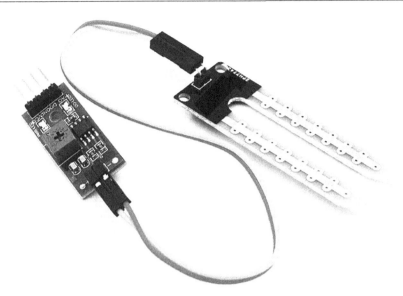

FIGURE 15.1   Soil moisture sensor

The two probes projecting out with exposed conductors (shown in Figure 15.2) act as a variable resistor. The resistance of these probes varies according to the water content in the soil and is inversely proportional to the soil moisture. Higher water in the soil produces lower resistance (higher conductivity) and lower water in the soil produces higher resistance (lower conductivity). Depending on the resistance, the sensor produces an output voltage that can be used to measure the soil moisture.

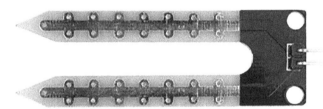

FIGURE 15.2   Probes

The soil moisture sensor also contains an electronic module (shown in Figure 15.3) that will help in connecting the sensor to an evaluation board or microcontroller. This module will help in producing an output voltage depending on the resistance of the probes which can be obtained from the Analog Output (AO) pin. The same voltage is sent through an LM393 comparator to convert it into a digital signal which can be obtained from the Digital Output (DO) pin on the module.

The potentiometer on the module will help in adjusting the sensitivity of Digital Output. It can be used to set a threshold at which the sensor should output LOW. If the output voltage exceeds the threshold (set by the potentiometer), the module will output LOW or else HIGH. Rotating the knob of the potentiometer clockwise will increase its sensitivity, and rotating counterclockwise will decrease its sensitivity. A screwdriver can be used to rotate the knob of the potentiometer.

Potentiometer ← → LM393 Comparator

FIGURE 15.3   Electronic module

15.3 DHT11 SENSOR

DHT11 is an inexpensive IoT sensor (shown in Figure 15.4) used to measure relative humidity and temperature. It is made up of two parts: a capacitive humidity sensor and a thermistor. The humidity sensing part has a thin-layer polymer sandwiched between two electrodes as shown in Figure 15.5. The thin-layer polymer is the moisture-holding substrate in the DHT11 sensor. The ions released by the polymer are absorbed by the electrodes which will increase the conductivity between them. This change in conductivity between the electrodes is inversely proportional to the relative humidity. So, higher relative humidity will decrease the conductivity between the electrodes and vice versa.

FIGURE 15.4   DHT11 sensor

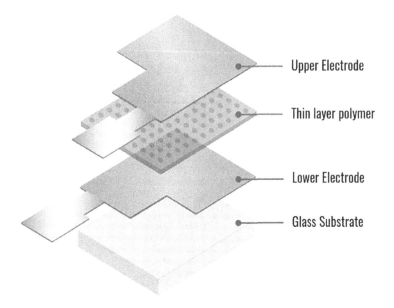

Upper Electrode

Thin layer polymer

Lower Electrode

Glass Substrate

FIGURE 15.5   Capacitive humidity sensing part

The second part of the DHT11 sensor is an NTC (negative temperature coefficient) thermistor to measure the temperature. A thermistor is nothing but a resistor whose resistance changes with temperature. The resistance of an NTC thermistor is inversely proportional to the temperature.

DHT11 sensor also has an IC at its back (inside the case) which helps in converting the analog signal with the stored calibration values to digital signal containing both temperature and relative humidity values. DHT11 sensor can measure relative humidity from 20% to 90% with an accuracy of 5% and temperature from 0°C to 50°C with an accuracy of 2°C.

DHT11 is a slow sensor whose sampling rate is 1 Hz (1 sample per second). So, an appropriate delay (at least 1,500 milliseconds) must be used between consecutive samples from the DHT11 sensor. This will ensure the consecutive samples obtained from the DHT11 sensor are valid and not duplicated. To know more about the DHT11 sensor, refer to its datasheet (https://www.mouser.com/datasheet/2/758/DHT11-Technical-Data-Sheet-Translated-Version-1143054.pdf).

15.4 RAIN SENSOR

Rain sensors help in the detection of rain at a particular place. Similar to soil moisture sensor, rain sensor is a combination of a sensing pad and an electronic module. The sensor shown in Figure 15.6 is the most commonly used rain sensor in IoT which detects rain based on electrical resistance.

The sensing pad of the rain sensor (shown in Figure 15.7) contains a series of copper lines spread across the pad that will act as a variable resistance. The resistance of the sensing pad is inversely proportional to the amount of water on its surface. More water on the surface of the sensing pad will result in lower resistance and less water on the surface of the sensing pad will result in higher resistance. Depending on the resistance, the sensor produces an output voltage that can be used to detect rain.

The rain sensor also contains an electronic module (shown in Figure 15.3) that will help in connecting the sensor to an evaluation board or microcontroller. The electronic module used in the rain sensor is the same electronic module as used in the soil moisture sensor. The usage of the electronic module in

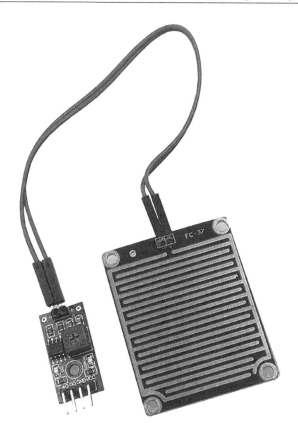

FIGURE 15.6   Rain sensor

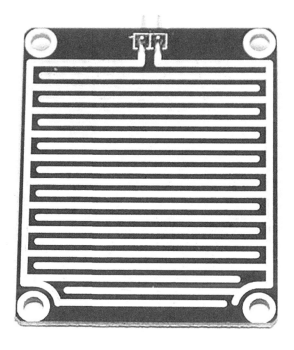

FIGURE 15.7   Sensing pad

rain sensor is for the same purpose as in soil moisture sensor. It will help in producing an output voltage depending on the resistance of the sensing pad which can be obtained from the Analog Output (AO) pin. The same voltage is sent through an LM393 comparator to convert it into a digital signal which can be obtained from the Digital Output (DO) pin on the module.

15.5 SUBMERSIBLE MINI WATER PUMP

Centrifugal water pumps are the most commonly used water pumps for agricultural purposes. They use high AC voltage for pumping water to the fields. It has a rotary part that rotates inside a casing to draw the liquid to its centre and throw out through an opening at the side of the casing using centrifugal force.

In this project, we'll use a mini DC submersible water pump shown in Figure 15.8 to pump the water. You can also use an AC water pump in place of a mini DC submersible water pump. For pumping the water out, a small tube must be connected to the projection of the mini submersible water pump. The specifications of the mini submersible water pump are tabulated in Table 15.1.

TABLE 15.1   Specifications of mini DC submersible water pump

Operating voltage	2.5–6 V
Operating current	130–220 mA
Flow rate	80–120 l/h
Maximum lift	40–110 mm

FIGURE 15.8   Mini DC submersible water pump

Every water pump (or motor) draws high current compared to any other normal sensor or module. Even the mini DC submersible water pump draws a minimum of 130 mA but the maximum DC current that can be drawn from a GPIO pin of Arduino Uno is 20 mA and for NodeMCU it is 12 mA. The same is the case for most of the evaluation boards and microcontrollers, and they cannot provide the high current required by a water pump.

If you directly connect a water pump that draws more current than the rated current of a GPIO pin, then there is a chance of damaging the evaluation board or microcontroller. If you connect a water pump directly to the 3.3 V or 5 V pins of an evaluation board which is connected to a USB port of a PC, then the water pump runs for few seconds and shuts down the USB port. Doing this repeatedly might damage both your evaluation board and the USB port of your PC. Rarely, it can damage your whole PC as well. So, be careful while dealing with high current devices.

To use the water pump with your evaluation board, you need to use an external power source that can provide the required current for it. But the power source should not always be connected to the water pump. It should only power the water pump when a signal is sent from the evaluation board. So, we need a switch to control the connection between the power source and the water pump depending on the signal from the evaluation board. For this purpose, you can use a transistor as a switch or a relay module. If used for a long time, transistors as a switch might cause heating issues. So, we'll use a 5 V DC 1 channel relay to control the connection between the power source and the water pump.

15.6 BATTERY

A battery is nothing but a power source consisting of one or more electrochemical cells with external connections for powering electronic or electrical devices. There are many types of batteries available in the market like nickel-cadmium batteries, lithium-ion batteries, nickel-metal hydride batteries, lead-acid batteries and many more. Each of the battery types is available in different voltages according to our usage.

In this project, we'll use a 9 V alkaline battery (shown in Figure 15.9) for powering the water pump. This is the most popular battery type available in the market for high DC voltage. If you have a power source that can provide a high current and has a voltage between 3 V and 6 V, then please use it and don't use a 9 V battery. These batteries are not rechargeable and need to be replaced after usage. You can also use a rechargeable battery instead of single use batteries.

FIGURE 15.9   9 V alkaline battery

Batteries can be conveniently used for prototypes and short-term usage projects. But long-term usage of batteries in high current consuming systems or devices is always troublesome as the batteries get drained very fast, and many batteries need to be replaced in a short span of time. To avoid this, you can use a 5 V DC power adapter with proper connectors instead of batteries for long-term usage projects.

15.7 LM7805 VOLTAGE REGULATOR

A voltage regulator is a device used to maintain a constant voltage. Both AC and DC voltage regulators are available in the market. Voltage regulators (either DC or AC) find widespread use in many places. You can find at least one voltage regulator in almost all complex electronic circuits.

In this project, we'll be using a 9 V DC power source (or battery) for powering the water pump but the maximum voltage that can be given to it is 6 V. So, directly connecting the water pump to a 9 V DC power source might damage or burn the water pump. To avoid this, we will use a voltage regulator which will convert the 9 V DC voltage to 5 V DC voltage.

LM7805 (or L7805) shown in Figure 15.10 is a widely used fixed 5 V voltage regulator. It outputs a constant 5 V DC voltage for any input voltage between 7 V and 25 V. There are also other voltage regulators in the same family of LM7805 like LM7809 (for 9 V DC), LM7812 (for 12 V DC), etc. For knowing more about the LM7805 voltage regulator and its use cases, refer to its datasheet (https://www.sparkfun.com/datasheets/Components/LM7805.pdf).

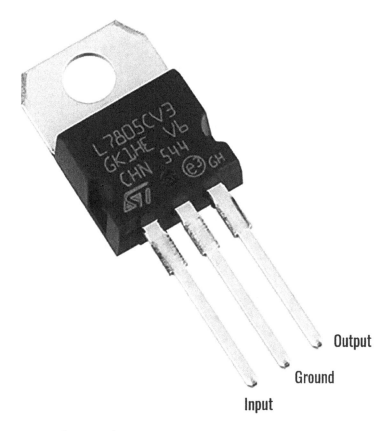

FIGURE 15.10   LM7805 voltage regulator

For having a consistent stable output voltage, you need to connect two capacitors each between the Input and Ground, Output and Ground of LM7805. These capacitors act as decoupling capacitors used for eliminating any AC components in the voltage supply as well as little voltage fluctuations that arise due to changes in load. Here, we have used a 0.33 µF capacitor between Input and Ground and a 0.1 µF capacitor between Output and Ground as shown in Figure 15.11. You can also use other capacitor combinations as well. Even without using the capacitors, LM7805 can be used for voltage regulation but with little fluctuations in voltage.

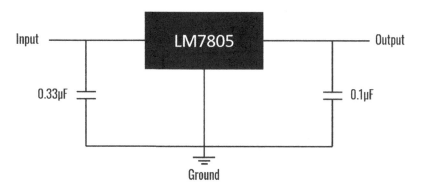

FIGURE 15.11   LM7805 capacitor connections

15.8 BLYNK

Blynk is a mobile application that is commonly used in Android and iOS for creating GUI dashboards that can control evaluation boards or microcontrollers over the Internet. You can create GUI dashboards for your application by dragging and dropping widgets. It is available for free on both Android and iOS platforms. For an Android smartphone, the Blynk application (shown in Figure 15.12) can be downloaded from the Google Play store (https://play.google.com/store/apps/details?id=cc.blynk). For an iOS smartphone, the Blynk application (shown in Figure 15.13) can be downloaded from the Apple App Store (https://itunes.apple.com/us/app/blynk-control-arduino-raspberry/id808760481?ls=1&mt=8).

In the project, we'll create a GUI dashboard for the farmers to remotely monitor their farm using Blynk mobile application. So, download and install the Blynk mobile application on your smartphone. As soon as you open the Blynk mobile application after installing it, you'll see the page as shown in Figure 15.14. Create an account, if you don't have one or else log in with your registered Email ID. If you have a Facebook account, you can also log in with it.

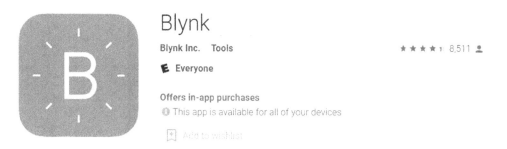

FIGURE 15.12   Android Blynk application

FIGURE 15.13   iOS Blynk application

FIGURE 15.14   Log in page

Click *Create New Account* and enter your Email and Password as shown in Figure 15.15 and click *Sign Up*. On clicking the *Sign Up* button, you'll be directly logged in to the application without verifying your Email ID. If you have an account in Blynk, then the same page can be used to log in as well.

After you successfully log in to the Blynk mobile application for the first time, you'll see a popup message as shown in Figure 15.16. Please read it carefully and click *Cool! Got It*. The popup message says that energy is required for adding each GUI widget in the project dashboard. If the added GUI widget is deleted, then the energy gets restored. If more energy is required than the provided free energy, you can buy it from their store. Mostly for the dashboards we build, the default energy is sufficient.

After you agree to the popup message, you'll see the main page of the Blynk mobile application as shown in Figure 15.17. From this page, you can create a new project or see the Apps which you have previously created or visit the evergrowing Blynk community to solve your doubts or help someone like you.

FIGURE 15.15   Create new account

FIGURE 15.16   Popup message

FIGURE 15.17   Main page

15.9 HARDWARE REQUIRED

TABLE 15.2   Hardware required for the project

HARDWARE	QUANTITY
NodeMCU	1
Soil Moisture Sensor	1
DHT11 Sensor	1
Rain Sensor	1
5 V DC Relay	1
Mini submersible water pump	1
LM7805 voltage Regulator	1
0.33 µF ceramic capacitor	1
0.1 µF ceramic capacitor	1
9 V battery	1
Breadboard	1
Jumper wires	As required

15.10 CONNECTIONS

The list of required components for the project is tabulated in Table 15.2, followed by Table 15.3, which presents the pin-wise connections. Also, the connection diagram is shown in Figure 15.18, which shall help get a visualization of the connections.

TABLE 15.3 Connection details

NODEMCU	SOIL MOISTURE SENSOR	DHT11 SENSOR	RAIN SENSOR	5 V DC RELAY	WATER PUMP	LM7805 VOLTAGE REGULATOR	0.33 µF CERAMIC CAPACITOR	0.1 µF CERAMIC CAPACITOR	9 V BATTERY
3V3	VCC	VCC	VCC	VCC					
A0	AO								
D3		SIG							
D2			DO						
D1				IN					
				Common	+ve pin				
						Input	1st pin		+ve pin
				NC		Output		1st pin	
					−ve pin	GND	2nd pin	2nd pin	−ve pin
GND	GND	GND	GND	GND					

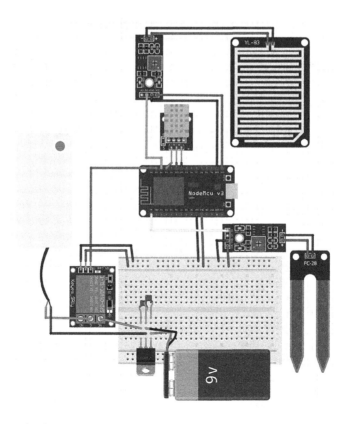

FIGURE 15.18 Connection layout

15.11 CREATE A NEW PROJECT USING BLYNK

Creating a new GUI dashboard (project) using Blynk is easy and straightforward. You can create as many projects as you want for free in Blynk mobile application, but the energy required to add GUI blocks provided for free is limited. To create a GUI dashboard, click *+New Project* on the main screen of the Blynk app. Enter the name of the project as *Smart Irrigation*, choose the device as *NodeMCU*, connection type as *WiFi* from the drop-down and click *Create* as shown in Figure 15.19. After successfully creating the project, you'll see a notification stating that Auth Token of the project is sent to your Email ID as shown in Figure 15.20. Click *OK* and open the email sent by Blynk. The email contains the Auth Token of the project and some important links as shown in Figure 15.21.

Initially, you'll see a plain dashboard as shown in Figure 15.22 where you can add different GUI widgets. For adding new GUI widgets, click the+symbol at the top-right corner of the page which will open the *Widget Box* from the side as shown in Figure 15.23. Blynk provides a lot of widgets, and it is difficult to

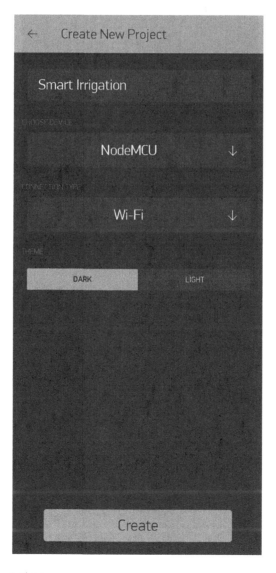

FIGURE 15.19 Create new project

mention all the widgets available in the Widget Box. With no exception, you'll get all the widgets required for your project. At the top of the Widget Box, you'll see the *Energy Balance* in your account. Initially, Blynk provides 2000 energy for free use. You can also purchase more energy by clicking + *Add* button at the top of the Widget Box. This energy is used to add a widget to your dashboard. Each widget takes a certain amount of energy, and the energy taken by a widget will be mentioned beneath it. For example, the *Button* widget takes 200 energy, *Joystick* widget takes 400 energy.

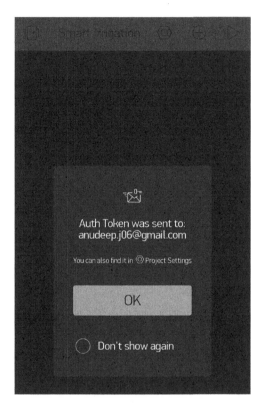

FIGURE 15.20   Auth token notification

Auth Token for Smart Irrigation project and device Smart Irrigation   Inbox ×

Blynk <dispatcher@blynk.io> Unsubscribe
to me ▾

Auth Token : Xwz-qx4jmAs1aWgo8ElnLbjGtdnPVr9r

Happy Blynking!

-

Getting Started Guide -> https://www.blynk.cc/getting-started
Documentation -> http://docs.blynk.cc/
Sketch generator -> https://examples.blynk.cc/

Latest Blynk library -> https://github.com/blynkkk/blynk-library/releases/download/v0.6.1/Blynk_Release_v0.6.1.zip
Latest Blynk server -> https://github.com/blynkkk/blynk-server/releases/download/v0.41.13/server-0.41.13.jar

https://www.blynk.cc
twitter.com/blynk_app
www.facebook.com/blynkapp

FIGURE 15.21   Auth token email

FIGURE 15.22   Plain dashboard

FIGURE 15.23   Widget box

First, add a *Gauge* widget for 300 energy to the *Smart Irrigation* dashboard and click it to configure its settings. Enter the name of the *Gauge* widget as *Soil Moisture*, select the Input as virtual pin *V0* from the drop-down, enter the range of input as 0–100, enter the label as */pin./ %*, reading rate as *2 sec* from the drop-down as shown in Figure 15.24 and click back button at the top for the settings to apply. This Gauge widget is used to display soil moisture in percentage depending on the output analog values from the soil moisture sensor.

Blynk provides different options for formatting the label in a widget. Some of the label formatting options supported by Blynk are listed below. For example, if your code sends 16.223 to the virtual pin, then

/pin/ - displays the value without formatting (16.223)

/pin./ - displays the value without decimal part (16)

/pin.#/ - displays the value with 1 decimal digit (16.2)

/pin.##/ - displays the value with 2 decimal digits (16.22)

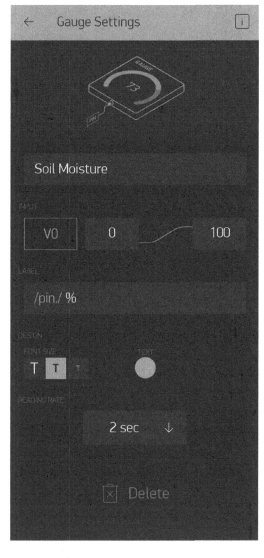

FIGURE 15.24   Soil moisture gauge settings

Next, add another *Gauge* widget for 300 energy to the *Smart Irrigation* dashboard and click it to configure its settings. Enter the name of the *Gauge* widget as *Humidity*, select the Input as virtual pin *V1* from the drop-down, enter the range of input as 0–100, enter the label as */pin./ %*, reading rate as *2 sec* from the drop-down as shown in Figure 15.25 and click back button at the top for the settings to apply. This Gauge widget is used to display relative humidity values obtained from the DHT11 sensor.

Next, add a *Labeled Value* widget for 400 energy to the *Smart Irrigation* dashboard and click it to configure its settings. Enter the name of the *Labeled Value* widget as *Temperature*, select the Input as virtual pin *V2* from the drop-down, select the middle label arrangement, enter the label as */pin.##/ °C*, reading rate as *2 sec* from the drop-down as shown in Figure 15.26 and click back button at the top for the settings to apply. This Labeled Value widget is used to display temperature values obtained from the DHT11 sensor.

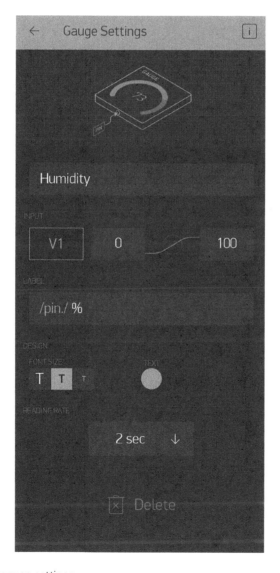

FIGURE 15.25   Humidity gauge settings

Next, add a *segmented switch* widget for 400 energy to the *Smart Irrigation* dashboard and click it to configure its settings. Select the output as virtual pin *V3* from the drop-down, enter option 1 as *Raining* and option 2 as *Not raining* as shown in Figure 15.27 and click the back button at the top for the settings to apply. This segmented switch widget is used to display whether it is raining or not depending on the values obtained from the rain sensor.

Next, add a *Button* widget for 200 energy to the *Smart Irrigation* dashboard and click it to configure its settings. Enter the name of the *Button* widget as *Water Pump*, select the output as virtual pin *V4* from the drop-down, enter the output values as *0* and *1*, select the mode as *Switch*, enter the ON/OFF labels as *Start* & *Stop* as shown in Figure 15.28 and click back button at the top for the settings to apply. This Button widget is used to manually switch ON or OFF the water pump using the Blynk app.

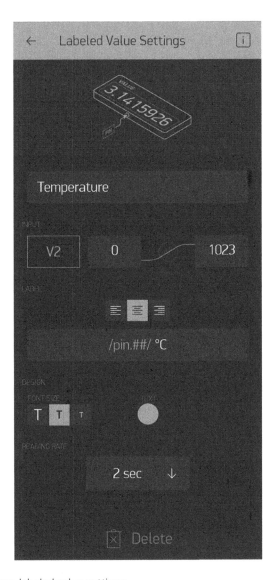

FIGURE 15.26   Temperature-labeled value settings

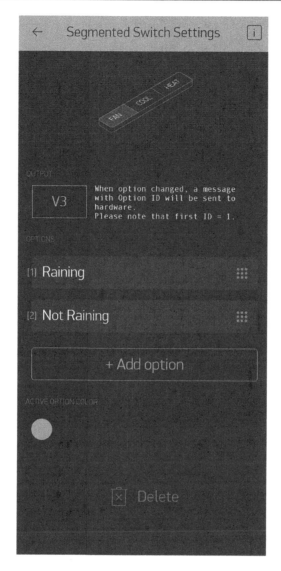

FIGURE 15.27   Segmented switch settings

Finally, add a *Text Input* widget for 400 energy to the *Smart Irrigation* dashboard and click it to configure its settings. Select the output as virtual pin *V5* from the drop-down, enter the character limit as *30* as shown in Figure 15.29 and click the back button at the top for the settings to apply. This Text Input widget is used for suggesting the farmer whether the farm needs to be watered or not depending on the values obtained from the soil moisture sensor, DHT11 sensor and rain sensor.

After adding all the widgets to the *Smart Irrigation* dashboard, arrange them as shown in Figure 15.30. The arrangement of widgets is optional, and it is only for our convenience. You can arrange the widgets in the dashboard as you wish.

If you are unable to build the dashboard in the Smart irrigation project or want to clone the same project build by us, then scan the QR code shown in Figure 15.31 using the QR code scanner at the top-right corner on the main page of the Blynk mobile application (shown in Figure 15.17). We recommend you to build the dashboard on your own and use the QR code as your last option.

FIGURE 15.28   Water pump button settings

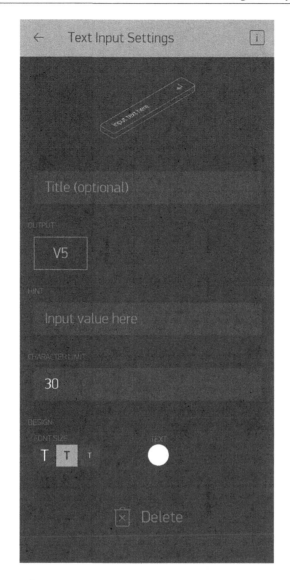

FIGURE 15.29 Text input settings

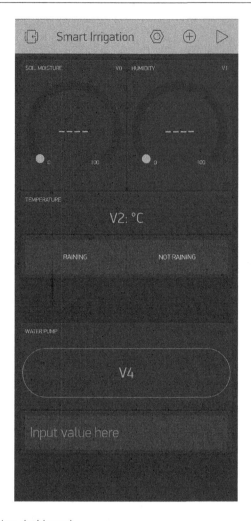

FIGURE 15.30 Smart irrigation dashboard

FIGURE 15.31 QR code for smart irrigation project

15.12 ARDUINO IDE CODE

If you remember the Adafruit projects, we have used Adafruit libraries, Adafruit feed and Adafruit dashboard for displaying sensor data on a GUI dashboard. Similar to Adafruit, Blynk also uses Blynk libraries, Blynk server and Blynk mobile application for displaying sensor data on its GUI dashboard. The best part of Blynk is that we need not worry about the server part. It'll be automatically configured for us and ready to use. We need to just use the Blynk libraries in the code and configure the GUI dashboard using widgets.

Blynk library can be easily downloaded from the Arduino IDE. First, go to *Tools* > *Manage Libraries* or directly Click *Ctrl+Shift+I* (in Windows) or *Cmd+Shift+I* (in macOS) to open the *Library Manager*. Search for *Blynk* in the search box and scroll down until you find *Blynk by Volodymyr Shymanskyy*. Select the latest stable version from the drop-down and click install as shown in Figure 15.32.

FIGURE 15.32 Installation of Blynk library

You can also directly download the latest Blynk library from this link (https://github.com/blynkkk/blynk-library) and extract it in the libraries folder of Arduino IDE.

`BlynkSimpleEsp8266.h` file is part of the Blynk library which is used to connect ESP8266-based devices to the WiFi and the Blynk server. `DHT.h` library is used to obtain the temperature, relative humidity and heat index values from the DHT11 sensor. Similar to the Blynk library, the DHT library can be downloaded from the *Library Manager*. Search for *DHT* in the search box and scroll down until you find *DHT sensor library by Adafruit*. Select the latest version from the drop-down and click Install as shown in Figure 15.33.

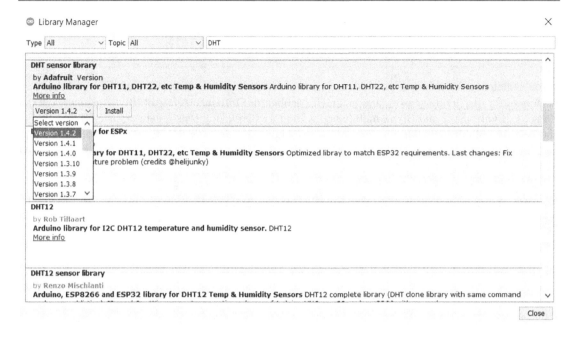

FIGURE 15.33   Installation of DHT library

You can also directly download the latest DHT library from this link (https://www.arduino.cc/reference/en/libraries/dht-sensor-library/) and extract it in the libraries folder of Arduino IDE.

```
#include <BlynkSimpleEsp8266.h> //Library for Blynk
#include <DHT.h> //Library for DHT sensor
```

`soilMoisturePin`, `DHTPin`, `rainPin` and `relayPin` are constant variables used to store the pin number to which soil moisture sensor, DHT11 sensor, rain sensor and relay are connected to the NodeMCU respectively.

Similar to digital and analog pins, the Blynk app has another set of pins known as virtual pins. These pins act as a virtual interface for sending and receiving data from an evaluation board or microcontroller over the Internet. `virtualSoilMoisturePin`, `virtualHumidityPin`, `virtualTemperaturePin`, `virtualRainPin`, `virtualWaterPumpPin` and `virtualTextBoxPin` are constant variables used to store the virtual pin number to which soil moisture gauge widget, humidity gauge widget, temperature-labeled value widget, rain segmented switch widget, water pump button widget and text box widget are connected respectively in the Blynk mobile application.

```
//Defining the Sensor & relay pins
#define soilMoisturePin A0
#define DHTPin D3
#define rainPin D2
#define relayPin D1

//Defining virtual pins for Blynk app
#define virtualSoilMoisturePin V0
#define virtualHumidityPin V1
#define virtualTemperaturePin V2
#define virtualRainPin V3
#define virtualWaterPumpPin V4
#define virtualTextBoxPin V5
```

If you remember, for Adafruit free account we can publish 30 data points per minute. Similarly, the Blynk server allows us to send not more than 10 data points per second. If you send more than 100 data points per second, you may cause flood error and your evaluation board or microcontroller will be automatically disconnected from the Blynk server. So, be careful and send data to the Blynk server in regular time intervals. Don't use `delay()` for sending data to the Blynk server, it might disconnect the evaluation board or microcontroller from the Blynk server and cause connectivity issues. To avoid this, Blynk documentation recommends using `BlynkTimer` which helps in sending data periodically to the Blynk server without interfering with Blynk library routines. `BlynkTimer` is built on top of the `SimpleTimer` library which is a widely used library to time multiple events on hardware.

`timer` is a `BlynkTimer` object used to call a function at regular intervals. `dht` is a `DHT` object used to obtain the temperature, relative humidity and heat index values from the DHT11 sensor. It accepts two arguments: the first one is the pin number to which the DHT11 sensor is connected, and the second argument is the type of DHT sensor.

```
BlynkTimer timer; //BlynkTimer object
DHT dht(DHTPin, DHT11); //DHT object
```

`authToken` stores the Auth Token for the Smart Irrigation project which you have received in your email from Blynk as shown in Figure 15.21. `ssid` and `password` store the name and password of your WiFi, respectively.

`soilMoisture`, `tempC` and `heatIndxC` are `float` variables used to store the soil moisture values, temperature values in °C and heat index values in °C, respectively. `relHum` and `waterPump` are `int` variables used to store relative humidity values and values from `virtualWaterPumpPin`, respectively. `rain` is a `bool` variable used to store the values from the rain sensor.

```
//Details for connecting to Blynk
char authToken[] = "YOUR AUTH TOKEN";
char ssid[] = "YOUR WIFI NAME";
char password[] = "YOUR WIFI PASSWORD";

//Variables for storing sensor data & controlling water pump
float soilMoisture;
int relHum;
float tempC;
float heatIndxC;
bool rain;
int waterPump;
```

NodeMCU's serial is initialized at 115200 using `Serial.begin()`. As we are reading values from the soil moisture sensor, DHT11 sensor and rain sensor, we'll declare the direction of `DHTPin`, `rainPin` and `relayPin` as INPUT. As we are writing voltages on the relay, we'll declare the `relayPin` as OUTPUT.

Previously, we used to write a lengthy code for connecting the NodeMCU to WiFi. Blynk library makes this process very simple. Using the Blynk library, we can connect the NodeMCU to WiFi and Blynk server in a single line of code. `Blynk.begin()` is used to connect the NodeMCU to WiFi and then to the Blynk server.

`timer.setInterval()` accepts two arguments that are interrelated to each other. The first argument is the time interval after which the second argument must be called. The second argument is the function that needs to called after every first argument interval. Here, `callFunc` function is called after every 2,000 milliseconds. `dht.begin()` is used to initialize the `dht` object.

```
void setup() {
  Serial.begin(115200); //Initialisation of NodeMCU Serial

  //Defining the direction of sensors & relay GPIO pins
```

```
pinMode(soilMoisturePin, INPUT);
pinMode(DHTPin, INPUT);
pinMode(rainPin, INPUT);
pinMode(relayPin, OUTPUT);

//Connecting to Blynk
Blynk.begin(authToken, ssid, password);
//Used to send data in intervals
timer.setInterval(2000, callFunc);

dht.begin(); //Initialisation of dht object
}
```

BLYNK_WRITE(<VIRTUAL PIN NUMBER>) function is used to read data from a particular virtual pin from the Blynk app where <VIRTUAL PIN NUMBER> must be replaced with the virtual pin number from which you want to read the data. In our project, we will switch ON or OFF the water pump depending on the state of water pump button in the Blynk mobile application. So, we need to read the virtual pin that is connected to the water pump button.

Whenever the water pump button in the Blynk mobile application is pressed, BLYNK_WRITE(virtualWaterPumpPin) function will be called. Inside the function, we'll use param.asInt() to read the data from the virtualWaterPumpPin as integer, you can also read the data as a string using asStr() and as float using asFloat(). The read data will be stored in the waterPump variable. If waterPump is 1, then the water pump will be switched ON or else the water pump will be switched OFF.

```
//Reading data from Blynk
BLYNK_WRITE(virtualWaterPumpPin){
  waterPump = param.asInt();
  if(waterPump == 1){
    Serial.println("Water Pump Switched ON");
    digitalWrite(relayPin, HIGH);
  } else {
    Serial.println("Water Pump Switched OFF");
    digitalWrite(relayPin, LOW);
  }
}
```

Inside void loop(), the code related to sending data to the Blynk server should not be written. If the code related to sending data to the Blynk server is written in the void loop() and proper delay is not used, the code might send 100s of requests within 1 second to the Blynk server which might disconnect the hardware from the Blynk server. To avoid this, it is advisable to write the code related to sending data to the Blynk server in a separate function and call it at regular intervals using BlynkTimer.

Blynk.run() is a single command which performs various functions. It takes care of all the background activity of the Blynk library. Even if the NodeMCU gets disconnected from the WiFi or Blynk server, Blynk.run() tries to reconnect it until it is connected. Similarly, timer.run() is used for initiating and running the BlynkTimer.

```
void loop() {
  Blynk.run(); //For managing Blynk
  timer.run(); //For managing BlynkTimer
}
```

All the code related to sending data to the Blynk server is written inside the callFunc() function, and it is called for every 2,000 milliseconds using BlynkTimer. Unlike setup() and loop(), callFunc() is a custom function that will not be present by default. Creating and calling new functions in Arduino IDE is very easy,

```
void <FUNCTION NAME>(<ARGUMENTS>) {
<WRITE YOUR CODE HERE>
}
```

will create a new function where <FUNCTION NAME> must be replaced with the name of the function and <ARGUMENTS> must be replaced with any arguments that need to be passed to the function. If you don't have any arguments, then leave it empty. The new function can be called using <FUNCTION NAME>(<ARGUMENTS>); anywhere in the code.

If void is written before the function name while declaring it, then the function returns nothing. Functions can also return int or float by writing int or float, respectively, before the function name while declaring it. For example,

```
int funcInt() {
      int valueInt;
      <WRITE YOUR CODE>
      return valueInt;
}
```

The function funcInt will return an integer value and can be called using the following lines of code anywhere in the code.

```
int recFuncValue;
recFuncValue = funcInt();
```

```
int funcFloat() {
      float valueFloat;
      <WRITE YOUR CODE>
      return valueFloat;
}
```

The function funcFloat will return a float value and can be called using the following lines of code anywhere in the code.

```
float recFuncValue;
recFuncValue = funcInt();
```

soilMoisture stores the analog value read from the soil moisture sensor. This value will be in the range of 0–1023. Higher values correspond to lower soil moisture, and lower values correspond to higher soil moisture. For anyone who doesn't know about this inverse relationship between soil moisture and analog values, it'll be difficult for them to interpret the soil moisture from analog values. So, for easy interpretation, we'll map the analog values read from the soil moisture sensor from the 1023–0 range to the 0–100 range using the map() function.

map() is a math function used in Arduino IDE to re-map a number from one range to another. It takes five arguments for mapping a number from one range to another. The first argument is the number to be mapped, the next two arguments are the initial range, and the last two numbers are the final range. By default, the map() function does not constrain the values to the given range. It can map out of the range values as well. For example, if we define map(input, 0, 100, 500, 600), then input can take any value and need not be restricted between 0 and 100. If input is –50, then the map function will output 451. Internally, map() is implemented using the following function:

```
long map(long inp, long ini_min, long ini_max, long fin_min, long fin_max)
{
  return (inp - ini_min) * (fin_max - fin_min) / (ini_max - ini_min) + fin_min;
}
```

where inp is the input value, ini_min and ini_max are the minimum and maximum values of the initial range and fin_min and fin_max are the minimum and maximum values of the final range. From the function implementation, we can see that the function accepts all the arguments in long type and returns a value in long type. So, the map() function cannot process the arguments in float type. But, if float type values are given as arguments to the map() function, then these will be converted to long type and then get processed. If your project requires mapping of float type values from one range to another range, then it is recommended to implement the function on your own by referring to the map() function implementation.

```
void callFunc() {
  //Read the soil moisture sensor values and map it between 0 to 100
  soilMoisture = analogRead(soilMoisturePin);
  soilMoisture = map(soilMoisture,1023,0,0,100);
```

tempC is used to store the temperature values (in °C) read from the DHT11 sensor using readTemperature() function. To read the temperature values in °F, readTemperature(true) can be used. relHum is used to store the relative humidity values read from the DHT11 sensor using readHumidity() function.

```
  //Read temperature and humidity values from DHT
  tempC = dht.readTemperature();
  relHum = dht.readHumidity();
```

Occasionally due to some problem, readTemperature() and readHumidity() functions cannot read the temperature and relative humidity values from the DHT11 sensor and store junk values in the tempC and relHum variables. So, we need to ensure the data obtained from DHT11 sensor is correct before using it. isnan() checks whether the data stored in a variable is not-a-number (NaN) or not. If the data stored in either tempC or relHum is not-a-number, then the callFunc() function will exit.

The heat index (or humiture) is a parameter that uses temperature and relative humidity to determine the equivalent temperature perceived by the human. computeHeatIndex() function is used to obtain the heat index from the temperature and relative humidity values. It accepts three arguments: the first one is the temperature value, the second one is the relative humidity value and the third one is used for determining the unit in which heat index (°C or °F) must be generated. If the third argument is false, then the heat index will be generated in °C or else in °F.

rain is used to store the values obtained from the rain sensor. It is used to determine whether it is raining or not. All the sensor values are printed on the serial monitor before writing them on their respective virtual pins.

```
  //Validating temeperature and humidity values
  if (isnan(tempC) || isnan(relHum))
    return;

  //Calculate heat index in °C
  heatIndxC = dht.computeHeatIndex(tempC, relHum, false);

  rain = digitalRead(rainPin); //Read data from rain sensor

  //Print the sensor values
  Serial.print("Mositure : ");
  Serial.print(soilMoisture);
  Serial.print("%  Humidity: ");
  Serial.print(relHum);
  Serial.print("%  Temperature: ");
  Serial.print(tempC);
```

```
Serial.print("°C  Heat Index: ");
Serial.print(heatIndxC);
Serial.print("°C  ");
```

`Blynk.virtualWrite()` is used to write data on a particular virtual pin. It accepts two arguments: the first is the virtual pin number, and the second one is the data to be written on the virtual pin. If you are using this command, then don't forget to use `BlynkTimer()` in your code. Read the documentation to know more about the Blynk library, Blynk server and Blynk mobile application (https://docs.blynk.cc/).

```
//Writing on Blynk Virtual Pins
Blynk.virtualWrite(virtualSoilMoisturePin, soilMoisture);
Blynk.virtualWrite(virtualHumidityPin, relHum);
Blynk.virtualWrite(virtualTemperaturePin, tempC);
```

The rain sensor will output 0 when rain is detected and 1 when there is no rain. Finally, using the soil moisture, relative humidity and rain sensor values, we'll suggest the farmer to water the farm or not. The suggestion will be displayed on the text box at the bottom of the GUI dashboard. The threshold values used for suggesting the farmer are not research based and can be adjusted as required.

```
if(rain == 0){
  Serial.print("Its Raining  ");
  Blynk.virtualWrite(virtualRainPin, 1);
} else {
  Serial.print("Not Raining  ");
  Blynk.virtualWrite(virtualRainPin, 2);
}

if((soilMoisture < 50 || relHum < 50) && rain == 1){
  Serial.println("Water the Farm");
  Blynk.virtualWrite(virtualTextBoxPin, "Water the Farm");
} else if((soilMoisture >= 50 && relHum >= 50) || rain == 0) {
  Serial.println("Don't Water the Farm");
  Blynk.virtualWrite(virtualTextBoxPin, "Don't Water the Farm");
}
}
```

Change the YOUR AUTH TOKEN, YOUR WIFI NAME and YOUR WIFI PASSWORD with the auth token of your project received in your email, WiFi name and password, respectively. Then select the appropriate board and COM port and upload the code to NodeMCU. After successfully uploading the code, open the serial monitor and change the baud rate to 115200 to see the sensor data.

Now, open the *Smart Irrigation* project in your Blynk mobile application and click the play button at the top-right corner (shown in Figure 15.30) to run the dashboard. If the NodeMCU is connected to WiFi, then the *Smart Irrigation* project will run and display the values as shown in Figure 15.34. If the NodeMCU is not connected to WiFi, then the Smart Irrigation project gets disconnected from NodeMCU and the values won't be updated in the dashboard as shown in Figure 15.35.

FIGURE 15.34   Connected smart irrigation project

FIGURE 15.35 Disconnected smart irrigation project

You can find all the resources of this project at this Github link (**https://github.com/anudeep-20/30IoTProjects/tree/main/Project%2015**).

In this project, you have learnt about various sensors like soil moisture sensor, DHT11 sensor and rain sensor. You have learnt about interfacing a water pump to the NodeMCU using a relay. You have also learnt to create a GUI dashboard using Blynk mobile application and then connect it to the NodeMCU. You also learnt to send and receive data from the virtual pins of Blynk mobile application. The next project will be an interesting one as you'll build a system that can control devices anywhere in the world using voice commands.

Let's build the next one!

Control Devices Using Google Assistant and IFTTT

16

Over the decades, smartphones have underwent many advances both in performance and in features. They have grown to such an extent that we can't even imagine to part with them even for a day. Smartphone applications used for email, messaging, video sharing, etc., help us a lot in our daily life. To make the device control even smoother, virtual assistants are developed. These can control the entire smartphone as well as smart home devices using simple commands. This project deals with one such application of controlling an LED using a voice-based virtual assistant.

16.1 MOTIVE OF THE PROJECT

In this project, you'll control an LED using Google Assistant and IFTTT. To make the project simple, we are controlling an LED but you can control any electronic or electrical device using the same method.

16.2 GOOGLE ASSISTANT

Google Assistant is an artificial intelligence (AI) powered virtual assistant developed by Google that is mainly available in mobile and smart home devices. Users primarily interacting with Google Assistant are interacted through voice commands using *Ok Google* or *Hey Google* wake words. If you are unable to convey the voice command properly or the Google Assistant cannot understand the command, then keyboard input can also be provided. Google Assistant can do almost any task on your smartphone like calling a friend, receiving a call, sending a message, reading a received message, playing songs or movies, booking a cab, scheduling a meeting, opening any app, searching online, unlocking devices, controlling smart home devices and many more. The Google Assistant Icon is depicted in Figure 16.1.

Hi, how can I help?

FIGURE 16.1 Google Assistant Icon

DOI: 10.1201/9781003147169-16

Not only Google Assistant there are many more intelligent virtual assistants developed by various other companies like Apple's Siri, Amazon's Alexa, Samsung's Bixby, Microsoft's Cortana and many more. Usage of virtual assistants is increasing day by day as these help in performing various tasks. Many smartphones, smartwatches and earphones provide additional physical buttons or dedicated actions for activating virtual assistants.

16.3 IFTTT

If This Then That (IFTTT) is a service that can be used to design response for an event. It has integrated APIs (Application Programming Interface) of various service providers to analyze events and execute responses. Using IFTTT you can integrate more than one service provider with each other to create event and response pairs. At first glance, you might think If This Then That is a little odd but the name reflects exactly the working of IFTTT service shown in Figure 16.2. In short, the IFTTT service monitors for a Trigger; if the Trigger (This) is detected, then an Action (That) is executed.

FIGURE 16.2   IFTTT

IFTTT can be used to integrate service providers like Adafruit, Amazon Alexa, Google Assistant, Google Sheets, Google Drive, Twitter, Facebook, Instagram, Telegram, YouTube and many more. Visit the IFTTT website (https://ifttt.com/) to know the complete list of service providers that can be used to create triggers or actions.

First, you need to login or sign up to IFTTT for using their services. Go to the IFTTT website and click *Log in* if you already have an account. If you are visiting the IFTTT website for the first time or if you don't have an account, then click *Get started* at the top-right corner of the page as shown in Figure 16.3. Now, you'll be redirected to a page where you can sign up with your Apple account or Google Account or Facebook account or sign up with any of your email address and password as shown in Figure 16.4.

FIGURE 16.3   IFTTT website

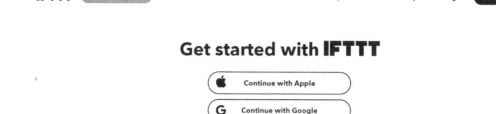

FIGURE 16.4   Getting started

As I'm having a Google Account and wanted to sign up with it, I'll click *Continue with Google*. But you can sign up with any account of your choice. Now, you'll be redirected to the login page of your respective account website as shown in Figure 16.5. After successfully logging in or signing up to IFTTT, you'll see the page as shown in Figure 16.6.

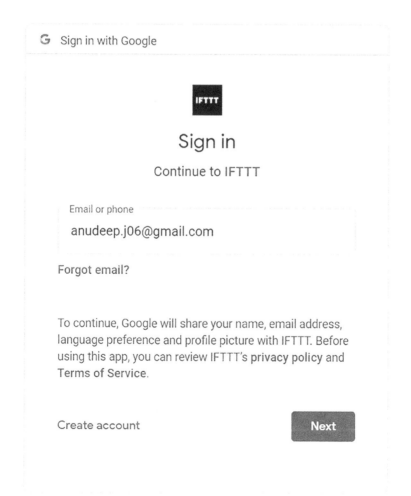

FIGURE 16.5   Sign in with Google

Start connecting your world.

Get more

FIGURE 16.6   IFTTT page after logging in

16.4 HARDWARE REQUIRED

TABLE 16.1   Hardware required for the project

HARDWARE	QUANTITY
NodeMCU	1
LED	1
Jumper wires	As required

16.5 CONNECTIONS

The list of required components for the project is tabulated in Table 16.1, followed by Table 16.2, which presents the pin-wise connections. Also, the connection diagram is shown in Figure 16.7, which shall help get a visualization of the connections.

TABLE 16.2   Connection details

NODEMCU	LED
D1	+ve pin (long pin)
GND	−ve pin (short pin)

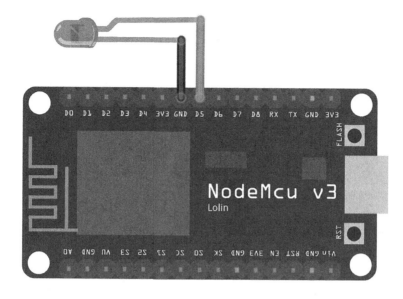

FIGURE 16.7   Connection layout

16.6 CREATE AN APPLET USING IFTTT

After successfully logging in or signing up to IFTTT, click *Create* at the top-right corner as shown in Figure 16.6 for creating an applet. Applets in IFTTT are used to integrate different services for creating trigger-action pairs. In the IFTTT free account, you can only create three applets for free and each applet can only contain one trigger-action pair. But you can create unlimited applets and unlimited trigger-action pairs in a single applet by opting for the subscription-based IFTTT pro service.

After you click *Create*, you'll be redirected to the applet creation page as shown in Figure 16.8. First, we need to create a trigger and then action must be created. For creating a trigger, click *If This Add* in the applet creation page. Now, you'll be presented with all the services as shown in Figure 16.9 that can be used for creating a trigger. Scroll down and select Google Assistant.

FIGURE 16.8   Trigger creation

FIGURE 16.9 Available services for trigger

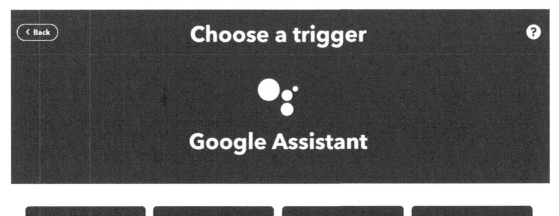

FIGURE 16.10 Choose a trigger

With the help of Google Assistant, you can create four types of triggers as shown in Figure 16.10. As we can control an LED using a simple phrase without any number or text ingredient, choose the *Say a simple phrase* trigger.

Click *Connect* as shown in Figure 16.11 for connecting IFTTT to the Google Assistant service. This is used to access the voice commands which are being told to your Google Assistant. Now, you'll be

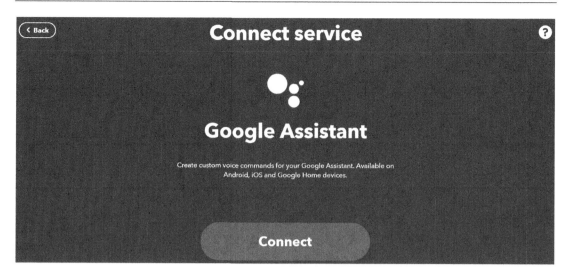

FIGURE 16.11   Connect Google Assistant service

FIGURE 16.12   Choose a Google Account

displayed with all the Google Accounts available in your browser as shown in Figure 16.12. Select the Google Account from which you want to control the LED. You need not choose the Google Account from which you have logged in to the IFTTT, it can be any other Google Account as well. After choosing the Google Account, you'll be asked to grant access to IFTTT for accessing your voice commands. Click *Allow* as shown in Figure 16.13 after completely reading the information displayed on that page.

After IFTTT gets connected to the Google Assistant service, all the required fields for creating the trigger will be displayed as shown in Figure 16.14. Enter the phrase which you want to tell the Google Assistant for switching ON the LED as *Switch On LED*. Enter *Turn On LED* and *On LED* for another way of saying the phrase. Enter *Switching On LED* as the response phrase from the Google Assistant. Select the language as *English* from the drop-down and click *Create trigger*.

If the trigger is created successfully, then you'll see the page as shown in Figure 16.15. You can also Edit or Delete the trigger from the same page. The next step is to create an Action.

Before creating an action, you need to create a feed named *LED* in Adafruit IO. Fill in the name and description details of the feed as shown in Figure 16.16 and click *Create*. After successfully creating the feed, come back to the IFTTT website and click *Then That Add* as shown in Figure 16.15 for

G   Sign in with Google

IFTTT wants to access your Google Account

👤 anudeep.j06@gmail.com

This will allow IFTTT to:

G   Manage your Google voice commands   ⓘ

Make sure that you trust IFTTT

You may be sharing sensitive info with this site or app. Find out how IFTTT will handle your data by reviewing its terms of service and privacy policies. You can always see or remove access in your Google Account.

Find out about the risks

Cancel   **Allow**

FIGURE 16.13   Grant access to IFTTT

creating an action. Now, you'll be presented with all the services as shown in Figure 16.17 that can be used for creating an action. Among them select Adafruit.

Using Adafruit you can only create one type of action as shown in Figure 16.18. Click the *Send data to Adafruit IO* action for sending data to Adafruit IO when the trigger is executed. Click *Connect* as shown in Figure 16.19 for connecting the IFTTT to Adafruit service. If you are logged in to Adafruit in your browser, you'll be directly shown the Authorization page as shown in Figure 16.20 or else log in to the Adafruit for seeing that page. Read the information displayed on the page completely and click *AUTHORIZE*.

After IFTTT gets connected to the Adafruit service, all the required fields for creating the action will be displayed as shown in Figure 16.21. The drop-down under the Feed name field shows all the available feeds in your Adafruit account. Among them, Select the *LED* feed. Enter the data to save as *ON* and click *Create action*. If the action is created successfully, then you'll see the page as shown in Figure 16.22. You can also Edit or Delete the action from the same page. If you are having an IFTTT pro account then you can create multiple such trigger-action pairs in the same applet by clicking the + symbol after the trigger and action. After successfully creating the trigger-action pair(s), click *Continue*.

Finally, review and edit the Applet Title and click *Finish* as shown in Figure 16.23. If your applet gets connected successfully, then you'll see the page as shown in Figure 16.24. From this page, you can connect or disconnect your Applet whenever you need or you can completely Archive the applet. You can also edit the trigger-action pair(s) by clicking the *Settings* at the top-right corner. Activity of the applet can also be reviewed by clicking the *View Activity* button.

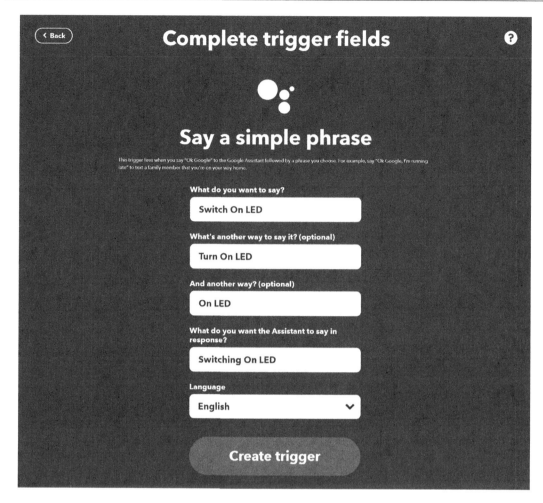

FIGURE 16.14   Trigger fields

FIGURE 16.15   Action creation

Create a new Feed   ×

Name

LED

Maximum length: 128 characters. Used: 3

Description

Feed for storing the ON and OFF information of LED.

Cancel   Create

FIGURE 16.16   LED feed creation

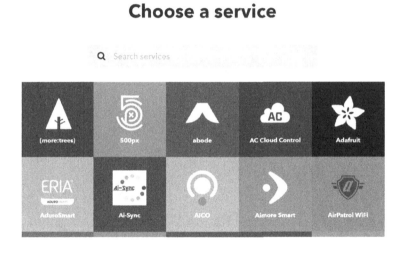

FIGURE 16.17   Available services for action

FIGURE 16.18   Choose an action

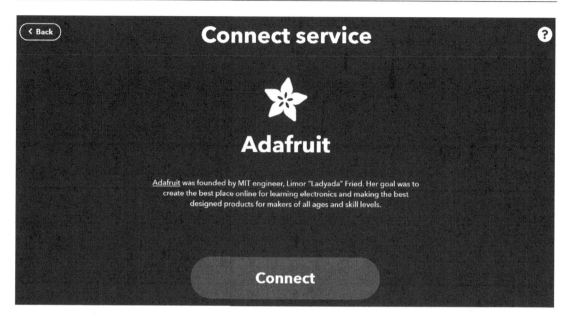

FIGURE 16.19 Connect Adafruit service

My Account

Profile

Addresses

Security & Privacy

Payment Methods

Subscriptions

Order History

My Wishlists

Gift Certificates and Coupons

Product Notifications

Newsletter Settings

Change Password

Services

The application IFTTT is requesting the following information for your account. Would you like to grant access?

- Name
- username
- Adafruit IO Dashboard URL
- Read and write to your feed data

By granting access, you'll be able to connect your Adafruit account to IFTTT, and enable any integrations that have been provided.

AUTHORIZE

FIGURE 16.20 Authorize access to IFTTT

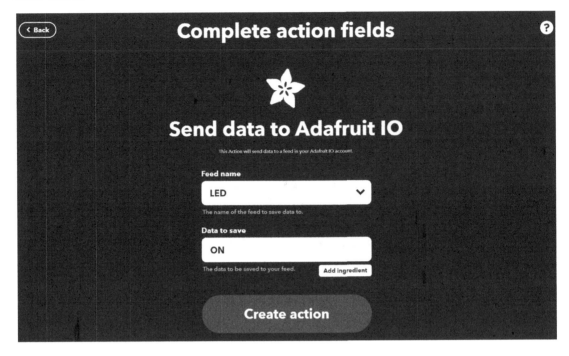

FIGURE 16.21 Action fields

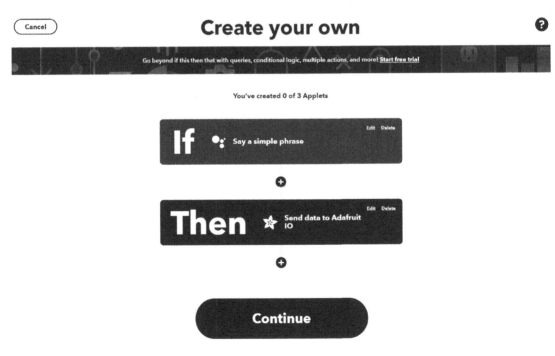

FIGURE 16.22 Trigger-action pair created

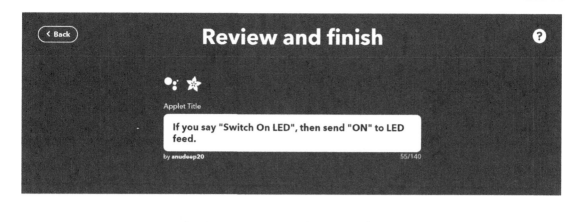

FIGURE 16.23   Review and finish

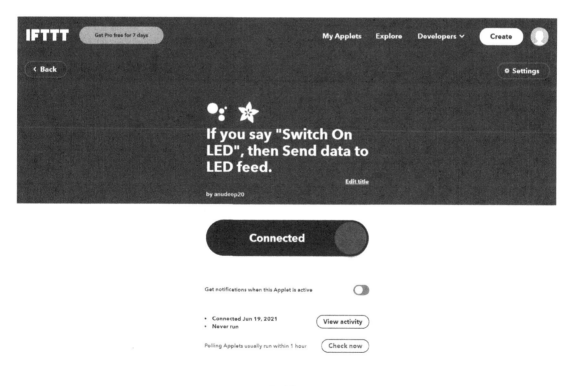

FIGURE 16.24   Connected applet

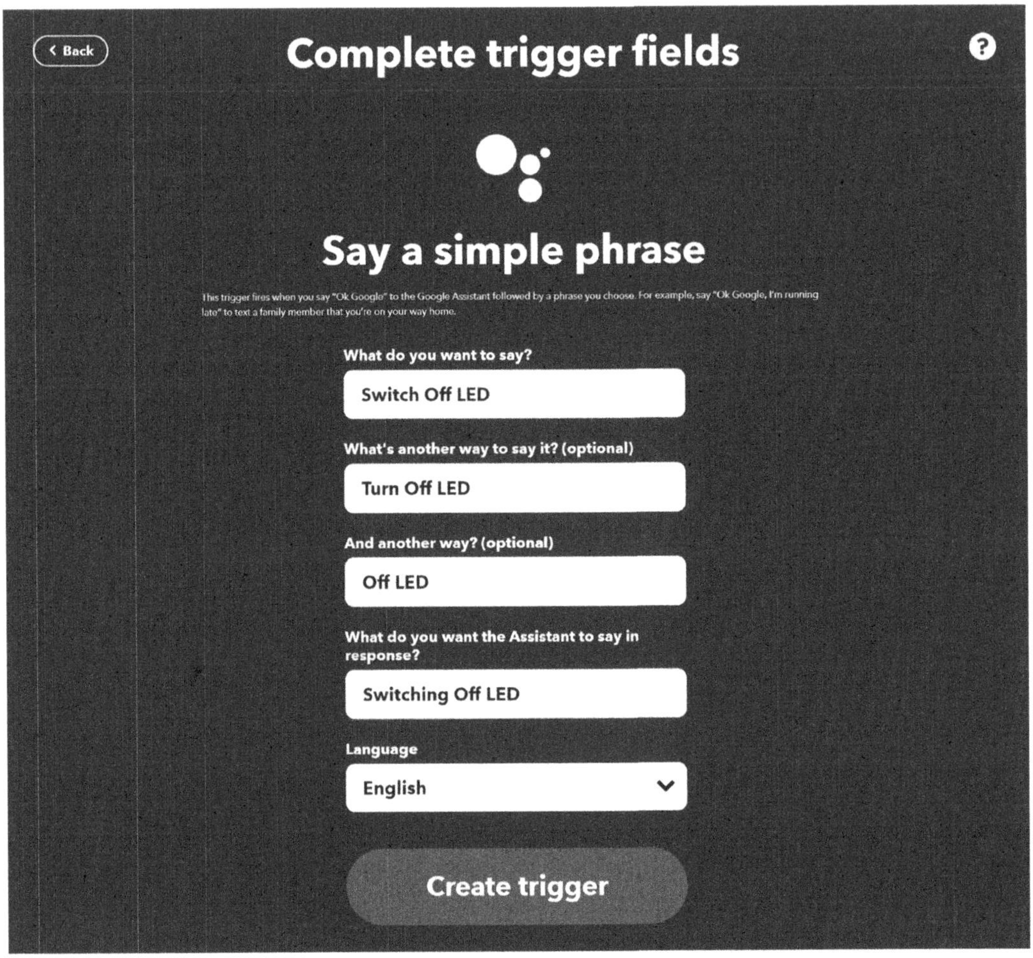

FIGURE 16.25   Trigger fields

Now, we have created a trigger-action pair for switching ON the LED. But, for switching OFF the LED we need to create another trigger-action pair. As discussed before, we cannot create multiple trigger-action pairs in a single applet of an IFTTT free account. So, we need to create another applet for switching OFF the LED in the same procedure as discussed before. Do this applet creation on your own as it'll be a practice for you. If you have an IFTTT pro account, then create another trigger-action pair in the same applet. No need of creating a new applet.

For creating another applet, click *Create* at the top-right corner of Figure 16.24. The process of creating an applet for switching OFF the LED is the same as discussed before, but the data entered in trigger and action fields must be changed. In the trigger fields, fill in the data as shown in Figure 16.25. Enter the phrase which you want to tell the Google Assistant for switching OFF the LED as *Switch Off LED*. Enter *Turn Off LED* and *Off LED* for another way of saying the phrase. Enter *Switching Off LED* as the response phrase from the Google Assistant. Select the language as *English* from the drop-down and click *Create trigger*. In the action fields, fill in the data as shown in Figure 16.26. Select the feed name as *LED* from the drop-down. Enter the data to save as *OFF* and click *Create action*. Finally, review and edit the Applet Title and click *Finish* as shown in Figure 16.27. If your applet gets connected successfully, then you'll see the page as shown in Figure 16.28.

After creating both applets, click on *My Applets* at the top-right corner. All the applets created in your account will be displayed as shown in Figure 16.29. Click any of these applets for editing or disconnecting

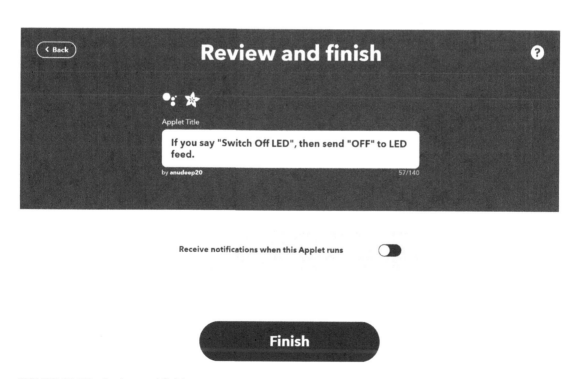

FIGURE 16.26   Action fields

FIGURE 16.27   Review and finish

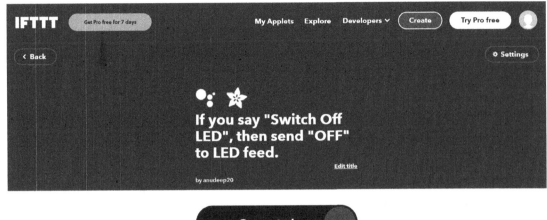

FIGURE 16.28   Connected applet

FIGURE 16.29   My applets

them. If you want to delete a service that is connected to IFTTT, click on the person image at the top-right corner and select *My services* as shown in Figure 16.30. All the services connected to your IFTTT account will be displayed as shown in Figure 16.31. Select the service which you want to delete and then click *Settings* at the top-right corner as shown in Figure 16.32. On the service settings page, you can see the activity log, see and edit the account info. You can also remove the service connection with IFTTT by clicking *Remove <SERVICE NAME>* at the bottom as shown in Figure 16.33.

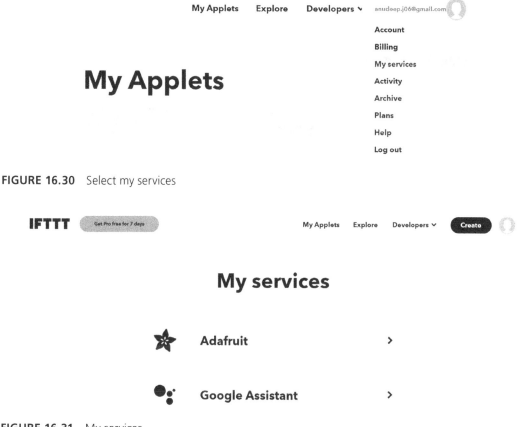

FIGURE 16.30   Select my services

FIGURE 16.31   My services

FIGURE 16.32   Google Assistant service

FIGURE 16.33   Remove service

16.7 ARDUINO IDE CODE

When we say *Switch On LED* in Google Assistant, IFTTT will upload *ON* to the *LED* feed. Similarly, when we say *Switch Off LED* in Google Assistant, IFTTT will upload *OFF* to the *LED* feed. So, we need to subscribe to the *LED* feed and read the data being added to it. If the data added to the *LED* feed is ON, then we need to switch ON the LED connected to NodeMCU, or if the data added to *LED* feed is OFF, then we need to switch OFF the LED connected to NodeMCU.

The code for subscribing to a feed and reading data added to it is already discussed in the previous project. We'll be using the same code with minor variable name changes. If you didn't follow it before, refer to Section 14.8 for understanding the following code.

```
#include <ESP8266WiFi.h> //Library for WiFi functionalities
#include <Adafruit_MQTT_Client.h> //Library for Adafruit MQTT Client

//Adafruit Details
#define server "io.adafruit.com"
#define port 1883
#define IO_USERNAME   "AIO_USERNAME"
#define IO_KEY        "AIO_KEY"

String ssid="WIFI_SSID"; //SSID of your WiFi
String password="WIFI_PASSWORD"; //Password of your WiFi

WiFiClient esp; //Create a WiFi Client
//Creation of Adafruit MQTT Client
Adafruit_MQTT_Client mqtt(&esp, server, port, IO_USERNAME, IO_KEY);
//Variable for subscribing data in an Adafruit Feed
```

```
Adafruit_MQTT_Subscribe LEDFeed=Adafruit_MQTT_Subscribe(&mqtt, IO_USERNAME"/
feeds/led");

int LEDPin=D5; //Initialize the LED pin
String dataFeed; //Stores data read from the feed

void setup()
{
  Serial.begin(115200); //Initialisation of NodeMCU Serial
  pinMode(LEDPin, OUTPUT); //Defining the direction of LED
  digitalWrite(LEDPin, LOW); //Initially the LED is set to OFF

  //Connecting NodeMCU to WiFi
  Serial.println("");
  Serial.print("Connecting to ");
  Serial.println(ssid);
  WiFi.begin(ssid, password);
  while (WiFi.status() != WL_CONNECTED)
  {
    delay(500);
    Serial.print(".");
  }
  Serial.println("");
  Serial.println("WiFi connected");
  Serial.println(WiFi.localIP()); //Print the local IP address of NodeMCU

  Serial.print("Connecting to MQTT");

  //Subscribing to LED Feed
  mqtt.subscribe(&LEDFeed);
}

void loop()
{
  //Connecting to MQTT
  if(!mqtt.connected()){
    while (mqtt.connect())
      Serial.print(".");
  }

  //Read data from the Feed
  Adafruit_MQTT_Subscribe *subscription;
  while ((subscription=mqtt.readSubscription(2000))) {
    if (subscription == &LEDFeed) {
      dataFeed=(char *)LEDFeed.lastread;

      Serial.print("Received Data: ");
      Serial.print(dataFeed);

      //Control the LED
      if(dataFeed == "ON")
        digitalWrite(LEDPin, HIGH);
      if(dataFeed == "OFF")
        digitalWrite(LEDPin, LOW);
    }
  }
}
```

Change the AIO_USERNAME and AIO_KEY with the Adafruit username and Adafruit IO key, respectively. Also, change the WIFI_SSID and WIFI_PASSWORD with your WiFi name and password, respectively. Then select the appropriate board and COM port and upload the code to NodeMCU. After successfully uploading the code, open the serial monitor and change the baud rate to 115200 to view the data received from the subscribed feed (*LED*).

Now, open Google Assistant from the Google Account which is connected to IFTTT and tell *Switch On LED*. Google Assistant will reply *Switching On LED* as shown in Figure 16.34 and simultaneously the LED connected to NodeMCU will switch on. Also, try telling other ways to Google Assistant like *Turn On LED* or *On LED* and see the same result happen.

Next, tell *Switch Off LED* to Google Assistant. It will reply Switching Off LED as shown in Figure 16.35, and simultaneously, the LED connected to NodeMCU will switch off. Also, try telling other ways to Google Assistant like *Turn Off LED* or *Off LED* and see the same result happen.

FIGURE 16.34   Google Assistant – switch on LED

FIGURE 16.35   Google Assistant – switch off LED

16.8 TRY IT

If you don't use Google Assistant much or if you are a fan of Amazon Alexa, you can try the same project using Amazon Alexa. The procedure and code remain the same. Try it on your own as it'll be a practice for you.

You can find all the resources of this project at this Github link (https://github.com/ anudeep-20/30IoTProjects/tree/main/Project%2016).

In this project, you have learnt to connect IFTTT to various services and build trigger-action pairs inside applets. You have learnt to control an LED using Google Assistant and IFTTT. Using the same concept you can control any electrical or electronic device using Google Assistant. In the next project, you'll build a home intrusion detection system that will automatically call the concerned person in case of an intrusion.

Let's GO to the next project!

Home Intrusion Detection System

17

The safety and security of our home should never be compromised. We live and store many of our valuables at home. With technological developments and recent innovations, security systems have advanced to a new level but are often costly and unaffordable. Using IoT and sensors, we can build efficient and simple intrusion detection systems. This project helps you build one such home intrusion detection system using a microcontroller and simple sensors.

17.1 MOTIVE OF THE PROJECT

In this project, you'll build a home intrusion detection system using LDR and laser that will alert the user using an automated VoIP call in case of an intrusion. The same concept can be used to detect intrusion in any room of a house or at any places that need surveillance.

As discussed in the previous project, the resistance of the LDR varies depending on the light incident on it. It'll output low resistance when light is incident on it and high resistance when no light is incident on it. This property of LDR can be used to detect an intrusion. When laser light is continuously incident on an LDR, it outputs low resistance. But if there is any obstruction to the laser beam falling on the LDR, it suddenly outputs high resistance. This change in resistance is detected as an intrusion.

17.2 LASERS

Laser is the acronym for Light Amplification by Stimulated Emission of Radiation. From the name itself you could understand that laser is nothing but light that is produced through a process of optical amplification based on the stimulated emission of electromagnetic radiation. In short, normal visible light which we see in our daily life spreads in the air as it moves forward whereas laser light travels as a single concentrated beam. The physics behind generating laser light is a bit complex and requires a deeper understanding. If you want to know more about it, you can read it on the Internet at this link (https://www.britannica.com/technology/laser).

In this project, we'll use a commonly used laser module in IoT as shown in Figure 17.1. Both laser modules shown in Figure 17.1 are the same, but the left side laser module is a breakout board built using the laser dot diode module. You can use any one of them depending on their availability. If both are available, then it is preferred to use the laser dot diode module as it only requires two pins to power it. Using the laser dot diode module is the same as using an LED. Its red wire needs to be connected to 5 V, and the blue wire needs to be connected to the Ground. You can also connect the red wire to any lower voltage (like 3.3 V) but the laser beam intensity will reduce.

DOI: 10.1201/9781003147169-17

Laser Module

Laser Dot Diode Module

FIGURE 17.1   Types of laser modules

17.3 VoIP CALL

VoIP is the acronym of Voice over Internet Protocol. It is similar to a normal cellular phone call, but the phone call is placed over the Internet connection. While normal cellular calls work with or without an Internet connection, VoIP calls work only with the help of an Internet connection.

Due to the rise of broadband and cheaper cellular data cost, almost every person even in remote areas are able to get Internet access. As a result, VoIP calls are getting popular day by day. Most of the daily used social media mobile applications like WhatsApp, Facebook, etc. also provide an option for VoIP calls. While cellular calls are chargeable, VoIP calls are actually free and you need to only pay for the Internet used. VoIP is also used to call between PC and mobile phones.

VoIP calling doesn't require any extra hardware. It uses the hardware used for connecting devices to the Internet. But cellular calls require extra hardware on all the devices between which the call is placed. So, to minimize the complexity and cost of the project, we'll use automated VoIP calls for alerting the user in case of an intrusion. Don't worry! You'll learn about the hardware required for cellular calls in future projects and you can also try this project using it.

17.4 HARDWARE REQUIRED

TABLE 17.1   Hardware required for the project

HARDWARE	QUANTITY
NodeMCU	1
LDR	1
Buzzer	1
Laser module	1
5 V power supply	1
Switch	1
Resistor (1 kΩ)	1
Breadboard	1
Jumper wires	As required

17.5 CONNECTIONS

TABLE 17.2 Connection details

NODEMCU	LDR	RESISTOR(1 KΩ)	BUZZER	LASER	SWITCH	5 V POWER SUPPLY
3V3	Pin 1					
A0	Pin 2	Pin 1				
D5			+ve pin			
				+ve pin	Pin 1	
					Pin 2	+ve pin
				−ve pin		−ve pin
GND		Pin 2	−ve pin			

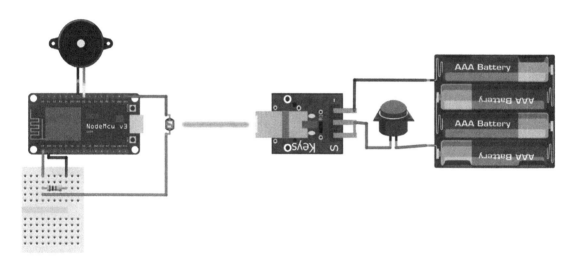

FIGURE 17.2 Connection layout

The list of required components for the project is tabulated in Table 17.1, followed by Table 17.2, which presents the pin-wise connections. Also, the connection diagram is shown in Figure 17.2, which shall help get a visualization of the connections.

As shown in Figure 17.2, if you have a laser module with three pins, then you can short the out pin and +ve pin using a wire. A 5 V power supply need not be a set of batteries connected in series as shown in Figure 17.2. You can also use a 5 V DC adapter using appropriate connectors or a 9 V battery with an LM7805 voltage regulator. You can use any type of SPST (single pole single throw) switch to ON and OFF the power supply to the laser module.

The NodeMCU setup (left side of Figure 17.2) and laser module setup (right side of Figure 17.2) must be placed on either side of the entrance of any location you want to monitor. They have to be placed such that when the laser light is switched ON, it has to directly fall on the LDR. When there is any obstruction to the laser light, the LDR immediately detects it as intrusion and an automated VoIP call will be generated. You can switch OFF the whole system when not needed as this will prevent false alarms.

17.6 IFTTT SETUP

As discussed in the previous project, IFTTT can be used to connect many services with each other. One such useful service provided by IFTTT is VoIP call that can be connected to any trigger. In this project, we'll create an applet for connecting an Adafruit trigger to the VoIP call action using the same procedure followed in the previous project. In IFTTT free account, you can only create three applets for free. Out of the three free applets, you have used two in the previous project. So, you need to use the remaining free applet in this project. If you have an IFTTT pro account then you need not worry about the number of applets being used.

First, log in to your IFTTT account and click + *Create* as shown in Figure 17.3 to create a new applet. Before creating a trigger, create a feed named *Intrusion* in Adafruit IO. Fill in the name and description details of the feed as shown in Figure 17.4 and click *Create*. After successfully creating the feed, come back to the IFTTT website and click *Add* beside *If This* to create a new trigger as shown in Figure 17.5. Among the listed services that can be used for creating a trigger, choose Adafruit service. Adafruit provides two types of triggers as shown in Figure 17.6 that can be used for triggering an action. It has to be remembered that the triggers provided by the Adafruit service are different from the actions provided by it. In the previous project, we have used the Adafruit service for action and it only provides a single action. But, in this project, we'll use the Adafruit service for trigger, and it provides two types of trigger. If you want to monitor a feed for a particular type of data added to it, then choose the first trigger, *Monitor a feed on Adafruit IO*. If you want to monitor a feed for any data added to it, then choose the second trigger, *Any new data*. In this project, we'll upload *YES* to the *Intrusion* feed when an intrusion is detected and no other data is uploaded to the *Intrusion* feed in any case. So, you can choose any type of Adafruit trigger among the two. We'll choose the *Monitor a feed on Adafruit IO* trigger. If IFTTT is already connected to

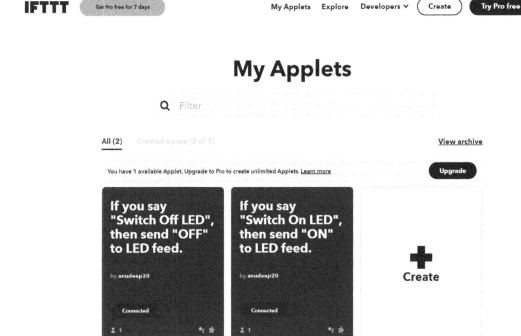

FIGURE 17.3   Creating a new applet

Create a new Feed   ✕

Name

Intrusion

Maximum length: 128 characters. Used: 9

Description

Feed used to store home intrusion data.

Cancel   Create

FIGURE 17.4 *Intrusion* feed creation

Cancel   **Create your own**   ❓

Go beyond if this then that with queries, conditional logic, multiple actions, and more! **Start free trial**

You've created 2 of 3 Applets

If This Add

Then That

FIGURE 17.5 Trigger creation

the Adafruit service, then you need not connect them again. But if the connection between them is deleted or if they are not connected before, then you need to connect them.

All the required fields for creating the *Monitor a feed on Adafruit IO* trigger will be displayed as shown in Figure 17.7. The drop-down under the feed name field shows all the available feeds in your Adafruit account. Among them, Select the *Intrusion* feed. The drop-down under the relationship field shows all the relationships with which the data in the Value field must be compared. As we'll monitor the feed for a particular value, the relationship needs to be selected as *equal to*. Enter the value to monitor as YES and click *Create trigger*. If the trigger is created successfully, then you'll see the page as shown in Figure 17.8.

Next, create an action by clicking *Add* beside *Then That* as shown in Figure 17.8. Among the listed services that can be used for creating an action, choose VoIP calls service. Using VoIP calls service, you can only create one type of action as shown in Figure 17.9. After selecting the *Call my device* action, you'll be redirected to a page as shown in Figure 17.10 for connecting IFTTT and VoIP calls service. Read the

FIGURE 17.6   Choose a trigger

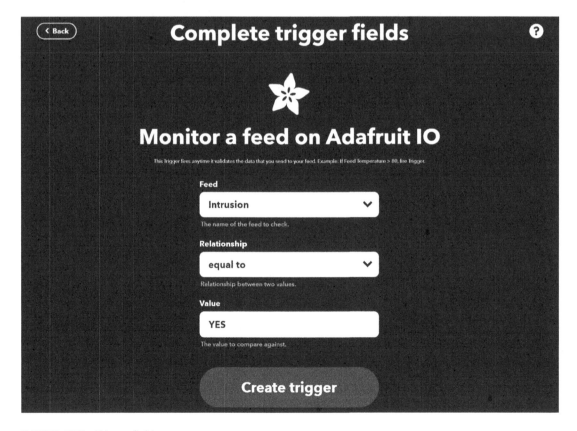

FIGURE 17.7   Trigger fields

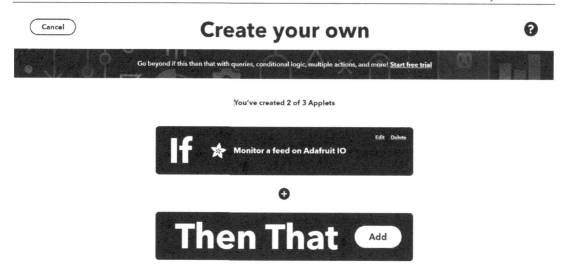

FIGURE 17.8   Trigger creation

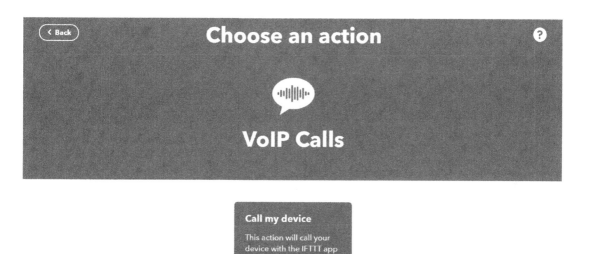

FIGURE 17.9   Choose an action

complete information displayed on the page and click *Connect*. A new window will open which automatically connects the IFTTT to VoIP calls service and you need not login or do anything.

After IFTTT gets connected to the VoIP calls service, all the required fields for creating the action will be displayed as shown in Figure 17.11. Write *Emergency! Someone broke into your house.* in the Voice message field and click *Create action*. If the action is created successfully, then you'll see the page as shown in Figure 17.12 and click *Continue*. Finally, review and edit the Applet Title and click *Finish* as shown in Figure 17.13. If your applet gets connected successfully, then you'll see the page as shown in Figure 17.14.

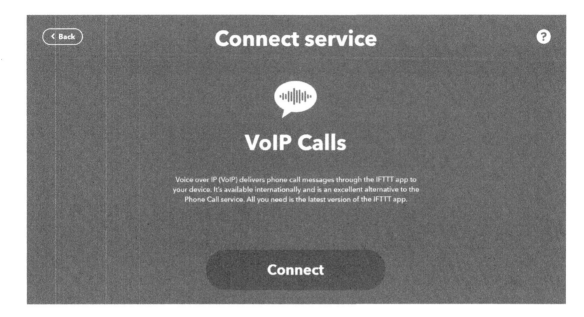

FIGURE 17.10   Connect VoIP calls service

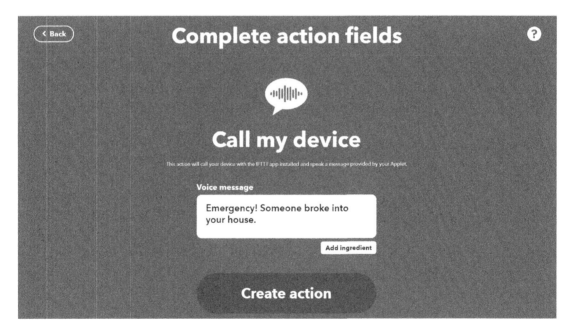

FIGURE 17.11   Action fields

FIGURE 17.12   Trigger-action pair created

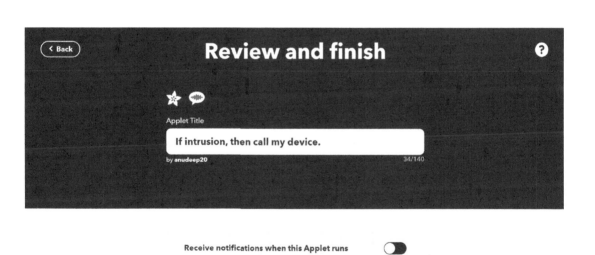

FIGURE 17.13   Review and finish

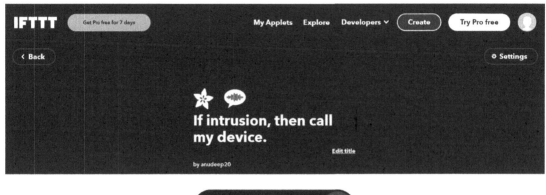

FIGURE 17.14   Connected applet

17.7 IFTTT MOBILE APPLICATION

Cellular calls don't require any specific mobile application to be installed for placing calls between devices. But VoIP calls generally require a specific mobile application installed on both the devices between which the VoIP call is placed. For example, if you want to do a VoIP call using WhatsApp, then the WhatsApp application needs to be installed on both devices whereas cellular calls don't need any such application to be installed. Similarly, the VoIP calls service provided by IFTTT requires IFTTT mobile application to be installed on your mobile phone for placing VoIP calls.

The latest version of the IFTTT mobile application is available for free and can be easily installed on both Android and iOS platform. For an Android-based mobile phone, the IFTTT application (shown in Figure 17.15) can be downloaded from the Google Play Store (https://play.google.com/store/apps/

FIGURE 17.15   Android IFTTT application

details?id=com.ifttt.ifttt&hl=en_IN&gl=US). For an iOS-based mobile phone, the IFTTT application (shown in Figure 17.16) can be downloaded from the Apple App Store (https://apps.apple.com/us/app/ifttt/id660944635).

The process of setting up the IFTTT mobile application is easy and straightforward. Open the IFTTT mobile application after installing it from Play Store or App Store. After opening the app, click *Continue* as shown in Figure 17.17 for logging in to the app. You'll be provided with a list of options as shown in Figure 17.18. The login options provided in the app are the same options provided on the website. So, log in to the app with the same account in which the applet was created. As we logged in to the website using Google and created the applets in it, we'll use Google for logging into the mobile app as well. Among the logged in Google Accounts on your mobile phone as shown in Figure 17.19, choose the account with

FIGURE 17.16   iOS IFTTT application

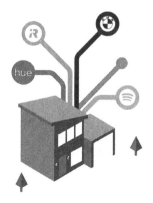

FIGURE 17.17   Landing page of IFTTT mobile application

FIGURE 17.18   List of login options

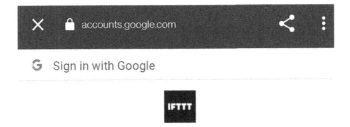

FIGURE 17.19   List of Google Accounts

which you want to login. After successfully logging in to the IFTTT mobile application, you'll see all the applets created in your account as shown in Figure 17.20. IFTTT mobile application can be used to do all the tasks that can be done using the website. You can create and delete an applet, edit and configure the trigger-action pairs and many more.

FIGURE 17.20   List of applets

17.8 ARDUINO IDE CODE

When an intrusion is detected by the NodeMCU, it publishes *YES* to the *Intrusion* feed or else no data will be published to the feed. If the data added to the *Intrusion* feed is *YES*, then a VoIP call will be immediately placed by the IFTTT mobile application. The code for connecting the NodeMCU to WiFi and publishing data to an Adafruit feed is already discussed in the previous projects. We'll be using the same code with minor variable name changes. If you didn't follow it before, refer to Section 9.6 for understanding the following code.

```
#include <ESP8266WiFi.h> //Library for WiFi functionalities
#include <Adafruit_MQTT_Client.h> //Library for Adafruit MQTT Client

//Adafruit Details
#define server "io.adafruit.com"
#define port 1883
#define IO_USERNAME   "AIO_USERNAME"
#define IO_KEY        "AIO_KEY"

String ssid = "WIFI_SSID"; //SSID of your WiFi
String password = "WIFI_PASSWORD"; //Password of your WiFi

WiFiClient esp; //Create a WiFi Client

//Creation of Adafruit MQTT Client
Adafruit_MQTT_Client mqtt(&esp, server, port, IO_USERNAME, IO_KEY);
//Variable for publishing data to Adafruit Feed
Adafruit_MQTT_Publish feed = Adafruit_MQTT_Publish(&mqtt, IO_USERNAME"/feeds/
intrusion");

int LDRPin = A0; //Initialize the LDR pin
int Buzzer = D5; //Initialize the Buzzer pin
int LDRvalue; //Stores LDR values
bool dataPublished = false; //Tracks whether the data is published to
Adafruit feed or not

void setup()
{
  Serial.begin(115200); //Initialisation of NodeMCU Serial
  pinMode(LDRPin, INPUT); //Defining the direction of LDR
  pinMode(Buzzer, OUTPUT); //Defining the direction of Buzzer

  //Connecting NodeMCU to WiFi
  Serial.println("");
  Serial.print("Connecting to ");
  Serial.println(ssid);
  WiFi.begin(ssid, password);
  while (WiFi.status() != WL_CONNECTED)
  {
    delay(500);
    Serial.print(".");
  }
  Serial.println("");
  Serial.println("WiFi connected");
  Serial.println(WiFi.localIP());  //Print the local IP address of NodeMCU
}
```

Previously, we have written the MQTT connection code in the `void setup()` function but now we have moved it to the `void loop()` function. As you already know `setup` function runs only once and the `loop` function runs infinitely. If the MQTT connection code is written in the `setup` function, the connection between NodeMCU and Adafruit will be established at the beginning and won't be reconnected if the connection is disconnected in between. But, if we move the MQTT connection code to `loop` function, the connection between NodeMCU and Adafruit will be checked every time the loop runs and the connection will also be reconnected if there is any disconnect. This will ensure successful publishing of data to the Adafruit feed. Extending this concept, you can also place the WiFi connection code in the `loop` function to ensure the NodeMCU is always connected to the WiFi.

```
void loop()
{
  //Connecting to MQTT
  if(!mqtt.connected()){
    Serial.print("Connecting to MQTT");
    while (mqtt.connect()){
      Serial.print(".");
    }
    Serial.println("");
  }
```

The resistance of LDR can be read using the analog pin of NodeMCU. You need to remember that the values read from the LDR need not be directly proportional to the resistance of LDR. Initially, upload the code without intrusion detection part and note down the LDR values when the laser light is falling on it and then with no light falling on it. According to the LDR values, determine a threshold value for which intrusion must be detected. In our case, when the laser light is falling on LDR, its values will be above 600 but when the laser light doesn't fall on LDR then its value drops below 500. So, we have arrived at a threshold of 500 for detecting an intrusion.

When an intrusion is detected, we'll continuously try to upload *YES* to the *Intrusion* feed until it is successfully uploaded. Also, the Buzzer will continuously ring until the NodeMCU is reset or the power supply to it is disconnected.

```
  //Reads and prints the LDR values
  LDRvalue = analogRead(LDRPin);
  Serial.print("LDR Value: ");
  Serial.println(LDRvalue); // Prints the value on serial monitor

  //If intrusion is detected
  if(LDRvalue < 500) {
    //Publish the data to Adafruit feed and continuously ring the buzzer
    while(1){
      if(!dataPublished && feed.publish("YES")){
        Serial.println("Successfully uploaded data to Adafruit Feed.");
        dataPublished = true;
      }
      digitalWrite(Buzzer, HIGH);
      delay(1000);
      digitalWrite(Buzzer, LOW);
      delay(1000);
    }
  }

  delay(1000);
}
```

Change the AIO _ USERNAME and AIO _ KEY with the Adafruit username and Adafruit IO key, respectively. Also, change the WIFI _ SSID and WIFI _ PASSWORD with your WiFi name and password, respectively. Then upload the code to NodeMCU after selecting the appropriate board and COM port. After successfully uploading the code, open the serial monitor and change the baud rate to 115200 to view the NodeMCU connection status and LDR values.

After NodeMCU gets connected to WiFi and Adafruit, try obstructing the laser light falling on the LDR. As soon as the laser light stop falling on LDR, NodeMCU uploads *YES* data to the *Intrusion* feed as shown in Figure 17.21. Wait for 10 seconds and you'll get an automated VoIP call from the IFTTT mobile application as shown in Figure 17.22. After lifting the call, you can see the applet name as shown in Figure 17.23 and hear an automated voice enunciating the sentence given in the action field (shown in Figure 17.11).

FIGURE 17.21   Data uploaded to *Intrusion* feed

FIGURE 17.22   VoIP call from IFTTT mobile application

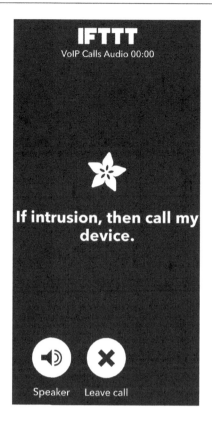

FIGURE 17.23 VoIP call audio

Generally, IFTTT applets work as expected most of the time. But sometimes if the applets are left unused for a few days, then they won't tend to work as expected. If you face any such problem refer to the common errors and troubleshooting tips provided by IFTTT (https://help.ifttt.com/hc/en-us/articles/115010194547-Common-errors-and-troubleshooting-tips). In short, if your applet doesn't work, then try disconnecting and reconnecting the applet or archive the applet and then restore it. If these don't work, then archive the applet and recreate a new one with the same trigger-action pair. You can also try reconnecting the services connected to your applet.

You can find all the resources of this project at this Github link (https://github.com/anudeep-20/30IoTProjects/tree/main/Project%2017).

In this project, you have built an intrusion detection system that can be used to protect any place. You have also learnt about lasers and VoIP calls. In the next project, you'll build a real-time alarm clock using an RTC module and OLED display.

Let's build the next one!

Real-Time Alarm Clock Using Arduino

18

Time is one of the important events we need to keep track of for meeting our deadlines. Everyone will have a digital wall clock at their home or a digital watch for knowing the time. But how do they build digital clocks? What's inside it? Can we also build a digital clock from scratch? Yes, it is easy to build an alarm clock from scratch using IoT. This project will help you know the contents of a digital clock and the procedure for building it from scratch.

18.1 MOTIVE OF THE PROJECT

In this project, you'll build a real-time alarm clock similar to a digital wall clock or digital wristwatch that can show the exact date-time and alert the user at a specific date-time. The real-time alarm clock is built from scratch using an RTC module, OLED and buzzer. It is designed to modify the alarm date-time using physical buttons without re-uploading the code.

18.2 I²C COMMUNICATION PROTOCOL

I²C (inter-integrated circuit) or IIC or I2C is one of the most important communication protocols used in electronics. The I²C communication protocol is also referred to as I²C communication bus by some authors and books. Even though it is an old protocol, it is widely used in short-distance, intra-board communications where low transmission speed is acceptable. I²C is a synchronous, multi-master, multi-slave, serial communication protocol developed by Philips Semiconductors.

The I²C protocol is mainly used in devices where low cost and simplicity are more important than speed. You'll see many IoT modules and sensors using the I²C protocol as most of the IoT applications need to be of low cost and don't require high-speed communication. One of the main advantages of the I²C protocol is that it can be used by microcontrollers to interface a large number of devices with the help of only two GPIO pins. Many other similar protocols like UART, SPI (serial peripheral interface) requires more pins to interface multiple devices.

I²C protocol uses only two bidirectional lines known as SDA (serial data) and SCL (serial clock) for data communication between multiple slaves and multiple masters as shown in Figure 18.1. The SDA line is used for transferring data between devices, and the SCL line carries the clock signal produced by the master device which is used for the synchronization of data transfer between devices on the I²C bus. Both the SDA and SCL lines are open-drain lines. So, pull-up resistors must be connected between VCC (Usually 5 V or 3.3 V) and SDA, SCL lines so that the lines are active high because the devices on the I²C bus are active low. Usually, 2 kΩ resistors are used for higher speeds (4 kb/s) and up to 10 kΩ resistors are used for lower speeds (100 kb/s). With recent modifications in the I²C protocol, much higher speeds up to 5 Mb/s can also be achieved.

DOI: 10.1201/9781003147169-18

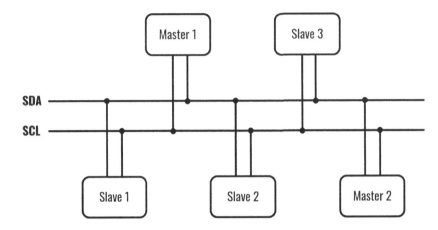

FIGURE 18.1   I²C communication bus

Every device in the I²C bus has a pre-defined unique address so that the master can choose the device to communicate. I²C protocol primarily uses 7-bit addressing for communication between devices through which 128 devices can be interfaced. But rarely I²C protocol is also used with 10-bit addressing through which 1028 devices can be interfaced. The data is transferred between master and slave in the form of blocks of 8-bit sequences. A complete data transfer includes a start bit, 7-bit device address, read or write bit, acknowledge bit (also called ACK or NACK bit), 8-bit internal register address, another acknowledge bit, 8-bit data, last acknowledge bit and finally a stop bit.

The data transfer between master and slave device begins with a start condition which occurs when the SDA line drops LOW while the SCL line is still HIGH. After the start condition, the clock begins and each data bit is transferred during each clock pulse. As you can observe from the timing diagram shown in Figure 18.2, a change in the logic level of SDA line occurs only when the SCL line is LOW, and the data bit to be transferred is the logic level of SDA line when the SCL line is HIGH. The data transfer of device address starts from MSB (most significant bit) to LSB (least significant bit). As an example, here we are transferring the data to an I²C device with 0x68 7-bit address. If you are new to different base systems of numbers and don't know the procedure for converting from one base to another, then please refer to this interesting tutorial provided by Tutorials point (https://www.tutorialspoint.com/computer_logical_organization/number_system_conversion.htm).

In the 8-bit sequence, the first 7-bits will be the device address bits and the 8th bit will be a Read or Write (R/W) bit. It is used for indicating whether the master will read data from the slave or write data to it. Logic HIGH of SDA line is considered as Read, while the logic LOW of SDA line is considered as Write. The next bit is the Acknowledge or Not Acknowledge (ACK/NACK) bit which is used for indicating whether the slave has successfully received the previous sequence of 8-bits or not. At this

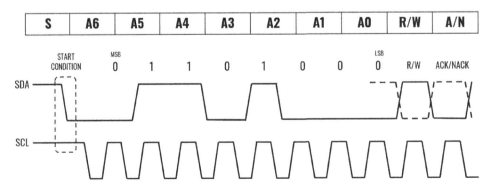

FIGURE 18.2   Timing diagram of the device address sequence

moment, the master device hands over the control of SDA line to the slave device. If the slave device has successfully received the previous sequence of bits, then the SDA line is pulled LOW, a condition known as Acknowledge (ACK). If the SDA line is not pulled to LOW, then the condition is known as Not Acknowledge (NACK) which means the slave device hasn't received the previous sequence of bits successfully. It may be caused due to various reasons like the slave might be busy or it might not have understood the sequence of bits or it may not be able to read any more data.

If the ACK/NACK bit is HIGH (Not Acknowledge), then the master decides the next procedure for dealing with NACK condition. If the slave device acknowledges, then the next set of 8-bit sequence containing the internal register address of the slave will be transferred. These are the internal memory locations in the slave device from which the data is read or written. As an example, here we are transferring 0x32 8-bit internal register address of the slave as shown in Figure 18.3. Similar to the first 8-bit sequence, after transferring the second 8-bit sequence, there'll be another ACK/NACK bit.

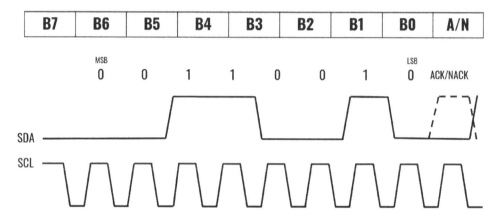

FIGURE 18.3   Timing diagram of the internal register address sequence

If the ACK/NACK bit is HIGH (Not Acknowledge), then the master decides the next procedure for dealing with NACK condition. If the slave device acknowledges, then the data transfer begins either from the slave or master device depending on the read or write mode selected using the R/W bit. As an example, here we are transferring 0xA2 8-bit data from slave to the master (Read is selected in the R/W bit). After the 8-bit data is transferred, there'll be another ACK/NACK bit for acknowledging whether the data is successfully sent or not. If the data is successfully sent, then the complete transfer will be terminated using a Stop condition which occurs when the SDA line goes from LOW to HIGH, while the SCL line is still HIGH as shown in Figure 18.4.

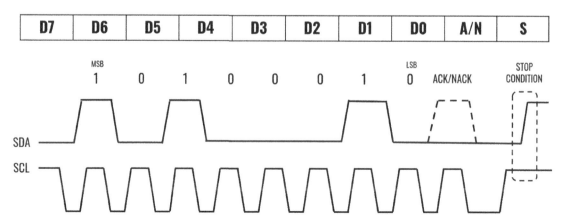

FIGURE 18.4   Timing diagram of data bits sequence

The explanation provided in this section is an overview of the complete features of the I²C communication bus which we'll be sufficient to understand most of the IoT projects. To know more about the I²C communication bus refer to the documentation provided by Texas Instruments (https://www.ti.com/lit/pdf/slva704). It has to be remembered that even this documentation doesn't explain every nuance of the I²C communication bus but gives you an in-depth understanding of it.

If you have an I²C device and don't know its I²C address. Don't worry! you can easily find it by using Arduino Uno and simple code. First, connect the VCC and GND of the I²C device to Arduino Uno according to its voltage rating and then connect the device's SDA and SCL pins to Arduino Uno's A4 and A5 pins, respectively. Not only Arduino Uno, but you can also use any evaluation board or microcontroller that has an Arduino bootloader for finding the I²C address of an I²C device. But the SDA and SCL pins of other evaluation boards or microcontrollers will differ from that of Arduino Uno. For example, the default SDA and SCL pins of NodeMCU are defined as GPIO 4 and GPIO 5, respectively, whereas Arduino Leonardo has SDA and SCL pins at digital pins 2 and 3, respectively.

For communicating with any I²C device, Wire.h library is a must. As Wire.h is one of the standard libraries of Arduino, it will be automatically installed during the installation of Arduino IDE. Upload the following code using Arduino IDE after selecting the appropriate board and COM port. It is taken from the official Arduino Playground website (https://playground.arduino.cc/Main/I2cScanner/). After successfully uploading the code, open the serial monitor and select the baud rate as 9600 to view the I²C addresses of all the connected I²C devices to the Arduino Uno.

```
#include <Wire.h>

void setup() {
  Wire.begin();

  Serial.begin(9600);
  while (!Serial);
  Serial.println("\nI2C Scanner");
}

void loop() {
  byte error, address;
  int nDevices;

  Serial.println("Scanning...");

  nDevices = 0;
  for(address = 1; address < 127; address++ ) {
    error = Wire.endTransmission();

    if (error == 0) {
      Serial.print("I2C device found at address 0x");
      if (address<16)
        Serial.print("0");
      Serial.print(address,HEX);
      Serial.println("  !");

      nDevices++;
    } else if (error==4) {
      Serial.print("Unknown error at address 0x");
      if (address<16)
        Serial.print("0");
      Serial.println(address,HEX);
    }
```

```
}
if (nDevices == 0)
  Serial.println("No I2C devices found\n");
else
  Serial.println("done\n");

delay(5000);
}
```

18.3 RTC MODULE

For building a real-time alarm clock, we need to know the exact date-time. But the evaluation boards or microcontrollers we use to build our projects can only provide the relative time from the start of program execution. To know the exact date-time, we need to use external modules known as RTC (real-time clock) modules.

There are many types of RTC modules available in the market that can be used in IoT projects. Among them, two of the most commonly used ones are the DS1307 RTC module and the DS3231 RTC module. The main difference between the two modules is the accuracy of time-keeping. The DS1307 RTC module has an external 32 kHz crystal oscillator whose oscillation frequency is easily affected by external temperature. So, when used continuously there will be a time delay of around 5 minutes per month. But the DS3231 RTC module has an internal 32 kHz temperature compensated crystal oscillator (TCXO) whose oscillation frequency is not affected by the change in external temperature. So, it can be used continuously and has an accuracy of ±2 minutes per year.

If your project requires time-keeping only for a few days and accuracy is not of the highest priority, then you can use the DS1307 RTC module. But, in this project, as we require utmost accuracy for continuously displaying the date-time and for ringing the alarm, the DS3231 RTC module is used.

The DS3231 RTC module (shown in Figure 18.5) has a large maxim manufactured DS3231 chip that is responsible for all the time-keeping functions of the RTC module. As the DS3231 chip supports the I²C protocol, it can be easily interfaced with any evaluation board or microcontroller. It can output year, month, date, day, hours, minutes and seconds information both in 12-hour and 24-hour format. Even the date corrections for a 28-day month or 31-day month or a leap year will be taken care of by the chip. It also provides two programmable time-of-day alarms which means we can program the DS3231 chip with two alarms at a particular time in a day but the year or month information cannot be provided.

For any real-time clock to work as expected, it has to run continuously without any interruption. All the RTC modules that are available in the market require power supply for uninterrupted working. But we cannot continuously supply power through an evaluation board for the uninterrupted working of the RTC module as it'll spoil the whole purpose. The DS3231 RTC module will work as expected when connected to an evaluation board using its VCC and GND pins. You can set the time of the RTC module according to your current time zone using a simple Arduino code. But, when there is any voltage fluctuation or if the connections are disconnected, then the date and time of the RTC module will be reset to 00:00:00 1st January 2000. Again we need to upload the code for setting the date and time. This process is not advisable, and it is not the correct way of using an RTC module.

So, to avoid all these procedures and to provide the RTC module with an uninterrupted power supply, the DS3231 RTC module offers a backup battery slot that can fit any 20 mm CR2032 3 V lithium coin cell. The RTC module has a built-in power sensing circuit that can sense the status of VCC power line to detect any power interruption and can automatically switch to the backup power supply. This will ensure the uninterrupted working of the RTC module even when it is not connected to any dedicated power supply.

The DS3231 RTC module usually consumes around 3 µA of current. So, a fully charged CR2032 lithium coincell can power a DS3231 RTC module for around 8 years without connecting to any dedicated power supply using the VCC pin.

The DS3231 RTC module also has 32 bytes 24C32 EEPROM chip from Atmel. This memory can be used for storing any information like settings, passwords, etc. The 24C32 EEPROM chip supports unlimited read-write cycles and uses I²C for data transfer. It shares the same I²C communication bus as the DS3231 chip. So, when you scan for the I²C address of the DS3231 RTC module using the I²C scanner code explained in the previous section, you'll see two I²C addresses printed on the Serial monitor. By default, the printed addresses will be 0x68 and 0x57 of which 0x68 is the I²C address of DS3231 chip and the 0x57 is the I²C address of 24C32 EEPROM. But the address of 24C32 can be changed using the A0, A1 and A2 unsoldered I²C address selection jumper pins.

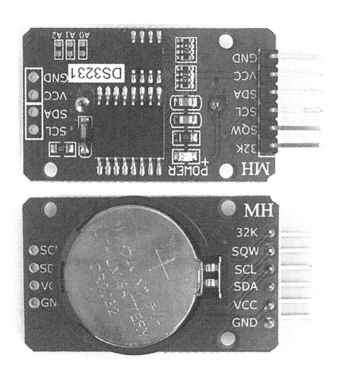

FIGURE 18.5   DS3231 RTC module

18.4 OLED

Any clock without an interface for displaying the date and time is as worse as a non-working clock. Daily used wall clocks show the time using mechanical hands whereas digital clocks use either 7-segment displays or LCD or OLED for displaying the time. The 7-segment displays are very much inexpensive. They can only display numbers and require more wires to operate. Hence, they are not advisable to use unless the system needs to display only a few numbers.

LCDs (liquid crystal displays) are very popular in electronics, and it is the first choice of display in most of the IoT projects. But even LCDs have their own advantages and disadvantages. LCDs are one of the less costly displays that can display not just numbers but more content on a screen with good resolution. They use a single backlight for the entire screen for displaying content. In an LCD screen, individual pixels don't emit their light but block the light emitting from the backlight. No matter how perfect an LCD is made its screen cannot block the entire light emitting from the backlight which means no LCD can display absolute black. But they can display cleaner white.

OLEDs (organic light emitting diode) are costly but the most efficient displays that can be used in electronics. They provide cleaner and high-resolution content on a screen. Unlike LCDs, they don't use any backlight for displaying the content but use independently illuminated pixels. So, each pixel in an OLED has its own backlight so that it can display light and colour independently. OLED screens can switch ON or OFF each pixel independently. So, to display absolute black, we need to just switch OFF all the pixels in an OLED screen. As OLEDs don't use a backlight, they cannot display absolute white and their brightness levels are a little less compared to LCD screens.

OLEDs are generally more power-efficient compared to LCDs. OLEDs consume power depending on the content displayed on the screen whereas LCDs consume almost the same power irrespective of the content displayed on the screen. When displaying content with more black colour, OLEDs consume the least power as they need to just switch OFF the pixels at the places where the content is black. But, when displaying content with more white colour, OLEDs require more power as they need to switch ON most of the pixels whereas LCDs consume constant power irrespective of the content as it uses the same backlight for any type of content. Each pixel in an OLED has an organic component whereas pixels in LCD don't. So, the life of OLEDs tends to be low compared to LCDs.

You can use any type of display in your projects. Select the correct type of display according to your use case, budget and the content displayed. The popularly used 16×2 character LCD display in IoT projects is shown in Figure 18.6. This LCD is less costly and can display up to 16 characters in each of the two rows. But the main disadvantage of using this LCD is that we require 16 wires for making it work

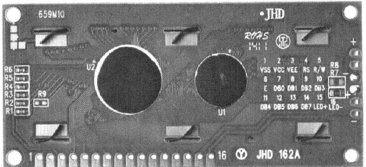

FIGURE 18.6 16×2 LCD with blue backlight

perfectly. It cannot display images and content of more than 32 characters. In our project, we'll display the content of more than 32 characters, and in future projects, we'll display images as well. So, for this reason, we'll only use OLEDs in our projects. But you can also implement the same project using LCD as well by reducing the number of characters to be displayed.

There are many types of OLEDs available in the market with different sizes and resolutions that can be used in our IoT projects. In this project, we'll use a commonly used 4-pin 1.3-inch 128×64 resolution OLED as shown in Figure 18.7. It consists of an SSD1306 CMOS OLED driver controller chip that can communicate with any microcontroller or evaluation board using I²C or SPI protocol. But the OLED used in this project supports only I²C communication protocol. There are also other types of OLED modules that use the same SSD1306 chip and can support SPI protocol and can display two different colours. The SSD1306 chip consists of inbuilt 1KB GDDRAM (Graphic Display Data RAM) for displaying the bit pattern on the screen. Each bit in the 1KB memory represents a particular pixel on the screen that can be switched ON or OFF programmatically. So, even complex images can be printed easily on the screen. If you are eager to know the complete specifications of the SSD1306 chip, refer to its datasheet (https://cdn-shop.adafruit.com/datasheets/SSD1306.pdf).

FIGURE 18.7   SSD1306 OLED display

18.5 TACTILE PUSH BUTTON

We cannot re-upload the code to the microcontroller or evaluation board every time for changing the alarm date and time. So, we need to design a system that can take input from the user to reset the alarm date and time without re-uploading the code. For that purpose, we use tactile push buttons (shown in Figure 18.8) for taking input from the user. These are normal push buttons with four ends among which both the parallel ends will be interconnected. When the push button is pressed, the two internally unconnected ends get connected, and when released, the same ends get disconnected as shown in Figure 18.9.

FIGURE 18.8   Tactile push button

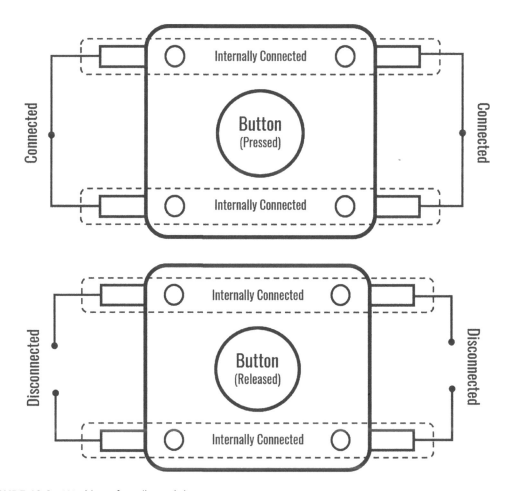

FIGURE 18.9   Working of tactile push button

18.6 HARDWARE REQUIRED

TABLE 18.1   Hardware required for the project

HARDWARE	QUANTITY
Arduino Uno	1
DS3231 RTC module	1
SSD1306 OLED module	1
Buzzer	1
Tactile Push Button	2
Resistor (1 kΩ)	2
Breadboard	1
Jumper wires	As required

18.7 CONNECTIONS

The list of required components for the project is tabulated in Table 18.1, followed by Table 18.2, which presents the pin-wise connections. Also, the connection diagram is shown in Figure 18.10, which shall help get a visualization of the connections.

TABLE 18.2   Connection details

ARDUINO UNO	DS3231 RTC	SSD1306 OLED	BUZZER	PUSH BUTTON 1	RESISTOR 1	PUSH BUTTON 2	RESISTOR 2
5V	VCC	VCC		Pin 1		Pin 1	
A4	SDA	SDA					
A5	SCL	SCL					
2				Pin 2	Pin 1		
3						Pin 2	Pin 1
4			+ve pin				
GND	GND	GND	−ve pin		Pin 2		Pin 2

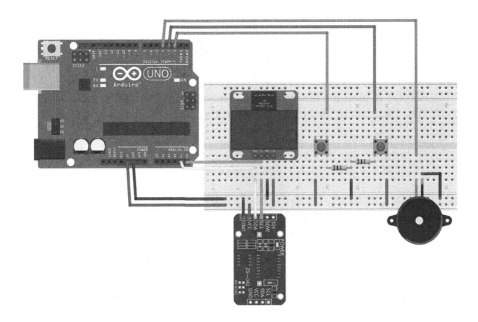

FIGURE 18.10 Connection layout

18.8 ARDUINO IDE CODE

As discussed earlier, Wire.h library helps to communicate with I²C supported devices. RTClib.h is a real-time clock library developed by Adafruit for DS3231, DS1307 and PCF8523 RTC modules. It is very useful for obtaining data from the RTC module through I²C protocol. You can easily install the RTClib.h library from the library manager (*Ctrl+Shift+I* or *Cmd+Shift+I* or *Tools > Manage Libraries*) in Arduino IDE or directly download it from Github at this link (https://github.com/adafruit/RTClib). RTClib.h library provides various functions that are useful for setting the time, alarm, extracting information and formatting the data obtained from the RTC module. But we have used only a limited number of functions provided by the RTClib.h library for implementing this project. If you want to know about all the functions provided by the RTClib.h library, refer to its documentation (https://adafruit.github.io/RTClib/html/index.html).

Adafruit _ SSD1306.h is an OLED library developed by Adafruit for SSD1306-based monochrome OLEDs. It is useful for controlling and performing various actions on the OLED. You can easily install the latest version of Adafruit _ SSD1306.h library from the library manager in Arduino IDE or can be directly downloaded from Github at this link (https://github.com/adafruit/Adafruit_SSD1306).

```
#include <Wire.h> //Library for I2C functionalities
#include <RTClib.h> //Library for RTC module
#include <Adafruit_SSD1306.h> //Library for OLED
```

display is an Adafruit _ SSD1306 constructor that accepts four arguments and is used for configuring an OLED screen. The first two arguments are the width and height of the OLED screen which is nothing but the resolution of the OLED. As discussed earlier, there are different types of OLEDs available in the market that work on different protocols like I²C, SPI, etc. But the OLED module used in this project supports only I²C protocol, and we'll use the same for controlling the OLED. So, &Wire is given as the third argument. Some OLED modules have a reset pin that can be connected to one of the evaluation board's pins. The fourth argument is the pin number to which the OLED module's reset pin is connected. If your OLED module doesn't have a reset pin or it is not connected to the evaluation board, then give the fourth argument as -1.

OLED _ ADDRESS is used for storing the I²C address of the OLED module. ALARM _ RING _ TIME stores the time (in minutes) for which the buzzer should ring from the alarm time.

```
#define SCREEN_WIDTH 128 //OLED screen width (in pixels)
#define SCREEN_HEIGHT 64 //OLED screen height (in pixels)
#define OLED_ADDRESS 0x3C //OLED I2C Address

#define ALARM_RING_TIME 1 //Alarm ring time (in minutes)

//Used for displaying content on OLED screen
Adafruit_SSD1306 display(SCREEN_WIDTH, SCREEN_HEIGHT, &Wire, -1);
```

RTC is an RTC _ DS3231 object used for performing various functions on the DS3231 RTC module. nowDT is a DateTime object that is used to store the current date and time obtained from the RTC module. numberChange and numberSet store the digital pins to which the first and second push buttons are connected respectively. Similarly, buzzer stores the digital pin to which the buzzer is connected.

nxtAlarm is a String type variable that is used to store the date and time at which the next alarm should ring. For manually setting the next alarm date and time, initialize the nxtAlarm with the next alarm date and time in "DD/MM/YYYY hh:mm" format and upload the code. All the remaining declared variables are used for a particular reason whose usage will be explained a little later in the code.

```
RTC_DS3231 RTC; //RTC_DS3231 object
DateTime nowDT; //Used for storing Date-Time obtained from RTC module

int numberChange = 2; //CHANGE push button pin
int numberSet = 3; //SET push button pin
int buzzer = 4; //Initialize Buzzer pin

//Stores the values read from the Buttons
bool numberChangeState;
bool numberSetState;

String nowDTStr=""; //Storing the Date-Time in String format
String nxtAlarm = "20/06/2025 00:00"; //Set your next alarm Date-Time
int alarmValues[5]; //Stores the Date-Time values individually

//Used to set Date-Time for new Alarm
int currNum = 0;
int currPos = 0;
String newAlarm = "0*/**/**** **:**";
char tempChar[2] = "";

long startTime; //Used for storing Alarm start time

void setup() {
  Serial.begin(9600); //Initialization of Arduino's Serial
  pinMode(numberChange, INPUT); //Direction of Number Change Button
  pinMode(numberSet, INPUT); //Direction of Number Set Button
  pinMode(buzzer, OUTPUT); //Direction of Buzzer
```

RTC.begin() is used for initializing the I²C communication on the RTC module. RTC.lostPower() is used to determine whether you are using the RTC module for the first time or the clock in the RTC module is reset due to any power disconnection. If the date and time of the RTC module gets reset, then RTC.adjust() is used for adjusting the date and time of the RTC module. During compilation of the code, _ _ DATE _ _ and _ _ TIME _ _ will be replaced with the date and time of your PC, respectively.

You can also create a custom DateTime object at any particular point in time using the DateTime() constructor. One of the ways of creating a custom DateTime object using the DateTime() constructor is by passing year, month, day, hour, minute and second arguments, respectively. Among these six arguments, year, month and day are the mandatory arguments.

```
RTC.begin(); //Initialize RTC

//Adjusts the time of RTC module when using it for the first time or when
its power is disconnected.
  if(RTC.lostPower()) {
    //DATE and TIME gets replaced with the date and time of the PC during
compilation.
    RTC.adjust(DateTime(F(__DATE__), F(__TIME__)));
  }
```

display.begin() is used for initializing the I²C communication between the OLED module and Arduino Uno. clearDisplay() is used for clearing any data on the OLED screen. setTextSize() is used to set the font size of the text displayed on the OLED screen. setTextColor() is used to set the text colour displayed on the OLED screen. For a monochrome display, you can choose between white and black text colours. SSD1306 _ WHITE considered as white colour, whereas SSD1306 _ BLACK is considered as black colour.

```
display.begin(SSD1306_SWITCHCAPVCC, OLED_ADDRESS); //Initialize OLED module

//Initial display configuration
display.clearDisplay();
display.setTextSize(1);
display.setTextColor(SSD1306_WHITE);
```

substring() is used for obtaining a part of a string. It accepts two arguments, starting and ending position of the part of the string to be obtained. If the second argument (ending position) is not given, then the complete string starting from the first argument will be returned. For example, the complete string is "HELLO WORLD" and we want to get "LLO W", then 2 and 7 have to be given as the starting and ending positions respectively. All the substrings are converted to integers using toInt() and are stored in alarmValues array.

```
//Obtaining the individual Date-Time values from Alarm string
alarmValues[0] = nxtAlarm.substring(0,2).toInt();
alarmValues[1] = nxtAlarm.substring(3,5).toInt();
alarmValues[2] = nxtAlarm.substring(6,10).toInt();
alarmValues[3] = nxtAlarm.substring(11,13).toInt();
alarmValues[4] = nxtAlarm.substring(14).toInt();
}
```

RTC.now() obtains the current date and time from the RTC module and returns it as a DateTime object. toString() is used to convert the DateTime object to a string in a particular format. DATE _ TIME _ FORMAT stores the format in which the DateTime object must be converted. Remember that the format should only contain D, M, Y, h, m and s. The format won't work if you use d instead of D or H instead of h. For example, "hh.mm YY-MM-DD" is the correct format whereas "HH:mm:s YY/M/dd" is the wrong format.

setCursor() is used to set the position from which the content should start printing on the OLED screen. It accepts the pixel coordinates from which the printing should start as an argument. Similar to Serial.print() and Serial.println(), display.print() and display.println() are used for printing text on the OLED screen. All the display functions used until now cannot display anything on the OLED screen and can only store it in a buffer. display.display() function is used to display all the content stored in the buffer on the OLED screen.

```
void loop() {
  //Used to obtain the current Date-Time
  nowDT = RTC.now();

  //Format of Date-Time to be displayed
  char DATE_TIME_FORMAT[] = "DD/MM/YYYY hh:mm:ss";
  Serial.print("Present Date-Time: ");
  Serial.println(nowDT.toString(DATE_TIME_FORMAT));

  //Displays the present Date-Time on the OLED
  display.setCursor(0,0);
  display.println("Clock: ");
  display.println(nowDT.toString(DATE_TIME_FORMAT));

  //Displays the next Alarm on the OLED
  display.setCursor(0,20);
  display.println("Next Alarm: ");
  display.println(nxtAlarm);
```

For changing the alarm date and time without re-uploading the code, we'll use two push buttons. One of them is used for changing the character at the cursor position, and the other is used for changing the cursor position in a string. We'll call the push button used for changing the character at the cursor position as CHANGE and the push button used for changing the cursor position as SET. For example, we need to modify the string from STR_ST="0*/**/**" to STR_END="27/08/25" using the CHANGE and SET push buttons.

- Initially, the cursor position will be at the first character of the STR_ST string. Press the CHANGE push button for changing the character at the cursor position. If the CHANGE push button is pressed once, then the STR_ST changes from "0*/**/**" to "1*/**/**". Press the CHANGE push button again to change the STR_ST to "2*/**/**". As the first character of STR_ST is the same as the first character of STR_END, press the SET push button to fix the first character of STR_ST and change the cursor position to the second character.
- After changing the cursor position, the STR_ST string changes from "2*/**/**" to "20/**/**". Now, press the CHANGE push button seven times to change the STR_ST from "20/**/**" to "27/**/**". As the second character of STR_ST is the same as the second character of STR_END, press the SET push button to fix the second character of STR_ST and change the cursor position to the fourth character. As the 3rd and 5th characters of STR_ST should not be modified, we'll code such that when the cursor is at 2nd or 4th character, and if the SET push button is pressed, then the cursor position skips a character.
- After changing the cursor position, the STR_ST string changes from "27/**/**" to "27/0*/**". As the fourth character of STR_ST is the same as the fourth character of STR_END, press the SET push button again to fix the fourth character of STR_ST and change the cursor position to the fifth character.
- After changing the cursor position, the STR_ST string changes from "27/0*/**" to "27/00/**". Press the CHANGE push button 8 times to change the STR_ST from "27/00/**" to "27/08/**".

As the fifth character of STR_ST is the same as the fifth character of STR_END, press the SET push button to fix the fifth character of STR_ST and change the cursor position to the sixth character. Continue the same procedure until STR_ST string is converted to STR_END string.

numberChangeState stores the digital reading read from the numberChange pin (connected to CHANGE push button). numberSetState stores the digital reading read from the numberSet pin (connected to SET push button). One of the common ways of modifying characters in a string is by using two pointers. One of them points to a particular character in a string and the other points to the position of characters in a string. currNum is used as a pointer to point to a particular character in the newAlarm string. currPos is another pointer used to point to the position of characters in the newAlarm string.

numberChangeState is used to change the currNum pointer and numberSetState is used to change the currPos pointer. Initially, currNum and currPos pointers will be pointing to the first character of newAlarm string. If the CHANGE push button is pressed once, then numberChangeState becomes HIGH and the currNum pointer will be incremented by 1. If the CHANGE push button is pressed continuously, then the currNum pointer also gets incremented continuously until it becomes 9. After that, the currNum pointer is reset to 0 and the increment starts again from 0. The character in newAlarm to which currPos is pointing must be replaced with the modified currNum pointer. But a character in newAlarm cannot be directly replaced with the currNum pointer as it is an int type variable. Only a character can replace a character in the newAlarm string. So, tempChar is used to convert the currNum pointer to a character array and then replace the character in newAlarm to which currPos is pointing.

If the SET push button is pressed once, then numberSetState becomes HIGH. This will set the character in the newAlarm string to which currPos is pointing and the currPos pointer will be incremented by 1. It has been coded such that the currPos pointer will not point to 3rd, 6th, 11th and 14th characters as those characters are fixed and should not be modified. If the SET push button is pressed continuously, then the currPos pointer will also be incremented continuously until it points to the last character of the newAlarm string. If the SET button is pressed when currPos is pointing to the last character of newAlarm string, then the values of newAlarm will be split into date, month, year, hour and minute values and stored in the alarmValues array. These values are used for validating whether the new alarm set by the user is valid or not. If it's a valid alarm, then the newAlarm string will be copied to the nxtAlarm string and the newAlarm string will be reset to 0*/**/**** **:**. If it's an invalid alarm, then INVALID is stored in the nxtAlarm and the newAlarm string will be reset to 0*/**/**** **:**.

While using tactile push buttons, many people face the problem of bounciness. It is nothing but the repeated pressing of push button even though it is pressed once. To avoid bounciness, you need to use either delay at the end of loop() function or use a debouncing code. So, we have used a delay of 500 milliseconds after each execution of the loop() function. You can also use the debounce code to avoid debouncing. The debounce code is one of the built-in examples of Arduino IDE which can be found at File > Examples > 02.Digital > Debounce in Arduino IDE.

```
//Reads the values from button pins
numberChangeState = digitalRead(numberChange);
numberSetState = digitalRead(numberSet);

//Used to set new Alarm Date-Time using Number Change and Number Set Buttons
display.setCursor(0,40);
display.println("New Alarm: ");
if(numberChangeState == HIGH){
  currNum += 1;

  if(currNum > 9)
```

```
      currNum = 0;
    String(currNum).toCharArray(tempChar, 2);
    newAlarm[currPos] = tempChar[0];
  }

  if(numberSetState == HIGH && currPos <= 15) {
    String(currNum).toCharArray(tempChar, 2);
    newAlarm[currPos] = tempChar[0];
    currPos += 1;

    if(currPos == 2 || currPos == 5 || currPos == 10 || currPos == 13){
      currPos += 1;
    }

    currNum = 0;
    newAlarm[currPos] = '0';
  }

  display.println(newAlarm);
  display.display();
  delay(500);
  display.clearDisplay();

  if(currPos > 15){
    alarmValues[0] = newAlarm.substring(0,2).toInt();
    alarmValues[1] = newAlarm.substring(3,5).toInt();
    alarmValues[2] = newAlarm.substring(6,10).toInt();
    alarmValues[3] = newAlarm.substring(11,13).toInt();
    alarmValues[4] = newAlarm.substring(14).toInt();

    if(!DateTime(alarmValues[2], alarmValues[1], alarmValues[0],
alarmValues[3], alarmValues[4]).isValid())
       nxtAlarm = "INVALID";
    else if(alarmValues[2] < nowDT.year())
       nxtAlarm = "INVALID";
    else if(alarmValues[2] == nowDT.year() && alarmValues[1] < nowDT.month())
       nxtAlarm = "INVALID";
    else if(alarmValues[2] == nowDT.year() && alarmValues[1] == nowDT.month()
&& alarmValues[0] < nowDT.day())
       nxtAlarm = "INVALID";
    else if (alarmValues[2] == nowDT.year() && alarmValues[1] == nowDT.month()
&& alarmValues[0] == nowDT.day() && alarmValues[3] < nowDT.hour())
       nxtAlarm = "INVALID";
    else if (alarmValues[2] == nowDT.year() && alarmValues[1] == nowDT.month()
&& alarmValues[0] == nowDT.day() && alarmValues[3] == nowDT.hour() &&
alarmValues[4] <= nowDT.minute())
       nxtAlarm = "INVALID";
    else {
      nxtAlarm = newAlarm;
      display.setCursor(10,20);
      display.println("Alarm successfully changed.");
      display.display();
      delay(2000);
      display.clearDisplay();
    }
```

```
    currNum = 0;
    currPos = 0;
    newAlarm = "0*/**/**** **:**";
  }
```

The alarmValues array is converted to a DateTime object and then compared to nowDT (which stores the current date and time). If they are equal, then the buzzer starts ringing until the numberSet button is pressed or until the ALARM_RING_TIME is elapsed.

```
    //Activate the Buzzer when its Alarm Date-Time
    if(DateTime(alarmValues[2], alarmValues[1], alarmValues[0], alarmValues[3],
alarmValues[4]) == nowDT && nowDT.second() < 3){
        display.setCursor(1,20);
        display.print("Alarm is ringing!");
        display.display();
        display.clearDisplay();

        startTime = millis();
        while(!numberSetState && millis() - startTime < ALARM_RING_TIME*60000){
          numberSetState = digitalRead(numberSet);
          digitalWrite(buzzer, HIGH);
          delay(500);
          digitalWrite(buzzer, LOW);
          delay(500);
        }
        nxtAlarm = "INVALID";

        display.setCursor(1,20);
        display.print("Alarm Ended.");
        display.display();
        delay(1000);
        display.clearDisplay();
    }
}
```

Finally, upload the code to Arduino Uno after selecting the appropriate board and COM port. After successfully uploading the code, open the serial monitor and change the baud rate to 9600 to view the data printed on the serial monitor. You'll also see the live clock, next alarm date and time and new alarm date and time on the OLED screen. Try changing the new alarm date and time using the two push buttons and then wait until the current time becomes the alarm time to listen to the alarm buzzer.

You can find all the resources of this project at this Github link (https://github.com/anudeep-20/30IoTProjects/tree/main/Project%2018).

In this project, you have learnt about the I²C communication protocol which is one of the most commonly used communication protocols in IoT. You have also learnt about different types of displays used in IoT and the differences between them. Next, you have learnt about commonly used RTC modules in IoT and tactile push buttons. Finally, you have built a real-time alarm clock whose alarm date and time can be changed without re-uploading the code. In the next project, you'll build a fruit classification system based on its colour using TCS3200 colour sensor.

Let's build the next project!

Classification of Fruits Using TCS3200 Colour Sensor

19

Eyes are one of the important sensory organs of humans. It is estimated that a healthy human eye can differentiate between 100,000 colours. A nearly equivalent electronic module for the human eye is a camera. But we cannot use a camera for differentiating the colours even along with the microcontrollers used in IoT as they don't have enough computing power for processing the images. So, we need to use a colour sensor for detecting colours instead of a camera in IoT. This project will use a low-cost colour sensor that can be easily interfaced with evaluation boards and microcontrollers for detecting the colours of objects.

19.1 MOTIVE OF THE PROJECT

In this project, you'll build a system to classify the fruits based on their colour using TCS3200 colour sensor and display the fruit images on the OLED screen.

Classification of items based on colour is a widely used industrial application. Depending on the colour, items are sorted into their respective containers. Automation of sorting helps industries save a lot of money and manpower as well. Colour sensors are also used in industries to detect defects in a uniform object, identify different colour marks on items, verify colours applied to the objects and many more.

19.2 TCS3200 OR TCS230 COLOUR SENSOR

The colour sensors used in industries are very expensive and accurate. They can even detect minute changes in colour. But, for prototype applications, we need not require high accuracy, and even inexpensive colour sensors like TCS3200 will serve the purpose. TCS3200 colour sensor shown in Figure 19.1 is the most commonly used colour sensor in IoT applications.

Before understanding the working of the TCS3200 colour sensor, we need to understand how are we able to see and detect the colours of objects. White light coming from the sun or the flashlight consists of seven different colours (Violet, Indigo, Blue, Green, Yellow, Orange, Red or in short VIBGYOR) which can be verified from the prism experiment. Each of the colours in the VIBGYOR has a different frequency and wavelength; among them, violet has the highest frequency and least wavelength, whereas red has the lowest frequency and highest wavelength.

When white light falls on any object, some wavelengths get absorbed and some get reflected depending on the material properties. The colour of the object which we see with our eyes is the wavelengths of white light that are reflected by the object. For example, if you are having orange jumper wires and some

DOI: 10.1201/9781003147169-19

FIGURE 19.1 TCS3200 colour sensor

white light is falling on them, then the orange colour on the jumper wires absorbs all the wavelengths except the orange colour wavelength which gets reflected and can be seen by your eyes.

The TCS3200 colour sensor is basically a breakout board for the TCS3200 chip (shown in Figure 19.2) manufactured by TAOS (Texas Advanced Optoelectronic Solutions). The board has four high-intensity white LEDs used to illuminate the object whose colour needs to be detected. These LEDs help in easier and accurate detection of colours even in low light conditions. If you have a closer look at the TCS3200 chip, you'll see an 8x8 grid of colour filters known as Bayer filter of which 16 have red filters, 16 have green filters, 16 have blue filters and another 16 have no filters. The Bayer filter consists of only red, green and blue colour filters because they are the primary colours that can be used to construct any colour. For example, purple colour can be obtained by mixing 50% red, 0% green and 50% blue colours. Each pixel in the Bayer filter is made up of a Bayer pattern which consists of four filters (one red, one green, one blue and one clear). Underneath each colour filter, there'll be a photodiode to detect the intensity of light passed to them.

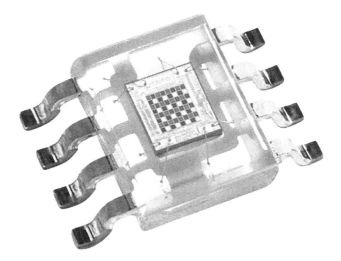

FIGURE 19.2 TCS3200 chip

After the TCS3200 colour sensor is powered, the four white LEDs project light onto the object whose colour needs to be detected. The object absorbs certain wavelengths of white light and reflects some wavelengths. These reflected wavelengths of white light will fall on the Bayer filter of the TCS3200 chip. The red filters in the Bayer filter will only pass the red wavelengths of the reflected white light to the photodiode underneath it. Similarly, green and blue filters will pass only the green and blue light wavelengths onto the photodiodes underneath them respectively. But the clear filter will not restrict any wavelength and pass the complete reflected white light to the photodiode underneath it. The intensity of light falling on each photodiode is measured and then each colour's light intensity is averaged and sent as output. The colour of the object is obtained by measuring the relative level of red, green and blue intensities from the output of the TCS3200 colour sensor.

As the photodiodes corresponding to each colour filter in the TCS3200 chip are connected in parallel, the intensity of light falling on each photodiode type can be easily obtained using two pins. The S2 and S3 pins on the TCS3200 colour sensor are used for this purpose. For example, to obtain the intensity of red colour both S2 and S3 must be set to LOW. Similarly, to obtain other colour intensities follow the combinations tabulated in Table 19.1.

TABLE 19.1   Configurations for choosing a photodiode type

S2	S3	PHOTODIODE TYPE
LOW	LOW	Red
LOW	HIGH	Blue
HIGH	LOW	Clear (no filter)
HIGH	HIGH	Green

The intensity of each colour is output as square waves from the OUT pin of the TCS3200 colour sensor. The frequency of the output square waves varies from 2 Hz to 500 kHz depending on the intensity of the colour. The TCS3200 colour sensor also provides the option of frequency scaling using S0 and S1 pins. It is used for optimizing the output and easy interfacing of the TCS3200 colour sensor with various evaluation boards and microcontrollers. Generally, 20% frequency scaling is used for Arduino boards. All the frequency scaling options in TCS3200 colour sensor are tabulated in Table 19.2. If you want to know more about the TCS3200 chip, refer to its datasheet (https://www.mouser.com/catalog/specsheets/tcs3200-e11.pdf).

TABLE 19.2   Configurations for choosing an output frequency scaling

S0	S1	OUTPUT FREQUENCY SCALING
LOW	LOW	Power down
LOW	HIGH	2%
HIGH	LOW	20%
HIGH	HIGH	100%

19.3 HARDWARE REQUIRED

TABLE 19.3   Hardware required for the project

HARDWARE	QUANTITY
Arduino Uno	1
TCS3200 colour sensor	1
SSD1306 OLED module	1
Breadboard	1
Jumper wires	As required

19.4 CONNECTIONS

The list of required components for the project is tabulated in Table 19.3, followed by Table 19.4, which presents the pin-wise connections. Also, the connection diagram is shown in Figure 19.3, which shall help get a visualization of the connections.

TABLE 19.4   Connection details

ARDUINO UNO	TCS3200 COLOUR SENSOR	SSD1306 OLED MODULE
5V	VCC	VCC
2	S0	
3	S1	
4	S2	
5	S3	
7	Out	
A4		SDA
A5		SCL
GND	GND	GND

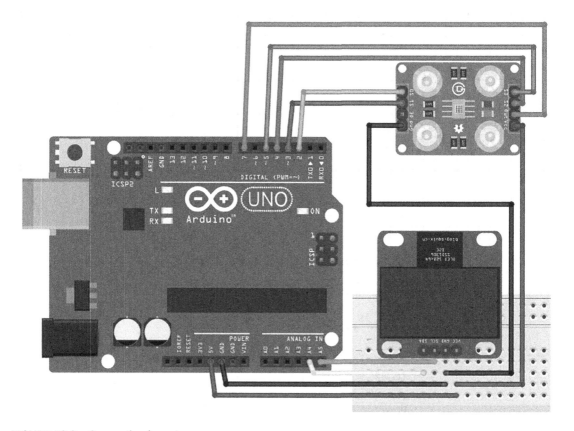

FIGURE 19.3   Connection layout

19.5 DISPLAYING IMAGES ON OLED

In the day-to-day used displays like TV or the monitor of a PC, images are displayed as soon as the device is powered. It is because all the information regarding the images to be displayed will be already present in the OS. Similarly, the SSD1306 OLED module can be configured to display images as soon as it is powered. As we cannot directly attach images in the Arduino IDE code, we need to convert the image to be loaded on the OLED display to a byte array. Don't worry! It's not a difficult process and can be easily done using a website (http://javl.github.io/image2cpp/) first developed by Javl. This website can even run locally without any Internet connection.

The resolution of the OLED used in this project is 128×64. So, we can completely display images only of resolution 128×64 or less. If the resolution of the images is greater than 128×64, then only a part of the image will be displayed on the OLED. In this project, we write the detected fruit name in the top 128×10 size, and in the remaining 128×54 size, we'll display the detected fruit image. Follow the steps one after another to convert your image to a byte array.

- The first step is to choose the image you want to convert into a byte array. Next, change the resolution of the image from its original resolution to any resolution less than or equal to the resolution of your display using any image editing software like paint, adobe photoshop, etc. Remember you can also change the resolution of your image using the *image2cpp* website mentioned earlier. Here, we'll change the resolution of the image shown in Figure 19.4 to 128×54 using the *image2cpp* website.

FIGURE 19.4 Image of banana

For converting the image resolution using the *image2cpp* website, open the website using any one of your favourite browsers and then upload the image to be converted to byte array as shown in Figure 19.5. Next, move to the *Image Settings* section and change the resolution of the image to 128×54 in the *Canvas size(s)* option as shown in Figure 19.6.

- Next, we need to change some settings for displaying the image properly on the OLED. Change the *Brightness/alpha threshold* to 190, *Scaling* to scale to fit, keeping proportions from the dropdown and finally check both the horizontally and vertically boxes of *Center.* You can also

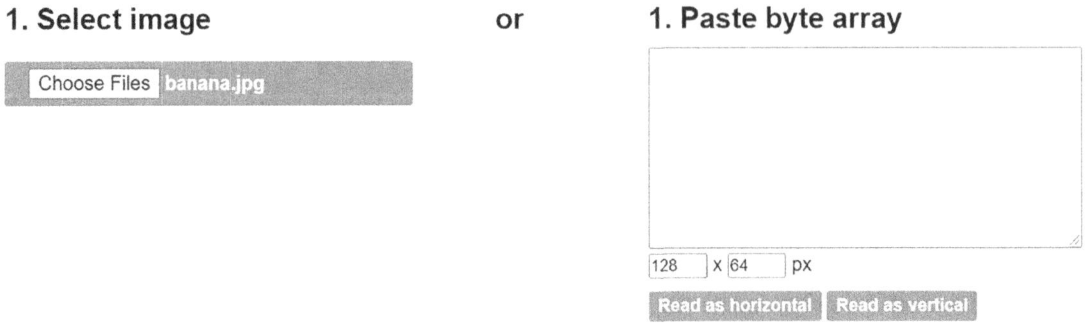

1. Select image   or   1. Paste byte array

Choose Files | **banana.jpg**

| 128 | x | 64 | px |

Read as horizontal | Read as vertical

FIGURE 19.5   Upload of image

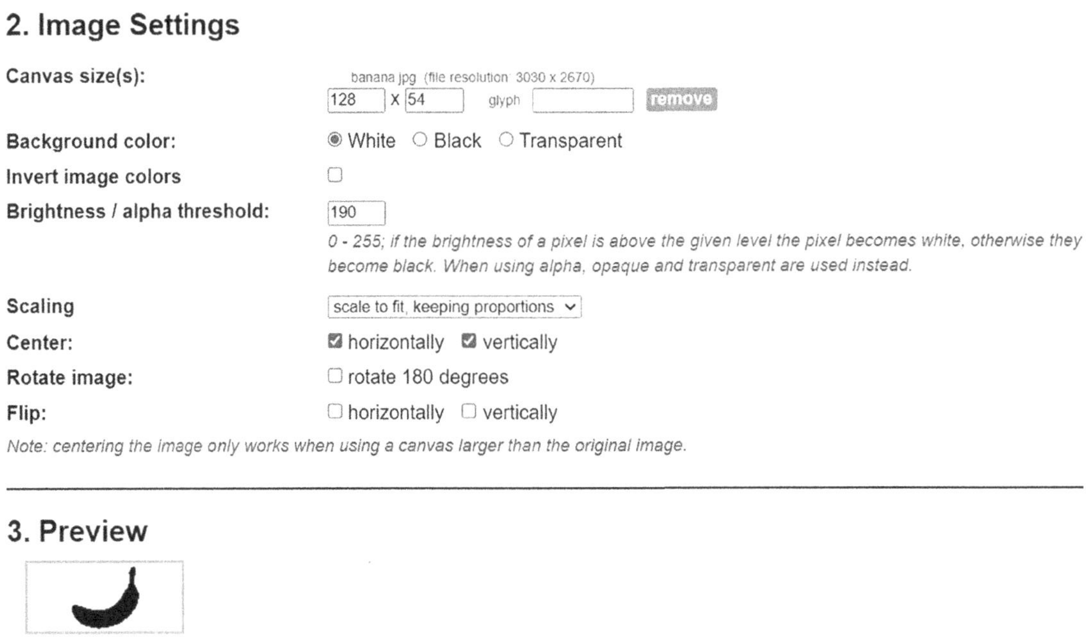

2. Image Settings

Canvas size(s):   banana.jpg (file resolution 3030 x 2670)

128 | x | 54 | glyph | | remove

Background color: ● White   ○ Black   ○ Transparent

Invert image colors   ☐

Brightness / alpha threshold: 190

0 - 255; if the brightness of a pixel is above the given level the pixel becomes white, otherwise they become black. When using alpha, opaque and transparent are used instead.

Scaling   scale to fit, keeping proportions ∨

Center: ☑ horizontally   ☑ vertically

Rotate image: ☐ rotate 180 degrees

Flip: ☐ horizontally   ☐ vertically

Note: centering the image only works when using a canvas larger than the original image.

3. Preview

FIGURE 19.6   Image settings and preview

change the background colour of the image, invert the image colours, rotate or flip the image using the *Image Settings* section. As you change the settings, the image in the *Preview* section will also be changed. It is the final image that will be displayed on the OLED.

- For converting the image shown in the *Preview* section, move to the *Output* section and change the *Code output format* option to plain bytes from the dropdown. Keep the *Draw mode* as Horizontal – 1 bit per pixel and don't change it unless you see a distorted image displayed on the OLED screen. Finally, click *Generate code* and copy the byte array generated in the box below (as shown in Figure 19.7) to the Arduino IDE code.

By the same process, convert the image of apple and grapes to byte array and copy it to the Arduino code. We have used open-source images from Google to display on the OLED which are available in the resources. If you are using the images provided in the resources, then it is recommended to use 169 brightness for apple and 145 brightness for grapes. Please remember that you can use any image to display on the OLED, but its brightness/alpha threshold must be adjusted accordingly for displaying it as expected on the OLED.

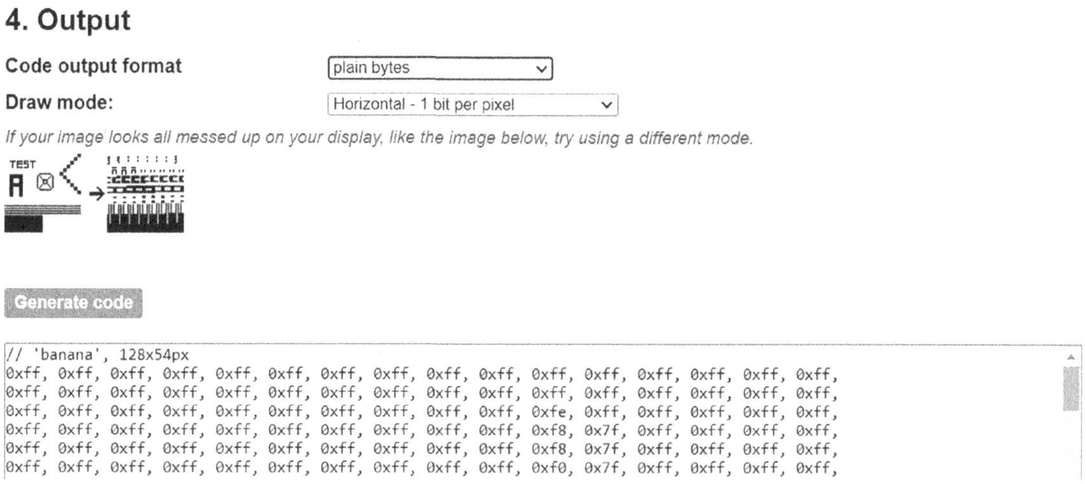

4. Output

Code output format   [plain bytes   ⌄]

Draw mode: [Horizontal - 1 bit per pixel   ⌄]

If your image looks all messed up on your display, like the image below, try using a different mode.

[Generate code]

```
// 'banana', 128x54px
0xff, 0xff, 0xff, 0xff, 0xff, 0xff, 0xff, 0xff, 0xff, 0xff, 0xff, 0xff, 0xff, 0xff, 0xff, 0xff,
0xff, 0xff, 0xff, 0xff, 0xff, 0xff, 0xff, 0xff, 0xff, 0xff, 0xff, 0xff, 0xff, 0xff, 0xff, 0xff,
0xff, 0xff, 0xff, 0xff, 0xff, 0xff, 0xff, 0xff, 0xff, 0xff, 0xfe, 0xff, 0xff, 0xff, 0xff, 0xff,
0xff, 0xff, 0xff, 0xff, 0xff, 0xff, 0xff, 0xff, 0xff, 0xff, 0xf8, 0x7f, 0xff, 0xff, 0xff, 0xff,
0xff, 0xff, 0xff, 0xff, 0xff, 0xff, 0xff, 0xff, 0xff, 0xff, 0xf8, 0x7f, 0xff, 0xff, 0xff, 0xff,
0xff, 0xff, 0xff, 0xff, 0xff, 0xff, 0xff, 0xff, 0xff, 0xff, 0xf0, 0x7f, 0xff, 0xff, 0xff, 0xff,
```

FIGURE 19.7   Bitmap generation

19.6 CALIBRATION OF TCS3200 COLOUR SENSOR

The sensors used until now are either precalibrated or don't require any calibration. But there are some sensors like the GY-61 accelerometer and TCS3200 colour sensor which need to be calibrated before usage. For calibrating the TCS3200 colour sensor, connect it to the Arduino Uno as shown in Figure 19.3 (with or without OLED module) and upload the following code for obtaining raw values from the TCS3200 colour sensor.

The preprocessors S0, S1, S2, S3 and out are used for defining the digital pins to which S0, S1, S2, S3 and OUT of TCS3200 colour sensor are connected respectively. Generally, the output from the TCS3200 colour sensor won't be stable and will oscillate even with a slight movement of the sensor. So, to obtain a more stable output from the TCS3200 colour sensor, we'll average a few output samples before printing the average. nSampleAvg is the number of consecutive samples to be averaged before printing the final output value. freqRed, freqGreen and freqBlue store the output frequency read from the red, green and blue filtered photodiodes, respectively.

```
//Used for controlling colour sensor
#define S0 2
#define S1 3
#define S2 4
#define S3 5
#define out 7
#define nSampleAvg 10

int freqRed = 0;
int freqGreen = 0;
int freqBlue = 0;
```

As the S0, S1, S2 and S3 pins are used to control the colour sensor by sending signals through them, the direction of operation of these GPIO pins is set as OUTPUT. Similarly, as the out pin is used to receive the output from the colour sensor, its direction of operation is set as INPUT. As discussed earlier, for Arduino boards the recommended frequency scaling is 20% which is set using the S0 and S1 pins.

```
void setup() {
  Serial.begin(9600); //Initialisation of Arduino's Serial

  //Direction of colour sensor pins
  pinMode(S0, OUTPUT);
  pinMode(S1, OUTPUT);
  pinMode(S2, OUTPUT);
  pinMode(S3, OUTPUT);
  pinMode(out, INPUT);

  //Frequency scaling of colour sensor set to 20%
  digitalWrite(S0,HIGH);
  digitalWrite(S1,LOW);
}
```

Initially, the colour sensor is set to read the output frequency from red filtered photodiodes by setting the S2 and S3 pins to LOW. If you remember, the pulseIn() function is already discussed in one of the previous projects about the ultrasonic sensor. It is used to obtain the time in microseconds for which the GPIO pin is either HIGH or LOW. The pulseIn(out, LOW) is used to obtain the time in microseconds for which the out pin is LOW. Next, the colour sensor is set to read the output frequency from green and blue filtered photodiodes by setting the S2 and S3 pins as mentioned in Table 19.1. The same process is repeated for nSampleAvg times and then printed on the serial monitor by averaging the final value.

```
void loop() {
  for(int i=0; i < nSampleAvg; i++){
    //Reading the output frequency of RED filtered photodiodes
    digitalWrite(S2,LOW);
    digitalWrite(S3,LOW);
    freqRed += pulseIn(out, LOW);
    delay(100);

    //Reading the output frequency of GREEN filtered photodiodes
    digitalWrite(S2,HIGH);
    digitalWrite(S3,HIGH);
    freqGreen += pulseIn(out, LOW);
    delay(100);

    //Reading the output frequency of BLUE filtered photodiodes
    digitalWrite(S2,LOW);
    digitalWrite(S3,HIGH);
    freqBlue += pulseIn(out, LOW);
    delay(100);
  }
  Serial.print("R= ");
  Serial.print(freqRed/nSampleAvg); //Printing RED colour frequency
  Serial.print("  G= ");
  Serial.print(freqGreen/nSampleAvg); //Printing GREEN colour frequency
  Serial.print("  B= ");
  Serial.println(freqBlue/nSampleAvg); //Printing BLUE colour frequency

  freqRed = 0;
  freqGreen = 0;
  freqBlue = 0;
}
```

After the code is successfully uploaded to Arduino Uno, open the serial monitor and set the baud rate to 9600. The RGB values printed on the serial monitor are raw precalibrated sensor values. The colour sensor needs to be calibrated using these values for obtaining the correct RGB values. To obtain the calibration values, take two reference sheets of white and black colour. You can also use any two reference objects with nearly pure black and pure white colour. The accuracy of measurement of RGB values using the colour sensor depends on these calibration values. So, choose the reference objects correctly.

First, place the black object in front of the colour sensor and note down the RGB values printed on the serial monitor. Similarly, place the white object in front of the colour sensor and note down the RGB values printed on the serial monitor. Using our TCS3200 colour sensor and reference objects, we have obtained the calibration RGB values as 240, 240, 180 for black and 37, 37, 27 for white. It has to be remembered that the calibration values will be different for different colour sensors, and it also varies for the same sensor depending on the reference objects used.

19.7 ARDUINO IDE CODE

The following Arduino code detects the fruit in front of the TCS3200 colour sensor by calculating the correct RGB values using the calibration values obtained in the previous section. The following code is a combination of the code used in the previous project and the code used in the previous section. Only the code required for displaying an image on OLED using byte array is added to it.

```
#include <Wire.h> //Used for I2C functionalities
#include <Adafruit_SSD1306.h> //Used for OLED module

#define SCREEN_WIDTH 128 //OLED screen width (in pixels)
#define SCREEN_HEIGHT 64 //OLED screen height (in pixels)
#define OLED_ADDRESS 0x3C //OLED I2C Address

//Used for controlling colour sensor
#define S0 2
#define S1 3
#define S2 4
#define S3 5
#define out 7
#define nSampleAvg 10
#define RGBTHRESH 10

//Used for displaying content on OLED screen
Adafruit_SSD1306 display(SCREEN_WIDTH, SCREEN_HEIGHT, &Wire, -1);

//Stores output from colour sensor
int freqRed = 0;
int freqGreen = 0;
int freqBlue = 0;
```

Copy the byte array of banana, apple and grapes images obtained in Section 19.5 in the place of <Byte Array of Banana image>, <Byte Array of Apple image>, <Byte Array of Grapes image> respectively.

```
//Bitmap of Banana image [190 brightness]
const unsigned char banana [] PROGMEM = {
  <Byte Array of Banana image>
};

//Bitmap of Apple image [169 brightness]
```

```
const unsigned char apple [] PROGMEM = {
  <Byte Array of Apple image>
};
//Bitmap of Grapes image [145 brightness]
const unsigned char grapes [] PROGMEM = {
  <Byte Array of Grapes image>
};

void setup() {
  Serial.begin(9600); //Initialisation of Arduino's Serial
  display.begin(SSD1306_SWITCHCAPVCC, OLED_ADDRESS); //Initialize OLED module

  //Direction of colour sensor pins
  pinMode(S0, OUTPUT);
  pinMode(S1, OUTPUT);
  pinMode(S2, OUTPUT);
  pinMode(S3, OUTPUT);
  pinMode(out, INPUT);

  //Frequency scaling of colour sensor set to 20%
  digitalWrite(S0,HIGH);
  digitalWrite(S1,LOW);

  //Initial display configuration
  display.setTextSize(1);
  display.setTextColor(SSD1306_WHITE);
  display.clearDisplay();
}
```

The correct RGB values of black colour are 0, 0, 0, whereas the correct RGB values of white colour are 255, 255, 255. But we have obtained the raw RGB values for black as 240, 240, 180 and for white as 37, 37, 27 using the TCS3200 colour sensor. So, for calibrating the TCS3200 colour sensor, we need to map the raw RGB values to the correct RGB values using a linear function. For mapping them, we can use the map() function which is an inbuilt linear mapping function in Arduino.

For calibrating the red frequency, map(pulseIn(out, LOW), 37,240, 255,0) is used where pulseIn(out, LOW) is used to obtain the raw output value. The second argument is the raw red value for white, the third argument is the raw red value for black, the fourth argument is the correct red value for white, and the fifth argument is the correct red value for black. So, the raw red values in the second and third arguments are mapped to correct red values in the fourth and fifth arguments. Similarly, the green and blue raw values are mapped to the correct green and blue values, respectively. As we have already discussed, the map() function can map the values not only in the given range but also outside the range. But the RGB values outside the range of 0 and 255 are considered invalid. So, to constrain the values between 0 and 255, constrain() function is used. The same process is repeated for nSampleAvg times and then averaged to obtain the final value.

```
void loop() {
  for(int i=0; i < nSampleAvg; i++){
    //Reading the output frequency of RED filtered photodiodes
    digitalWrite(S2,LOW);
    digitalWrite(S3,LOW);
    freqRed += constrain(map(pulseIn(out, LOW), 37,240, 255,0), 0, 255);
    delay(100);

    //Reading the output frequency of GREEN filtered photodiodes
    digitalWrite(S2,HIGH);
```

```
    digitalWrite(S3,HIGH);
    freqGreen += constrain(map(pulseIn(out, LOW), 37,240, 255,0), 0, 255);
    delay(100);

    //Reading the output frequency of BLUE filtered photodiodes
    digitalWrite(S2,LOW);
    digitalWrite(S3,HIGH);
    freqBlue += constrain(map(pulseIn(out, LOW), 27,180, 255,0), 0, 255);
    delay(100);
}

//Averaging the output from colour sensor
freqRed /= nSampleAvg;
freqGreen /= nSampleAvg;
freqBlue /= nSampleAvg;

//Printing the RGB values detected by colour sensor
Serial.print("R= ");
Serial.print(freqRed); //Printing RED colour frequency
Serial.print("  G= ");
Serial.print(freqGreen); //Printing GREEN colour frequency
Serial.print("  B= ");
Serial.println(freqBlue); //Printing BLUE colour frequency
```

Now, the averaged final RGB value is compared with the RGB value of banana, apple and grapes. If you don't know the RGB values of the fruits you want to detect, upload the code by commenting out the following detection part. Open the serial monitor and change the baud rate to 9600. Next, place the fruits you want to detect in front of the colour sensor and note down the RGB values. We have obtained the RGB values for banana as 248, 199, 66 similarly for apple and grapes as 150, 5, 10 and 203, 220, 97, respectively. Please remember that the RGB values of fruits used by us need not be the same as the fruits used by you. So, the RGB values of fruits detected by our colour sensor may be different from yours.

Even though the colour of all the bananas might look the same, the RGB value of a particular banana will slightly differ from the other. For example, the RGB values of the banana used for comparison are 248, 199, 66, but the RGB values of another similar type of banana that needs to be detected are 250, 195, 72. So, a threshold of RGBTHRESH is used to detect most of the bananas that have RGB values in the range of the banana used for comparison. Similarly, the same concept is used for detecting most of the apples and grapes. You can also use different thresholds for different fruits and different colour frequencies to further improve the accuracy.

If a fruit is detected, the name of the fruit is written at the top of the OLED and its image is displayed beneath it. drawBitmap() function is used to display an image on the OLED screen. The first two arguments of the drawBitmap() function accept the pixel coordinates from which the image should be displayed. The third argument accepts the byte array of the image. The fourth and fifth arguments accept the resolution of the image to be displayed. The sixth argument accepts the background colour of the image.

```
//Checking whether the fruit is Banana [RGB of Banana - 248,199,66]
 if(freqRed >= 248-RGBTHRESH && freqRed <= 248+RGBTHRESH && freqGreen >=
199-RGBTHRESH && freqGreen <= 199+RGBTHRESH && freqBlue >= 66-RGBTHRESH &&
freqBlue <= 66+RGBTHRESH){
    display.clearDisplay();
    display.setCursor(0,0);
    display.print("Detected Fruit:BANANA");
    display.drawBitmap(0, 10, banana, 128, 54, WHITE);
    display.display();
}
```

```
  //Checking whether the fruit is Apple [RGB of Apple - 150,5,10]
  else if(freqRed >= 150-RGBTHRESH && freqRed <= 150+RGBTHRESH && freqGreen >=
5-RGBTHRESH && freqGreen <= 5+RGBTHRESH && freqBlue >= 10-RGBTHRESH &&
freqBlue <= 10+RGBTHRESH){
    display.clearDisplay();
    display.setCursor(0,0);
    display.print("Detected Fruit:APPLE");
    display.drawBitmap(0, 10, apple, 128, 54, WHITE);
    display.display();
  }

  //Checking whether the fruit is Grapes [RGB of Grapes - 203,220,97]
  else if(freqRed >= 203-RGBTHRESH && freqRed <= 203+RGBTHRESH && freqGreen >=
220-RGBTHRESH && freqGreen <= 220+RGBTHRESH && freqBlue >= 97-RGBTHRESH &&
freqBlue <= 97+RGBTHRESH){
    display.clearDisplay();
    display.setCursor(0,0);
    display.print("Detected Fruit:GRAPES");
    display.drawBitmap(0, 10, grapes, 128, 54, WHITE);
    display.display();
  }

  //If the detected colour doesn't match any fruit range
  else {
    display.clearDisplay();
    display.setCursor(0,0);
    display.print("This fruit is not in the list of detectable fruits.");
    display.display();
  }

  freqRed = 0;
  freqGreen = 0;
  freqBlue = 0;
}
```

Finally, upload the code to Arduino Uno after selecting the appropriate board and COM port. After successfully uploading the code, open the serial monitor and change the baud rate to 9600 to view the RGB values of the colour in front of the TCS3200 colour sensor. Now, place any fruit in front of the TCS3200 colour sensor and wait for 5 seconds for detecting the fruit. As soon as the fruit is detected, its name and image will be displayed on the OLED screen. If the colour sensor is unable to detect the fruit, you'll see an unable to detect message on the OLED screen.

You can find all the resources of this project at this Github link (https://github.com/ anudeep-20/30IoTProjects/tree/main/Project%2019).

In this project, you have learnt to interface and calibrate the TCS3200 colour sensor for obtaining correct RGB values. You have also learnt to convert an image into a byte array for displaying it on the OLED screen. Using these concepts, you have built a system to detect and display different types of fruits using the TCS3200 colour sensor and OLED. In the next project, you'll build a device to continuously track heart rate and oxygen saturation ($\%SpO_2$) of a person.

Let's build the next project!

Build Your Own Pulse Oximeter

20

During the coronavirus pandemic, the most important vital parameter to be monitored happened to be the level of oxygen in the blood also known as oxygen saturation (% SpO_2). This can be easily monitored using a reliable pulse oximeter even at home. It is also one of the widely sold healthcare devices in the world during the pandemic. There is a time when people were unable to get these devices and some people used to buy fake products due to lack of awareness about the quality. Using faulty pulse oximeters is very dangerous as these might show wrong readings leading to serious consequences such as unnecessary hospitalization, etc. This project will help you understand the internal working of a pulse oximeter and will also help in building an affordable pulse oximeter right from scratch.

20.1 MOTIVE OF THE PROJECT

In this project, you'll build a device to continuously track the heart rate and oxygen saturation (% SpO_2) of a person using the MAX30100/02 pulse oximeter and OLED module.

Disclaimer: The pulse oximeter built in this project should only be used for learning purposes and for a better understanding of the device. Its readings should not be considered for any medical purposes or diagnosis. Doing so is solely under the discretion of the user. The readings of a pulse oximeter depend largely on the quality and calibration of the sensors used. So, for any medical purpose or diagnosis, use a reliable accurate pulse oximeter.

20.2 WORKING OF PULSE OXIMETER

When a person inhales air through the nose, the air goes to the lungs and the oxygen in the air gets transferred to the blood through alveoli in the lungs. The haemoglobin in the blood carries the oxygen required for the organs in the body. The haemoglobin or blood with oxygen is known as oxygenated haemoglobin/blood. Similarly, the haemoglobin or blood without oxygen is known as deoxygenated haemoglobin/blood. The oxygenated and deoxygenated blood can also be differentiated visually as the oxygenated blood will be in bright red colour whereas deoxygenated blood will be dark red in colour. When the heart expands and contracts, the deoxygenated blood gets oxygenated and it is transferred to various parts of the body through arteries.

A pulse oximeter is a device that can non-invasively measure the heart rate and level of oxygen (SpO_2) in a person's blood using light. It has two LEDs: one of them emits monochromatic red light at a wavelength of 660 nm, and the other emits monochromatic infrared light at a wavelength of 940 nm. There is a specific reason for choosing these wavelengths. At these wavelengths, the oxygenated and deoxygenated haemoglobin have different absorptive properties as shown in Figure 20.1, which will help us in knowing the proportion of oxygenated haemoglobin and deoxygenated haemoglobin in the blood. The pulse

DOI: 10.1201/9781003147169-20

oximeter also has two photodiodes for detecting the amount of red and infrared light passed through the tissues. These are placed inside a case to minimize the surrounding ambient light falling on it.

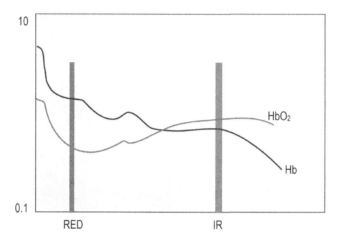

FIGURE 20.1   Absorptive properties of oxygenated and deoxygenated haemoglobin at different wavelengths

When a person places any part of the body (generally finger, toe or ear) on the LEDs of pulse oximeter, some amount of light is absorbed by the body and the remaining passes through the body and falls on the photodetector. Experimentally, it is determined that oxygenated blood absorbs more infrared light and passes more red light, whereas deoxygenated blood passes more infrared light and absorbs more red light. The oxygen saturation (level of oxygen in the blood) of a person is determined using the level of red light and infrared light absorbed by the person's body. When the heart expands and contracts, the volume of oxygenated blood increases and then decreases. Using the time difference between the increase and decrease of oxygenated blood, the heart rate is determined.

20.3 PROBLEMS WITH MAX30100/02 PULSE OXIMETER

The most commonly used low-cost pulse oximeters in IoT are MAX30100 and MAX30102. There are different types of MAX30100/02 modules available in the market as shown in Figures 20.2 and 20.3 developed by different manufacturers. The hardware design of each module is different from the others.

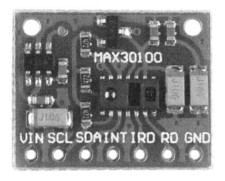

FIGURE 20.2   Types of MAX30100 breakout boards

FIGURE 20.3   Types of MAX30102 breakout boards

The MAX30100/02 modules on the right side of Figures 20.2 and 20.3 have good hardware design and works perfectly without any modification of the hardware on the breakout board. But the modules on the left side of Figures 20.2 and 20.3 have serious hardware design problem and cannot be used without modifying the hardware circuitry on the breakout board.

The MAX30100/02 modules shown on the left side of Figures 20.2 and 20.3 are assembled on a similar breakout board with hardware circuit design shown in Figure 20.4. As you can see from the circuit design, the module can support voltage levels from 1.8 V to 5.5 V. First, the U2 voltage regulator takes 5 V input and converts it to 3.3 V. Next, the 3.3 V is passed to the U3 voltage regulator for converting it to 1.8 V. The 3.3 V voltage level is used to control the LEDs of the MAX30100/02 sensor, whereas 1.8 V is used to control the ADC and sensor logic.

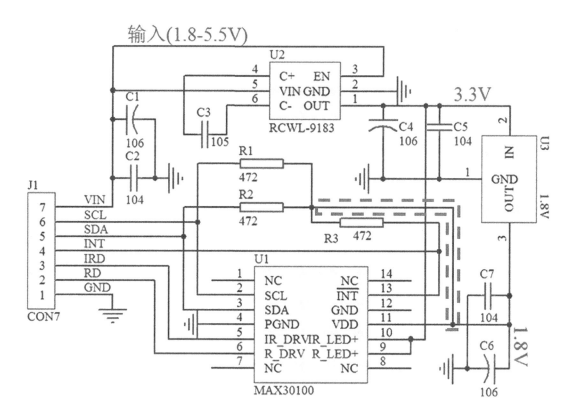

FIGURE 20.4   Hardware circuit design of MAX30100/02 module

If you carefully look at the circuit, the SDA, SCL and INT pins of MAX30100/02 sensor are pulled up to 1.8 V through 4.7 kΩ resistors (R1, R2, R3). The pull-up line connected between 4.7 kΩ resistors and 1.8 V is indicated by a dashed box as shown in Figure 20.4. If this module is directly interfaced to a 5 V logic-based microcontroller or evaluation board (like Arduino Uno), it won't be detected on the I²C bus because logic levels of SDA and SCL lines are too low. The minimum voltage for detecting HIGH for a 5 V logic-based microcontroller or evaluation board is 2 V. Even with 3.3 V logic-based microcontrollers or evaluation boards, this module won't work properly. You can also run the I²C scanner code used in the previous project to make sure whether the module is detected over the I²C bus or not.

This error cannot be solved without modifying the hardware on the breakout board. Don't worry! Modifying the hardware on the breakout board is not that difficult and can be easily done by following any one of the two methods.

20.3.1 Removing pull-up resistors

This error can be solved by removing the three 4.7 kΩ pull-up resistors connected to SDA, SCL, INT pins of MAX30100/02 sensor on the breakout board and then externally connecting 4.7 kΩ pull-up resistors between SDA, SCL, INT pins and 3.3 V. The 4.7 kΩ pull-up resistors on the MAX30100/02 module are indicated by dashed boxes in Figure 20.5.

Removing the 4.7 kΩ pull-up resistors from the breakout board is not that difficult. Initially, we too thought that removing resistors from the breakout board is very difficult, and we might end up in damaging the complete board. But let us tell you, what you are thinking is absolutely wrong. These resistors and other components on the breakout board are not soldered strongly and can be easily removed by little physical pressure or heat. If you have a soldering rod, then heat it without any soldering lead and carefully place it on one resistor after another you want to remove for 15–30 seconds. Remove the soldering rod and try to remove the resistor with your hand. If you are not able to remove the resistor, then repeat the process again. If you don't have a soldering rod, then you can also use a sharp cutter or pliers for removing the resistors.

FIGURE 20.5   Removal of pull-up resistors

In the new hardware circuit after removing the 4.7 kΩ pull-up resistors as shown in Figure 20.6, the SDA and SCL lines of MAX30100/02 sensor are not connected to any pull-up resistors. If you recall the explanation of I²C protocol, it is mentioned that the SDA and SCL lines must be pulled up so that they are active high because the devices on the I²C bus are active low. So, externally 4.7 kΩ pull-up resistors must be connected between the SDA, SCL pins of MAX30100/02 module and 3.3 V for detecting it over the I²C bus. In this project, we have followed this method for interfacing MAX30100/02 module to the evaluation board.

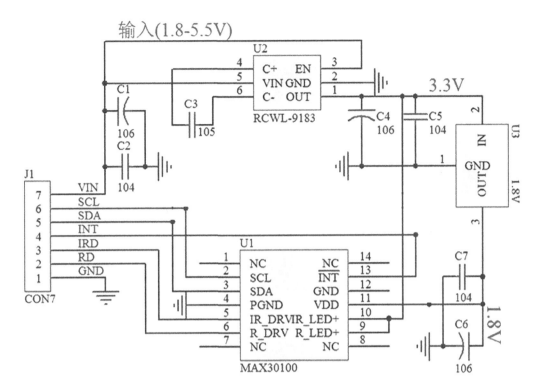

FIGURE 20.6 Hardware circuit after removing pull-up resistors

20.3.2 Reconnection of pull-up resistors

This is a little difficult method compared to the previous one as you need to remove a connection and make a connection on the breakout board. Removing a connection on the breakout board is the easiest of all but connecting two components on the breakout board is the hardest as you need some kind of special equipment or specialized skill. Don't try this method unless you are very good at soldering.

First, we need to remove the connection between the 4.7 kΩ pull-up resistor and the OUT pin of 1.8 V voltage regulator which is indicated by a dashed line in Figure 20.7. For removing a connection on a

FIGURE 20.7 Reconnection of pull-up resistors

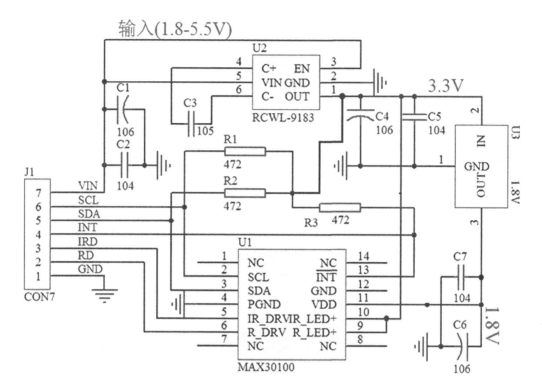

FIGURE 20.8   Hardware circuit after reconnecting pull-up resistors

breakout board, scratch the connecting light green line with any sharp end instrument like a screwdriver or knife. Next, we need to connect the disconnected end of 4.7 kΩ pull-up resistor and the OUT pin of 3.3 V voltage regulator which is indicated by a straight line in Figure 20.7. For connecting the disconnected end of 4.7 kΩ pull-up resistor and the OUT pin of 3.3 V voltage regulator, take a small piece of single strand wire and solder it between them. As the components on the breakout board are very small, be careful while soldering a single strand wire between them.

In the new hardware circuit after reconnecting the pull-up resistors as shown in Figure 20.8, the SCL, SDA and INT pins of MAX30100/02 sensor are pulled up through 4.7 kΩ resistors to 3.3 V. So, there is no need of connecting any external pull-up resistors for making the module detectable over the I²C bus. For knowing more about the specifications and features of MAX30100, refer to its datasheet (https://datasheets.maximintegrated.com/en/ds/MAX30100.pdf). Similarly, refer to the MAX30102 datasheet (https://datasheets.maximintegrated.com/en/ds/MAX30102.pdf) for knowing more about its configurations and specifications.

20.4 HARDWARE REQUIRED

If you have modified the MAX30100/02 module as explained in Section 20.3.1, then you need to compulsorily use 4.7 kΩ resistors for pulling up the SDA and SCL pins. If you don't have exact 4.7 kΩ resistors, then you can also use any resistor with a nearby resistance value like 5.1 kΩ or 4.2 kΩ, but the resistor

TABLE 20.1 Hardware required for the project

HARDWARE	QUANTITY
Arduino Uno	1
MAX30100/02 Pulse oximeter	1
SSD1306 OLED module	1
Buzzer	1
4.7 kΩ resistor	2
Breadboard	1
Jumper wires	As required

value should not be too high or too low. If you have followed the method explained in Section 20.3.2, then you don't require any pull-up resistors. The list of required components for the project is tabulated in Table 20.1.

20.5 CONNECTIONS

If you have modified the MAX30100/02 module as explained in Section 20.3.1, then follow the connection diagram shown in Figure 20.9. If you have followed the method explained in 20.3.2, then remove both the pull-up resistors and follow the connection diagram shown in Figure 20.9. The pin-wise connection details are tabulated in Table 20.2.

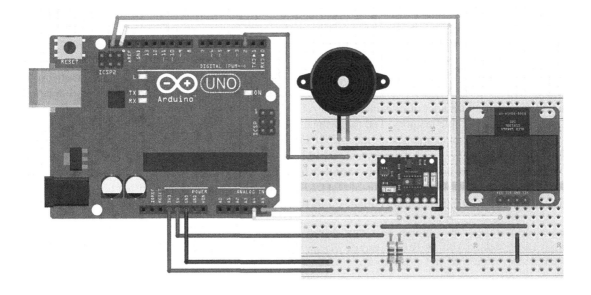

FIGURE 20.9 Connection layout

TABLE 20.2 Connection details

ARDUINO UNO	MAX30100/02 PULSE OXIMETER	4.7 KΩ RESISTOR 1	4.7 KΩ RESISTOR 2	SSD1306 OLED MODULE	BUZZER
5V	VIN			VCC	
3.3V		Pin 1	Pin 1		
A4	SDA		Pin 2		
A5	SCL	Pin 2			
SDA				SDA	
SCL				SCL	
2					+ve pin
GND	GND			GND	−ve pin

20.6 ARDUINO IDE CODE

As we are interfacing the MAX30100 pulse oximeter with Arduino Uno using I²C protocol, `Wire.h` library is used. For calculating the heart rate and oxygen saturation values from the raw MAX30100 pulse oximeter data, `MAX30100_PulseOximeter.h` library is used. For installing it, open the library manager in Arduino IDE by going to Tools > Manage Libraries or by pressing Ctrl+Shift+I in Windows or Cmd+Shift+I in macOS. Search for *MAX30100lib* and install the latest version of *MAX30100lib by OXullo Intersecans* among the listed libraries. You can also directly download the library from Github at this link (https://github.com/oxullo/Arduino-MAX30100).

Initially, we have tried using the `Adafruit_SSD1306.h` library that is used in the previous projects for displaying content on the OLED. But, due to some reasons, it failed to work properly with the MAX30100 pulse oximeter. We have also tried the famous `U8g2lib.h` library, but the MAX30100 pulse oximeter did not work with it as well. Similarly, we have tried some other OLED libraries but none of them worked properly along with the MAX30100 pulse oximeter. At last, we have found out `OakOLED.h` library that worked perfectly along with the MAX30100 pulse oximeter. OakOLED library is written on top of the Adafruit_GFX library. It is mainly used for SSD1306 chips without a reset line. The main problem with the OakOLED library is that it cannot be used with all types of microcontrollers and evaluation boards as it consumes a lot of dynamic memory. Even with Arduino Uno, it consumes almost 50–60% of the available dynamic memory. The following code consumes 92% of Arduino's dynamic memory so you might observe some glitches on the OLED screen. If you can reduce the program memory below 90% by removing the Serial print statements, then you won't see any glitches on the OLED screen. The `OakOLED.h` library can be easily installed from the library manager in Arduino IDE. Search for *OakOLED* and install the latest version of *OakOLED by Brian Taylor* among the listed libraries. You can directly download the library from Github at this link (https://github.com/netguy204/OakOLED).

```
#include <Wire.h> //Used for I2C functionalities
#include <MAX30100_PulseOximeter.h> //Used for MAX30100 Pulse oximeter
#include <OakOLED.h> //Used for OLED module
```

We need to continuously get the raw data from the MAX30100 pulse oximeter for accurately calculating heart rate and oxygen saturation. So, `delay()` cannot be used anywhere in the `void loop()` for pausing the execution of the code. Instead of using `delay()`, we'll execute part of the code where `delay()` is required once in every `TIME_PERIOD` second(s). If the oxygen saturation of the person using the pulse oximeter is below `SpO2_THRES`, then the buzzer will be switched ON to indicate his oxygen saturation is not normal and need to consult a doctor. buzzer stores the digital pin of Arduino to which the positive pin of buzzer is connected.

```
#define TIME_PERIOD 1 //Delay in seconds
#define SpO2_THRES 94 //Threshold for safe oxygen saturation
#define buzzer 2 //Digital pin to which buzzer is connected
```

display is an OakOLED object used to display content on the OLED screen. It is similar to the display (Adafruit _ SSD1306 constructor) that is used in the previous projects. Similarly, pox is a PulseOximeter object used for setting various parameters and obtaining data from the MAX30100 pulse oximeter. Bpm and SpO2 are used to store the heart rate and oxygen saturation values calculated from the raw data obtained from the MAX30100 pulse oximeter.

```
OakOLED display; //OakOLED object
PulseOximeter pox; //PulseOximeter object

//Variables used for controlling the pulse oximeter & buzzer
unsigned long prevTime = 0;
int bpm;
int SpO2;
bool buzzerState = false;
```

We'll display the heart rate image and oxygen saturation image each in one-quarter of the OLED screen. So, convert the heart rate image provided in the resources to a byte array with a canvas size of 64×34 pixels and brightness of 160 and paste it in the place of <Byte Array of heart rate image>. Similarly, convert the oxygen saturation image provided in the resources to a byte array with a canvas size of 64×34 pixels and brightness of 120 and paste it in the place of <Byte Array of oxygen saturation image>. onBeatDetected() function is executed whenever a heartbeat is detected by the pulse oximeter. So, you can write the code that needs to be executed when a heartbeat is detected in the onBeatDetected() function.

```
//Bitmap of heart rate image [64x34px, 160 brightness]
const unsigned char heart [] PROGMEM = {
  <Byte Array of heart rate image>
};

//Bitmap of oxygen saturation image [64x34px, 120 brightness]
const unsigned char oxysat [] PROGMEM = {
  <Byte Array of oxygen saturation image>
};

//Function executed when a heart beat is detected
void onBeatDetected() {
  Serial.println("Beat! ♥");
}
```

As we are writing voltage on the buzzer for controlling the buzzer, we'll set its direction of operation as OUTPUT. Almost all the functions supported for the Adafruit_SSD1306 constructor used in the previous project are supported for the OakOLED object. display.begin() is used for initializing the I²C communication between the OLED module and Arduino Uno. If you remember, in the previous project, we have passed the OLED I²C address as an argument to the begin() function. But we need not pass it for the OakOLED object because the OLED I²C address is already mentioned in the library header file. If you want to use an OLED with a different I²C address, then change it in the OakOLED.h file located in the Arduino libraries folder (Arduino>libraries>OakOLED>OakOLED.h). setTextColor() is used for setting the text colour displayed on the OLED. 1 is used for white colour and 0 is used for black colour.

pox.begin() is used for initializing the I²C communication between MAX30100 pulse oximeter and Arduino Uno. By default, 50 mA current is supplied to the IR LED of the MAX30100 pulse oximeter. But sometimes the MAX30100 pulse oximeter does not work as expected due to more current supplied to the IR LED or the evaluation board to which the MAX30100 pulse oximeter is connected might not provide the required current. So, setIRLedCurrent() is used to change the current supplied to IR LED. The current supply can be changed to default 50 mA by commenting out the pox.setIRLedCurrent(MAX30100_LED_CURR_7_6MA). The current required for the IR LED of pulse oximeter depends on various factors. So, the current supply that worked for us need not work for you. So, if 50 mA or 7.6 mA didn't work, then change the LED current configuration with all the available options one after another until your pulse oximeter works as expected. For knowing all the available LED current configurations, check the MAX30100_Registers.h file in Arduino libraries folder (Arduino > libraries > libraries > Arduino-MAX30100 > src > MAX30100_Registers.h). setOnBeatDetectedCallback() is used to set the function to be called when a heartbeat is detected.

```
void setup() {
  Serial.begin(9600); //Initialisation of Arduino's Serial
  pinMode(buzzer, OUTPUT); //Direction of buzzer

  display.begin(); //Initialisation of OLED module
  //Initial display configurations
  display.clearDisplay();
  display.setTextSize(1);
  display.setTextColor(1);

  if (!pox.begin()){ //Initialisation of pulse oximeter module
    Serial.println("Initialisation of MAX30100 FAILED.");
    while(1);
  } else
    Serial.println("MAX30100 Successfully Initialised.");

  //Initial pulse oximeter configurations
  pox.setIRLedCurrent(MAX30100_LED_CURR_7_6MA);
  pox.setOnBeatDetectedCallback(onBeatDetected);
}
```

MAX30100 pulse oximeter sends a lot of raw data like the level of RED light and IR light absorbed that can be used to calculate the heart rate and oxygen saturation. pox.update() is used to update the raw data obtained from the MAX30100 pulse oximeter. So, it must be called as fast as possible for accurate calculation of the heart rate and oxygen saturation.

millis() will return the time (in milliseconds) passed since the Arduino board began execution of the current program. Dividing it by 1,000 will give us the time in seconds. prevTime records the time at the end of each previous execution. So, we'll check whether TIME_PERIOD second(s) have passed or not since the previous execution. In short, the code in the if statement executes once for every TIME_PERIOD second(s).

getHeartRate() returns the heart rate of the person by calculating it from the raw values obtained from the pulse oximeter. Similarly, getSpO2() returns the oxygen saturation of the person by calculating it from the raw values obtained from the pulse oximeter. As oxygen saturation is in percentage, its value will never be negative or exceed 100. So, we'll use constrain() to constrain the oxygen saturation value between 0 and 100. It has to be remembered that a return value of 0 by any one of the functions mean it is invalid.

```
void loop() {
  pox.update(); //Run the update as fast as possible
```

```
//Run once for every TIME_PERIOD seconds
if (millis()/1000 - prevTime > TIME_PERIOD) {
  bpm = pox.getHeartRate(); //Calculate the heart rate
  //Calculate the oxygen saturation
  SpO2 = constrain(pox.getSpO2(), 0, 100);

  Serial.print("Heart rate: ");
  Serial.print(bpm);
  Serial.print("bpm  SpO2: ");
  Serial.print(SpO2);
  Serial.println("%");
  prevTime = millis()/1000;
```

When the MAX30100 pulse oximeter detects an oxygen saturation value below SpO2 _ THRES, then the person must be alerted using a buzzer. As switching ON the buzzer continuously will not generate a buzzing effect, we need to switch ON and OFF the buzzer periodically for creating a buzzing effect. For periodically switching ON and OFF the buzzer, we should not use delay(). So, we have used buzzer-State and the following if-else statements for periodically switching ON and OFF the buzzer without using delay().

```
//Switch ON the buzzer if oxygen saturation is below safe threshold
if(SpO2 != 0 && SpO2 < SpO2_THRES && !buzzerState){
  digitalWrite(buzzer, HIGH);
  buzzerState = true;
} else if(SpO2 != 0 && SpO2 < SpO2_THRES && buzzerState) {
  digitalWrite(buzzer, LOW);
  buzzerState = false;
} else
  digitalWrite(buzzer, LOW);
```

Finally, the heart rate and oxygen saturation data obtained from the MAX30100 pulse oximeter along with the heart rate and oxygen saturation images are printed on the OLED. The same functions used in the previous project are used for printing data and drawing bitmap on the OLED.

```
//Display the heart rate and oxygen saturation on the OLED
display.clearDisplay();

display.setCursor(0,0);
display.print("Heart Rate:");
display.setCursor(0,10);
display.print(bpm);
display.print(" bpm");

display.setCursor(74,0);
display.print("SpO2:");
display.setCursor(74,10);
display.print(SpO2);
display.print(" %");

display.drawBitmap(0, 25, heart, 64, 34, 1);
display.drawBitmap(64, 25, oxysat, 64, 34, 1);
display.display();
  }
}
```

Please remember that the Arduino code used for interfacing different pulse oximeters will be different. The Arduino code explained until now can only be used for interfacing MAX30100 pulse oximeters (Figure 20.2). It cannot be used for interfacing other types of pulse oximeters. If you are using a MAX30102 or MAX30105 pulse oximeter, then open Arduino IDE's library manager and search for *MAX3010x*. Install the latest version of *SparkFun MAX3010x Pulse and Proximity Sensor Library by SparkFun Electronics* among the listed libraries. After successfully installing the library, go to *File > Examples > SparkFun MAX3010x Pulse and Proximity Sensor Library* in Arduino IDE and open *Example8_SpO2*. This example provides the complete code for obtaining the heart rate and oxygen saturation of a person using MAX30102 or MAX30105 pulse oximeters. You can also display these values on an OLED using the OakOLED library.

Finally, upload the code to Arduino Uno after selecting the appropriate board and COM port. After successfully uploading the code, open the serial monitor and change the baud rate to 9600 to view the data printed on the serial monitor. If the MAX30100 pulse oximeter is initialized successfully, then place your finger over the red light of MAX30100 pulse oximeter. Wait for some time and see your heart rate and oxygen saturation values on the OLED screen.

You can find all the resources of this project at this Github link (https://github.com/anudeep-20/30IoTProjects/tree/main/Project%2020).

In this project, you have learnt about the working of a pulse oximeter and the types of pulse oximeters available in the market. You have also learnt to rectify the design problems in MAX30100/02 pulse oximeters for making them usable. Finally, you have interfaced the MAX30100 pulse oximeter and OLED with Arduino Uno for continuously tracking a person's heart rate and oxygen saturation. In the next project, you'll build an Arduino board from scratch.

Let's build the next one!

Build Your Own Arduino Board from Scratch

21

Have you ever thought how Arduino Uno is built? and why is it so easy to use? To answer these questions, you need to understand the components used in the board and the procedure for building it from scratch. This project helps you build an Arduino Uno board cheaper than the one available in the market without compromising the features.

21.1 MOTIVE OF THE PROJECT

In this project, you'll build an Arduino board on a breadboard from scratch. As Arduino is an open-source project, anyone can access their design and reproduce them without any restrictions.

21.2 ATmega328P

ATmega328P is a single-chip 8-bit AVR RISC-based microcontroller (shown in Figure 21.1) developed by Atmel (later acquired by Microchip). It is widely used in many Arduino boards like Arduino Uno,

FIGURE 21.1 ATmega328P microcontroller

DOI: 10.1201/9781003147169-21

Arduino Nano, Arduino Pro Mini and many more. The Arduino Uno boards used in our projects are not useful for performing any computational operations without the ATmega328P microcontroller. It is the main computational component of an Arduino Uno board. The Arduino Uno board acts as a breakout board for easily programming and using the ATmega328P microcontroller. The pin configurations of the ATmega328P microcontroller and its mapping to the Arduino Uno pins are shown in Figure 21.2.

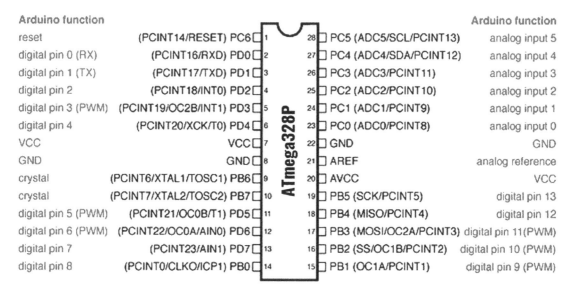

FIGURE 21.2   Pinout of ATmega328P microcontroller

ATmega328P has 32KB ISP (In-system Programming) flash memory with read-while-write capabilities, 1KB EEPROM and 2KB SRAM (Static Random Access Memory). It can only operate in the voltage range of 1.8–5.5 V. Any input voltage higher than 5.5 V might damage the board. All the digital communication peripherals like UART, SPI and I²C that are supported in Arduino Uno are supported in ATmega328P. Some of the important specifications of ATmega328P are tabulated in Table 21.1. To know more about the ATmega328P, refer to its datasheet (https://ww1.microchip.com/downloads/en/DeviceDoc/Atmel-7810-Automotive-Microcontrollers-ATmega328P_Datasheet.pdf).

TABLE 21.1   Specifications of ATmega328P

Program memory type	Flash
Program memory size	32 KB
CPU speed	20 MIPS/DMIPS
SRAM	2,048 bytes
Data EEPROM/HEF	1,024 bytes
Timers	2×8-bit, 1×16-bit
Number of comparators	1
Temperature range	−40–85 °C
Operating voltage range	1.8–5.5 V

21.3 CRYSTAL OSCILLATOR

An electronic oscillator is nothing but an electronic circuit that produces a periodic, oscillating electronic signal. Some of the commonly used electronic oscillators are RC oscillator and LC oscillator. These oscillators can be used to generate both low and high frequencies, but the main problem with these oscillators is that their oscillation frequency varies with temperature, power supply voltage or even with a slight change in component values. So, crystal oscillators are used in applications where high accuracy and stability are required. These are used in clocks, microcontrollers and in most of the ICs. These oscillators can generate frequencies from a few kilo Hz to 100s of mega Hz. A crystal oscillator is also an electronic oscillator that uses a vibrating crystal of piezoelectric material for generating an electrical signal with constant frequency. There are many naturally occurring crystals with the piezoelectric property like Rochelle salt, Quartz, Tourmaline, etc. Among them Quartz is the most commonly used crystal because of its easy availability, low cost, mechanical strength and piezoelectricity compared to other crystals. Crystal oscillators work on the principle of inverse piezoelectric effect, a property in which the shape of the material slightly changes under an electric field. The Arduino Uno board uses a 16 MHz crystal oscillator as shown in Figure 21.3 for providing clock signals to the microcontroller.

FIGURE 21.3   16 MHz crystal oscillator

21.4 HARDWARE REQUIRED

The list of components required for this project is presented in Table 21.2. The components marked with a single asterisk (*) are only used for burning the Arduino bootloader to the ATmega328P microcontroller. The components marked with a double asterisk (**) are not compulsorily required for completing the project but their inclusion will improve the quality of the output. The 220 Ω resistors are used for protecting the LEDs against high voltages. Even without the resistors LEDs will work as expected. The 0.33 μF ceramic capacitor and 0.1 μF ceramic capacitor are used for producing a stable output voltage from the LM7805 voltage regulator. Even without the capacitors, the LM7805 voltage regulator produces a stable 5 V output but with little fluctuations.

TABLE 21.2   Hardware required for the project

HARDWARE	QUANTITY
Arduino Uno*	1
ATmega328P microcontroller	1
CP2102 USB to TTL serial converter (with DTR pin)	1
16 MHz crystal oscillator	1
LM7805 voltage regulator	1
LEDs	2
Tactile push button	1
22 pF ceramic capacitor	2
100 pF ceramic capacitor	1
0.33 µF ceramic capacitor**	1
0.1 µF ceramic capacitor**	1
10 kΩ resistor*	1
220 Ω resistor**	2
9 V Battery	1
Breadboard	1
Jumper wires	As required

21.5 BURNING ARDUINO BOOTLOADER

The bootloader is simply a small piece of executable code stored inside the memory of a microcontroller that is executed as soon as the microcontroller is powered. Generally, microcontrollers are programmed using an external programmer unless it has a bootloader for programming without the need for it. The bootloader in the ATmega328P microcontroller of the Arduino board will allow us to program the Arduino board over the serial port using a USB cable. It will also help in placing the code uploaded from the PC in the memory of the microcontroller. So, for building an Arduino board from scratch, we need to have an ATmega328P microcontroller that has the Arduino bootloader burnt in it.

If you have bought an ATmega328P microcontroller that already has the Arduino bootloader or if you are using an ATmega328P microcontroller that is taken out from an Arduino board, then skip the following explanation and move to the next section. If you have bought a new ATmega328P microcontroller without the Arduino bootloader, then follow the below explanation for burning the bootloader into the ATmega328P microcontroller.

- Arduino bootloader can be burnt to the ATmega328P microcontroller by using an Arduino board as an ISP. For making an Arduino board as an ISP, upload the ArduinoISP code which is available at *File > Examples > ArduinoISP > ArduinoISP* in the Arduino IDE. Don't forget to select the appropriate board and COM port before uploading the code.
- Follow the connection diagram shown in Figure 21.4 and the pin-wise connection details in Table 21.3 for connecting the ATmega328P microcontroller to the Arduino Uno for burning the bootloader.

If you observe any IC carefully, you'll see a small dot at one corner of the IC. The pin exactly beside the dot is the pin number 1 of that IC. The remaining pins can be numbered in counterclockwise direction starting from that pin.

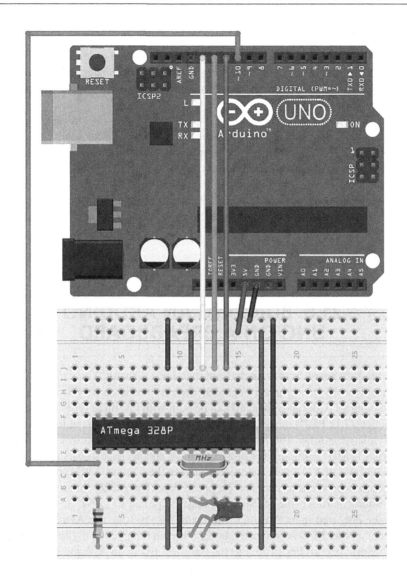

FIGURE 21.4   Connection layout for uploading bootloader

TABLE 21.3   Connection details for uploading bootloader

ARDUINO UNO	ATMEGA328P MICROCONTROLLER	16 MHZ CRYSTAL OSCILLATOR	22 PF CERAMIC CAPACITOR 1	22 PF CERAMIC CAPACITOR 2	10 KΩ RESISTOR
5 V	Pin 7 (VCC), Pin 20 (AVCC)				Pin 1
10	Pin 1 (RESET)				Pin 2
11	Pin 17 (MOSI)				
12	Pin 18 (MISO)				
13	Pin 19 (SCK)				
	Pin 9 (XTAL1)	Pin 1	Pin 1		
	Pin 10 (XTAL2)	Pin 2		Pin 1	
GND	Pin 8, Pin 22 (GND)		Pin 2	Pin 2	

- If you are building an Arduino Uno board from scratch, then select the board as *Arduino Uno* from the *Tools > Board > Arduino AVR Boards* in Arduino IDE and don't select any other board. You might come across some books or blogs suggesting to select Arduino Duemilanove or Arduino Nano with ATmega328P for burning Arduino bootloader to ATmega328P. Selecting any other board and burning the bootloader might cause problems while uploading new codes to the Arduino board built from scratch.
- Select the programmer as *Arduino as ISP* from the *Tools > Programmer* in Arduino IDE. Finally, click *Burn Bootloader* from the *Tools* section and wait for some time for burning the bootloader to your ATmega328P microcontroller. If burning the bootloader fails or if you get any error, then check the connections and try again.

You need to burn the bootloader to the ATmega328P microcontroller only once. After successfully burning the bootloader, you can remove the ATmega328P microcontroller and use it to build an Arduino board from scratch.

21.6 BUILDING AN ARDUINO UNO ON A BREADBOARD

Building an Arduino Uno board from scratch is not that much difficult. You can easily build an Arduino Uno board on a breadboard by using an ATmega328P microcontroller, a 16MHz oscillator and a couple of capacitors. Even though an Arduino Uno board can be built with just three types of components, we'll use some other components like LEDs, resistors and a tactile push button to make it more versatile.

ATmega328P microcontroller also has an internal 8MHz RC oscillator but due to its low reliability, we'll use an external 16MHz oscillator for providing more stable clock signals to the microcontroller. A green LED is used as power LED that is connected between the 5 V and GND through a resistor for knowing whether the board is powered or not. The 1st pin of the ATmega328P microcontroller is the RESET pin. To reset the microcontroller, the RESET pin must be pulled to LOW. So, we'll use a tactile push button similar to the one used in an Arduino Uno board for resetting the ATmega328P microcontroller whenever required. Another LED is connected to the 19th pin (mapped to digital pin 13 of Arduino Uno) of the ATmega328P microcontroller similar to the built-in LED in an Arduino Uno board. You can also add some more components according to your use case and improve the design of the Arduino Uno built in this project. For improving the design, you can also refer to the original Arduino Uno schematics and design at this link (https://www.arduino.cc/en/uploads/Main/arduino-uno-schematic.pdf).

For uploading code to the ATmega328P microcontroller, we'll use a CP2102 USB to TTL serial converter (with DTR pin). The RX of CP2102 USB to TTL serial converter is connected to the 3rd pin (TX) of the ATmega328P microcontroller. Similarly, the TX of CP2102 USB to TTL serial converter is connected to the 2nd pin (RX) of the ATmega328P microcontroller. The DTR (Data Terminal Ready) pin is connected to the RESET pin through a 100 pF capacitor for resetting the ATmega328p microcontroller to ensure that it synchronizes with the data sent through the CP2102 USB to TTL serial converter. Finally, connect all the components as shown in Figure 21.5. For your easier understanding, the connection details are also tabulated in Table 21.4.

After connecting all the components, we need to test whether our Arduino Uno is working similar to the original one or not. For testing our Arduino Uno, connect the CP2102 USB to TTL serial converter to your PC and upload the following Blink code. You can also get the following Blink code from *File > Examples > 01.Basics* in Arduino IDE. Before uploading the code, don't forget to select the board as Arduino Uno and the port as the COM port assigned to your CP2102 USB to TTL serial converter connected to the PC.

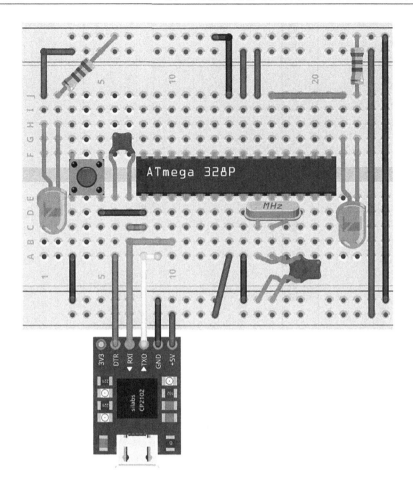

FIGURE 21.5 Connection layout of Arduino Uno on a breadboard (with CP2102 USB to serial converter)

```
void setup() {
  Serial.begin(9600); //Initialize Arduino's Serial
  //Initialize digital pin LED_BUILTIN as an output
  pinMode(LED_BUILTIN, OUTPUT);
}

void loop() {
  //Turn the LED on (HIGH is the voltage level)
  digitalWrite(LED_BUILTIN, HIGH);
  Serial.println("LED is switched ON");
  delay(1000); //Wait for a second
  //Turn the LED off by making the voltage LOW
  digitalWrite(LED_BUILTIN, LOW);
  Serial.println("LED is switched OFF");
  delay(1000); //Wait for a second
}
```

After successfully uploading the code, you can see the LED connected to the 19th pin of ATmega328P microcontroller blinking. If uploading the code failed or if the LED is not blinking, then the Arduino Uno which you have built is not working properly. Check the connections once again and re-upload the code. Also, open the serial monitor and change the baud rate to 9600 to view the data printed on it.

TABLE 21.4 Connection details of Arduino Uno on a breadboard (with CP2102 USB to serial converter)

ATMEGA328P	CP2102	100 PF CAPACITOR	16MHZ OSCILLATOR	22 PF CAPACITOR 1	22 PF CAPACITOR 2	PUSH BUTTON	LED 1	220 RESISTOR	LED 2	220 RESISTOR
Pin 7 (VCC), Pin 20 (AVCC), Pin 21 (AREF)	5V						+ve Pin			
	DTR	Pin 1								
Pin 1 (RESET)		Pin 2				Pin 1				
Pin 2 (RX)	TX									
Pin 3 (TX)	RX									
Pin 9 (XTAL1)			Pin 1	Pin 1						
Pin 10 (XTAL2)			Pin 2		Pin 1					
Pin 19 (D13)							−ve Pin	Pin 1	+ve Pin	Pin 1
Pin 8, Pin 22 (GND)	GND			Pin 2	Pin 2	Pin 2		Pin 2	−ve Pin	Pin 2

TABLE 21.5 Connection details of Arduino Uno on a breadboard (without CP2102 USB to serial converter)

ATMEGA328P	LM7805	0.33 MF CAPACITOR	0.1 MF CAPACITOR	16MHZ OSCILLATOR	22 PF CAPACITOR 1	22 PF CAPACITOR 2	PUSH BUTTON	LED 1	220 RESISTOR	LED 2	220 RESISTOR	9V BATTERY
	Input	Pin 1										+ve Pin
Pin 7 (VCC), Pin 20 (AVCC), Pin 21 (AREF)	Output		Pin 1					+ve Pin				
Pin 1 (RESET)							Pin 1					
Pin 9 (XTAL1)				Pin 1	Pin 1							
Pin 10 (XTAL2)				Pin 2		Pin 1						
								-ve Pin	Pin 1			
Pin 19 (D13)										+ve Pin		
										-ve Pin	Pin 1	
Pin 8, Pin 22 (GND)	GND	Pin 2	Pin 2		Pin 2	Pin 2	Pin 2		Pin 2		Pin 2	-ve Pin

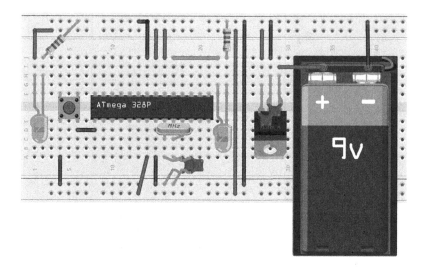

FIGURE 21.6   Connection layout of Arduino Uno on a breadboard (without CP2102 USB to serial converter)

You can use this newly built Arduino Uno same as the original Arduino Uno. You can upload any code that is supported on an Arduino Uno to it. If you want to use the new Arduino Uno independently, then you can remove the CP2102 USB to TTL serial converter and connect an independent power source. Please remember not to connect a voltage supply more than 5.5 V or else you might damage the complete board. If you have a 5 V voltage supply, then directly connect it to the VCC and GND lines on the bread-board. If you don't have a 5 V voltage supply, then you can use a 9 V battery along with an LM7805 voltage regulator and a couple of resistors for generating a stable 5 V voltage supply as shown in Figure 21.6. For independently using the new Arduino Uno with a 9 V battery, connect the components as mentioned in Table 21.5.

After connecting the new Arduino Uno with a 9 V battery, you can see the LED connected to the 19th pin of ATmega328P microcontroller blinking. It means the previously uploaded blink code to the ATmega328P microcontroller is being executed.

You can find all the resources of this project at this Github link (https://github.com/anudeep-20/30IoTProjects/tree/main/Project%2021).

In this project, you have learnt about the components used in building an Arduino Uno and also built an actual working Arduino Uno board from scratch. In the next project, you'll build a smart attendance system using RFID and Firebase.

Let's build the next project!

Smart Attendance System Using RFID and Firebase

22

Attendance is an important part of most of our academic and professional life. Even though there is a lot of advancement in technology, there are still many places where attendance is taken manually. You might have experienced this! How boring and waste of time it is to manually enter the attendance? Won't it be great if we have an automated system for recording our entry as soon as we enter the classroom or any premises for that matter and also record our exit after we leave the place. This project will help you build an automated attendance system along with a cloud database that can be readily implemented anywhere.

22.1 MOTIVE OF THE PROJECT

In this project, you'll build an automated attendance system that can track individual attendance, their entry time, exit time and the total number of people currently inside an entry using RFID, RTC module and Firebase.

If a person wants to enter a place, he/she needs to scan their RFID tag with the RFID reader placed at the entrance. If the scanned RFID tag is one of the verified RFID tags, then the person is permitted to enter the place and the time of entry obtained from the RTC module is entered into the Firebase. The person's attendance and the total number of people inside the entry are also updated in the Firebase. Similarly, when a person wants to leave the place, they need to scan their RFID tag with the RFID reader placed at the exit. After their exit, the time of exit and the total number of people inside the entry are updated in the Firebase. In this project, we'll use a single RFID reader at both the entry and exit for minimizing the cost of the system. The same system can also be implemented using different RFID readers at the entry and exit, but the cost of the system might double.

22.2 SPI

SPI (Serial Peripheral Interface) is a synchronous serial interface specification used for short-distance communication between a microcontroller and allied sensors. It is a full-duplex, master-slave-based interface first developed by Motorola. SPI bus specifies four signals:

- Serial Clock (SCLK)
- Master Out, Serial In (MOSI)
- Master In, Serial Out (MISO)
- Chip select or Slave Select (CS or SS)

DOI: 10.1201/9781003147169-22

The clock signal generated by the master is transferred to the slaves using SCLK line. The clock signal is used to synchronize the data transfer between master and slave devices on the SPI bus. The clock frequencies on the SPI bus are much higher compared to the I²C protocol; hence, the data transfer speed is also higher compared to I²C. Unlike I²C protocol, SPI cannot support multi-master multi-slave configuration but can support single-master multiple slaves. Multiple slaves can be connected to master in either regular configuration or daisy-chained configuration. In the regular configuration, each slave requires an individual SS line from the master as shown in Figure 22.1. A slave on the SPI bus can be selected by the master by pulling its SS line to active LOW and can be disconnected from the SPI bus by pulling its SS line to active HIGH. After the slave is selected, the clock signal from the master and the MOSI/MISO lines are available only for the selected slave for data transmission. If multiple slaves are selected, then the MISO line will be corrupted as the master cannot know from which slave the data is being transmitted. For a lesser number of slaves, regular SPI configuration is the best. But, as the number of slaves increases, the number of SS lines required from the master also increases. But the master cannot provide any number of SS lines to the slaves as the number of GPIO pins on the master is also limited. So, to increase the number of slaves connected to the master in regular SPI configuration, mux can be used to generate multiple SS lines.

In daisy-chain SPI configuration, a single SS line from the master is connected to all the slaves as shown in Figure 22.2. All the slaves receive the same serial clock signal from the master at the same time. The MOSI line from the master is directly connected to the first slave, and the MISO line is directly connected from the last slave to the master. In this configuration, the data is not transferred directly to the master but propagated from one slave to another. So, the number of clock cycles required to transfer the data is directly proportional to the position of the slave in the daisy chain. Data transfer to the slaves at the end of the daisy chain requires a large number of clock cycles compared to the slaves near to the master. Daisy-chain configuration is not supported in all the master and slave devices. To use the daisy-chain configuration, it must be supported on all the master and slave devices used.

To begin the SPI communication, the master must send a clock signal with a frequency supported by the slaves and select the slaves to communicate by pulling its SS line to active LOW. During each clock cycle, when the master sends a bit on the MOSI line, the slave reads it, and while the slave sends

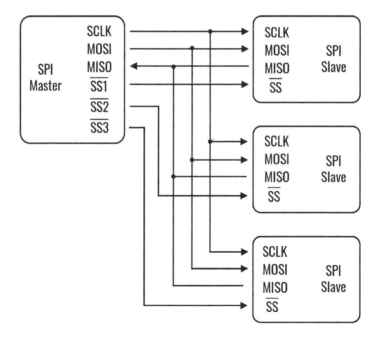

FIGURE 22.1   Regular multi-slave SPI configuration

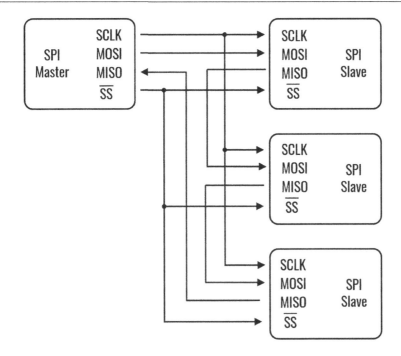

FIGURE 22.2   Daisy-chained multi-slave SPI configuration

a bit on the MISO line, the master reads it (full-duplex communication). Transmission of data between master and slave generally involves two shift registers of fixed word length (like 8 bits, 12 bits or 16 bits), one in master and one in slave apiece. The two shift registers are connected to each other in virtual ring topology as shown in Figure 22.3. Usually, the most significant bit in the shift registers is shifted out first. On the clock edge, both the master and slave shift out a bit and output it on the transmission for its counterpart to receive. On the next clock edge, at each receiver the bit is sampled from the transmission line and is set as the least significant bit of its shift register. For example, on the 2nd clock edge, the master and slave devices on the SPI bus shift out the 7th bit on their shift register and output it on the MOSI and MISO lines, respectively. On the 3rd clock edge, the master and slave samples the bit on the MISO and MOSI lines, respectively, and sets them as the new 0th bit of their shift register. This process continues until all the bits in the shift registers of master and slave are exchanged. If more data needs to be exchanged, then the shift registers are reloaded and the process is repeated. After the whole data is transferred, the master stops the clock signal on the SCLK line and deselects the slave by pulling its SS line to active HIGH.

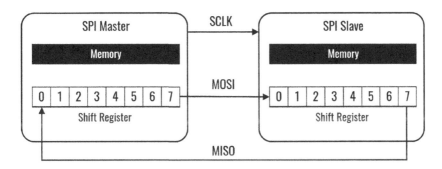

FIGURE 22.3   Shift register connection between master and slave in SPI bus

For the transmission of data, the master should also select clock polarity (CPOL) and clock phase (CPHA) as per the requirement of the slave. Depending on the selection of CPOL and CPHA bits, there are four different SPI modes as shown in Table 22.1.

TABLE 22.1   SPI modes depending on CPOL and CPHA

SPI MODE	CLOCK POLARITY (CPOL)	CLOCK PHASE (CPHA)
0	0	0
1	0	1
2	1	0
3	1	1

CPOL is used to set the polarity of the clock during the idle state. The idle state is the period when SS is high and transitioning to low at the start of the transmission and also when SS is low and transitioning to high at the end of the transmission. If the CPOL bit is 0, then the leading edge is a rising edge and the trailing edge is a falling edge. If the CPOL bit is 1, then the leading edge is a falling edge and the trailing edge is a rising edge. CPHA is used to select the clock phase. Depending on the CPHA bit, the rising or falling clock edge is used to sample or shift the data. If the CPHA bit is 0, then the data is sampled on the rising edge and shifted on the falling edge. If the CPHA bit is 1, then the data is sampled on the falling edge and shifted out on the rising edge. The timing diagram of SPI for different values of CPOL and CPHA is shown in Figure 22.4. The odd vertical lines denote leading clock edges, and the even vertical lines denote trailing clock edges.

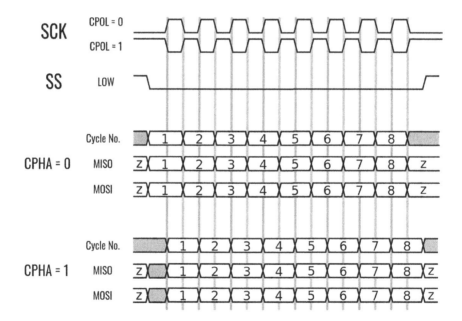

FIGURE 22.4   Timing diagram of SPI bus

22.3 RFID

Radio Frequency IDentification (RFID) is one of the widely used technology in IoT that uses high-frequency electromagnetic fields to identify an object. It has many use cases like inventory management, logistics, supply chain, tracking of people, access management and many more. An RFID system mainly consists of two components, an RFID reader and an RFID tag/card.

An RFID reader is made up of an RF module and an antenna which are used to generate a high-frequency electromagnetic field whereas an RFID tag/card consists of a microchip that stores identification information and an antenna to transmit the information. All the RFID tags/cards have a factory prere-corded UID (Unique Identifier) of either 4 bytes or 7 bytes. The 4 byte UIDs (single-size UIDs) have been stopped because the maximum number of usable UIDs with 4 bytes of size is approximately 3.7 billion. So, a 4 byte UID is no longer unique and can be assigned to more than one RFID tag/card during production. So, the 4 byte UID is known as NUID (Non-Unique Identifier). But the probability of having two RFID tags/cards with the same NUIDs in an RFID system is extremely low. So, the NUIDs of the RFID tags/cards used in this project can be considered almost unique. It is recommended to use 7 byte UID (double-size UID) RFID tags/cards for having guaranteed uniqueness. RFID tags/cards are usually passive devices that don't require any power source. When an RFID reader is powered using a microcontroller, it generates a high-frequency electromagnetic field in its close proximity. If an RFID tag/card is placed in this electro-magnetic field, current will be generated in the RFID tag/card's antenna for powering the microchip. The powered microchip sends back the information stored in it to the RFID reader in the form of a radio signal.

In this project, we'll use an RC522 RFID module shown in Figure 22.5 as an RFID reader. It is one of the commonly used low-cost RFID modules in IoT that is based on MFRC522 IC developed by NXP semiconductors. It is preferred to operate the module from 2.5 to 3.3 V, but the module can also be used with 5 V. The RC522 RFID module can create an electromagnetic field of 13.56 MHz that can be used to

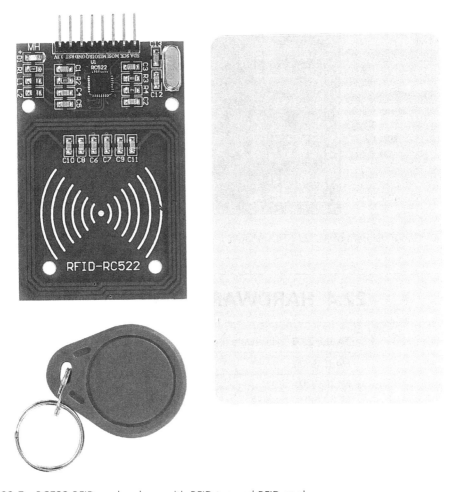

FIGURE 22.5   RC522 RFID reader along with RFID tag and RFID card

read the information in RFID tags/cards. Some of the important specifications of the RC522 RFID reader are tabulated in Table 22.2. For knowing the complete specifications and features of MFRC522 IC, refer to its datasheet (https://www.nxp.com/docs/en/data-sheet/MFRC522.pdf).

TABLE 22.2   Specifications of RC522 RFID module

Frequency range	13.56MHz
Operating voltage	2.5–3.3 V
Maximum data rate	10 Mbps
Read range	5 cm
Operating current	13–26 mA
Min. current (power-down mode)	10 μA

By default, the RC522 RFID module can be interfaced with a microcontroller using SPI alone. But by making little modifications to the board, the RC522 RFID module can also be interfaced using I²C and UART. This modification procedure is clearly explained in the MFRC522 IC datasheet, and for additional information, you can also refer to this article in the Arduino forum (https://forum.arduino.cc/index. php?topic=442750.0). The pinout diagram of the RC522 RFID module showing SPI, I²C and UART interfaces is shown in Figure 22.6. In this project, we'll use SPI for interfacing the RC522 RFID reader and NodeMCU.

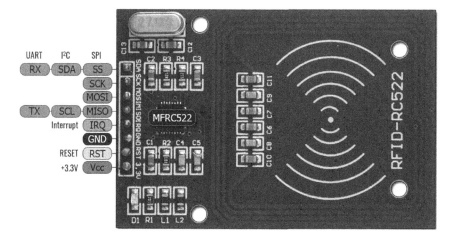

FIGURE 22.6   Pinout diagram of RC522 RFID module

22.4 HARDWARE REQUIRED

TABLE 22.3   Hardware required for the project

HARDWARE	QUANTITY
NodeMCU	1
RC522 RFID module	1
DS3231 RTC module	1
Buzzer	1
LEDs	2 (1*Green, 1*Red)
Breadboard	1
Jumper wires	As required

22.5 CONNECTIONS

The list of required components for the project is tabulated in Table 22.3, followed by Table 22.4, which presents the pin-wise connections. Also, the connection diagram is shown in Figure 22.7, which shall help get a visualization of the connections. Due to shortage of digital pins, we have connected the green LED and buzzer to default RX and TX pins of NodeMCU, respectively. So, remember to disconnect these connections before uploading the code and reconnect them after uploading.

TABLE 22.4   Connection details

NODEMCU	RC522 RFID	DS3231 RTC	RED LED	GREEN LED	BUZZER
3V3	3.3V	VCC			
D4	RST				
D6	MISO				
D7	MOSI				
D5	SCK				
D3	SDA (SS)				
D2		SDA			
D1		SCL			
D8			+ve pin (long pin)		
D9 (RX)				+ve pin (long pin)	
D10 (TX)					+ve pin
GND	GND	GND	−ve pin (short pin)	−ve pin (short pin)	−ve pin

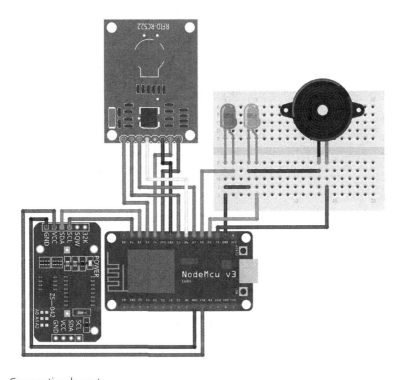

FIGURE 22.7   Connection layout

22.6 FIREBASE

In all the previous chapters wherever cloud is required, we have used Adafruit IO for storing time-stamped data points and for displaying the data. Even though Adafruit is very useful for storing simple data with little setup, it cannot be used for storing complex data. It also has security, data type and uploading speed limitations. To overcome these limitations, we'll use Firebase that provides high customization for security and storage of data. Firebase is a Backend-as-a-Service (BaaS) platform developed by Google primarily for web and mobile app creation. It has products like Firebase Realtime Database and Firebase Cloud Firestore that are useful for storing data with high security. In this project, we'll use Firebase Realtime Database (RTDB) for storing data and writing rules such that only authenticated users can modify the stored data. Follow the below steps for setting up your Firebase project.

- Go to Firebase Console (https://console.firebase.google.com/) and log in with your Google Account credentials. If you are visiting the Firebase for the first time or if you haven't created any Firebase projects in the past, you'll see the Firebase console as shown in Figure 22.8. Click *Create a project* for creating a new Firebase project.

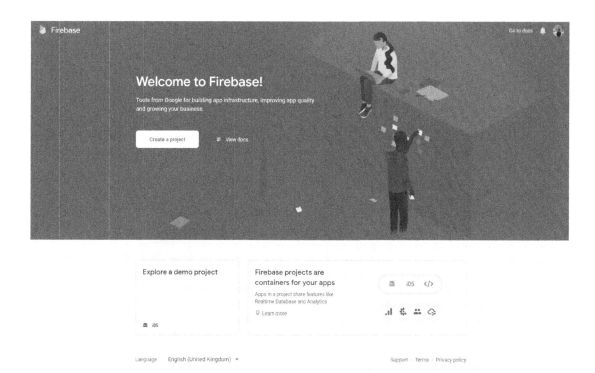

FIGURE 22.8   Creation of new Firebase project

- The creation of a Firebase project is easy and straightforward. Choose a name for your project and enter it in the *Project name* column as shown in Figure 22.9. Each project is assigned with a unique project ID which is displayed beneath the project name with a small edit symbol using which you can edit the project ID. In our case, the project name is *Smart Attendance RFID*, and the unique project ID is *smart-attendance-rfid*. After entering the project name and editing the project ID, click *Continue* for moving to the next step.

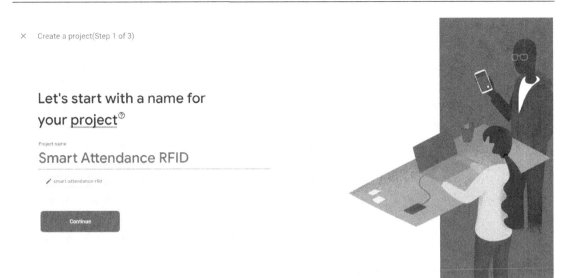

FIGURE 22.9 Naming the new Firebase project

- In the next step, you can enable or disable Google Analytics for your Firebase project. For our use case, enabling or disabling Google Analytics doesn't make any difference. But it is recommended to enable Google Analytics as it provides useful analytics about the functions in our Firebase project. After enabling/disabling the Google Analytics, click *Continue* as shown in Figure 22.10 for moving to the last step.

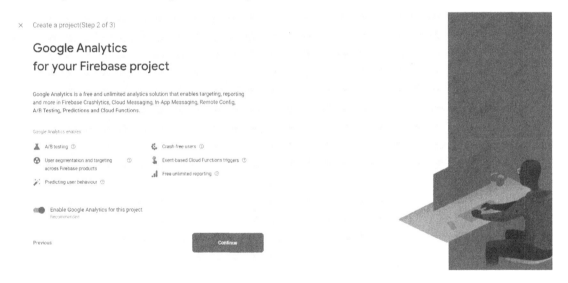

FIGURE 22.10 Enable/disable Google analytics

- If you have enabled Google Analytics in the 2nd step, then you'll see the 3rd step in which you need to configure the Google Analytics. Select the Google Account in which the analytics data is to be stored from the drop-down under *Choose or create a Google Analytics account* column. You can choose the *Default Account for Firebase* or link any other Google Account as shown in Figure 22.11. After choosing the Google Account, you can also edit the *Analytics property* using the edit button. In our case, we have chosen the *Default Account for Firebase* for storing the analytics data, and the *Analytics property* is left as default. After completing the configuration, click *Create project* and wait for at least 30 seconds for creating your new Firebase project.

FIGURE 22.11   Configure Google analytics

- If you have completed all the previous steps successfully without any errors, then the landing page of your Firebase project will open as shown in Figure 22.12. If it is not opened automatically, you'll see the list of projects created in your Firebase account. Among them, select your newly created Firebase project.

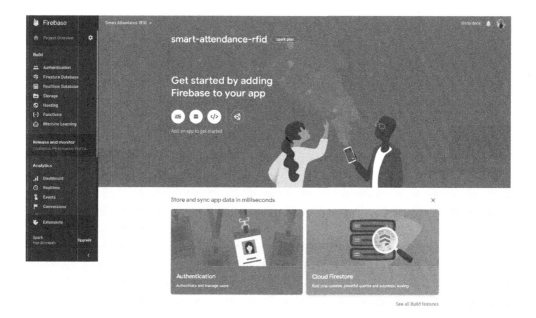

FIGURE 22.12   Landing page of Firebase project

- On the left side, you can see all the available features that can be used to configure your Firebase project. Among them, click *Authentication* for defining the authorized users who can access the data stored in the Firebase project. Click *Get started* on the landing page of *Authentication* as shown in Figure 22.13.

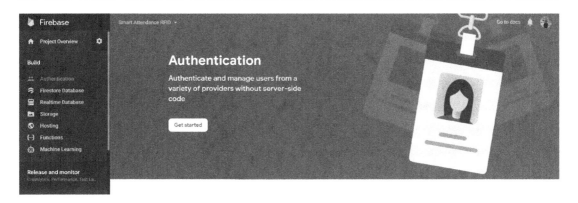

FIGURE 22.13   Landing page of Firebase authentication

* In the *Authentication > Sign-in method* column, you can see all the *Sign-in providers* (Figure 22.14) that are supported for authenticating a user. You can choose any of them for authenticating a user, but the easiest one with minimal setup is the *Email/Password* service. For enabling the *Email/Password* Sign-in provider, click on it and you'll see a pop-up window as shown in Figure 22.15. Click the top-right toggle button to *Enable* and click *Save* for enabling the *Email/Password* Sign-in provider.

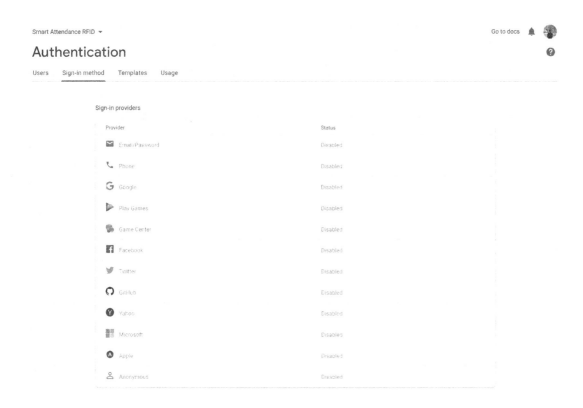

FIGURE 22.14   Sign-in providers

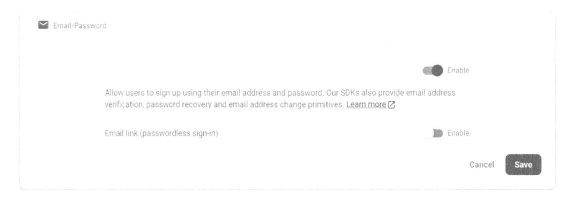

FIGURE 22.15   Configuration of email/password

- Now, go to the *Users* column and click *Add user* as shown in Figure 22.16 for creating a new user. Enter any custom *Email* and *Password* with which the new user can access the data stored in the Firebase project and click *Add User* as shown in Figure 22.17. The email and password can be anything and need not be an existing or verified email id. In our case, the *Email* and *Password* of the new user are admin@smartattendance.rfid and $ECRETPASSWORD@ADMIN, respectively. If the user is added successfully, then the newly created user will be listed in the *Users* column of *Authentication* as shown in Figure 22.18. You can also reset the password of the user, disable the account or delete the account by clicking the three horizontal dots at the right corner (*View more options*). You can add as many users as you require; there is no limit to it.

FIGURE 22.16   Users

FIGURE 22.17   Adding a new user

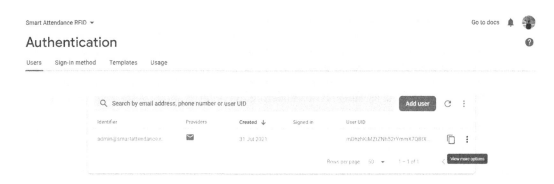

FIGURE 22.18   Addition of a new user

- After creating the users to whom you wanted to give the access, go to *Project settings* by clicking the gear button beside the *Project Overview* as shown in Figure 22.19. In the *General* column of *Project settings*, you can find most of the project-specific details like project ID, Project number, Web API key, etc. as shown in Figure 22.20. The *Public-facing name* in the *Public settings* is the name that is visible to the public. You can change it using the edit button beside the name. Also, configure the *Support email* by choosing an email id from the drop-down.

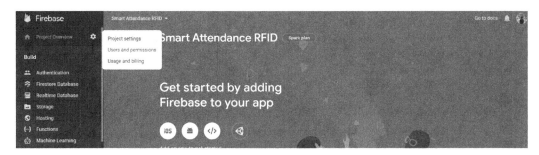

FIGURE 22.19   Go to project settings

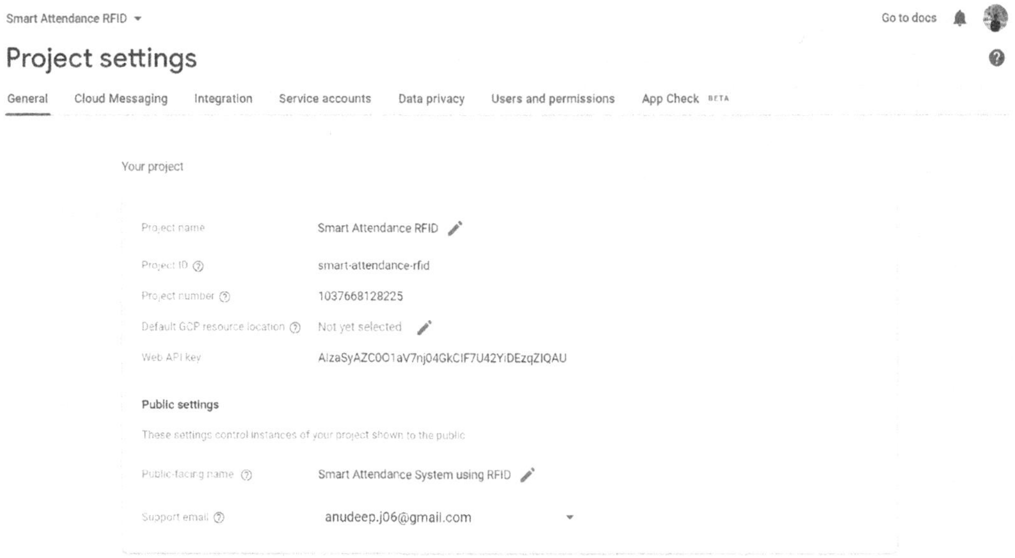

FIGURE 22.20   Project settings (general)

- Next, go to *Realtime Database* and click *Create Database* on its landing page as shown in Figure 22.21. You'll see a pop-up window for setting up the initial configurations of the database. Select the *Realtime Database location* as one of the nearest locations to your location from the drop-down and click *Next* as shown in Figure 22.22. Select the security mode with which you want to create the database. You can select either one of them as you can always edit the security rules. After selecting the security mode, click *Enable* as shown in Figure 22.23.

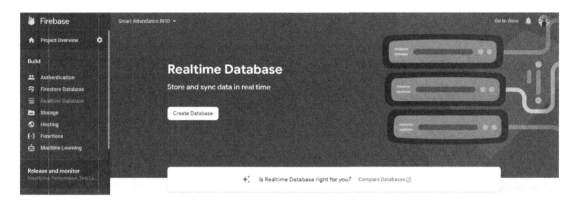

FIGURE 22.21   Realtime Database

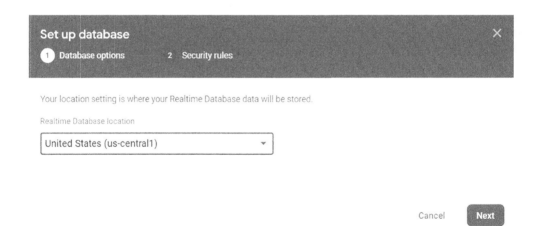

FIGURE 22.22   Realtime Database location

- If the Realtime Database is created successfully, you'll see an empty database in the *Data* column as shown in Figure 22.24. Next, go to the *Rules* column to edit the security rules of your database. Remove the previously existing rules and paste the following rules and click *Publish* as shown in Figure 22.25. The new security rules will allow anyone to read the data in your Realtime Database but allows only the authenticated users to edit the data. To know more about writing security rules for the Realtime Database, refer to the Firebase official documentation (https://firebase.google.com/docs/database/security).

FIGURE 22.23   Security of your database

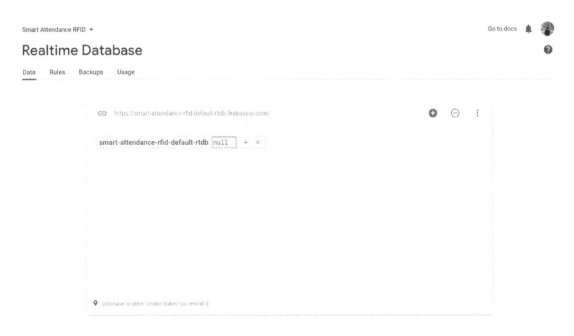

FIGURE 22.24   Empty database

```
{
  "rules": {
    ".read": true,
    ".write": "auth != null"
  }
}
```

FIGURE 22.25 Editing security rules of database

Any data uploaded to Realtime Database will be stored in JSON format. The information about the verified persons entering or leaving the place is stored in the Realtime Database in the following format.

```
<ROOT>: {
    <NUID>: {
        Attendance: <INT VALUE>
        IN_OUT: <STRING VALUE (IN or OUT)>
        IN_TIME: {
            <RANDOM ID>: <TIME STAMP>
        }
        OUT_TIME: {
            <RANDOM ID>: <TIME STAMP>
        }
    }
    Total_Inside: <INT VALUE>
}
```

22.7 ARDUINO IDE CODE

Most of the libraries used in this project are new and can be easily installed from the Library Manager of Arduino IDE. Similar to `Wire.h` which is used for I²C functionalities, `SPI.h` is used for SPI functionalities. As `SPI.h` is one of the standard libraries of Arduino, it will be automatically installed during the installation of Arduino IDE. `MFRC522.h` is an Arduino library for MFRC522 IC-based RFID modules. For installing the `MFRC522.h` library, open the library manager in Arduino IDE by going to Tools > Manage Libraries or by pressing Ctrl+Shift+I in Windows or Cmd+Shift+I in macOS. Search *MFRC522* in the search box and install the latest version of *MFRC522 by Github Community* among the listed libraries. You can also directly download the library from Github at this link (https://github.com/miguelbalboa/rfid). `Firebase _ ESP _ Client.h` is a complete Firebase-Arduino library for ESP8266 and ESP32 based evaluation boards. It is one of the few Firebase libraries that is being updated frequently and has good community support. It can be used not only for Realtime Database but also for Firestore, Storage, Authentication and many other services. `Firebase _ ESP _ Client.h` library can be directly installed from the library manager in Arduino IDE. Search *Firebase* in the search box and install the latest version of *Firebase-Arduino Client Library for ESP8266 and ESP32 by Mobizt* among the listed libraries. You can also directly download the library from Github at this link (https://github.

com/mobizt/Firebase-ESP-Client). The helper header files, `TokenHelper.h` and `RTDBHelper.h`, are automatically downloaded during the installation of *Firebase-Arduino Client Library*.

```
#include <ESP8266WiFi.h> //Library for WiFi functionalities
#include <SPI.h> //Library for I2C functionalities
#include <MFRC522.h> //Library for RC522 RFID
#include <RTClib.h> //Library for RTC module
#include <Firebase_ESP_Client.h> //Library for Firebase
//Helper header files
#include "addons/TokenHelper.h"
#include "addons/RTDBHelper.h"
```

`WIFI_SSID` and `WIFI_PASSWORD` store your WiFi credentials that are useful for connecting NodeMCU to the Internet. `RFID_SS` and `RFID_RST` store the NodeMCU digital pins to which the SDA and RST pins of the RC522 RFID reader are connected respectively. `redLED`, `greenLED` and `buzzer` store the NodeMCU digital pins to which the red LED, green LED and buzzer are connected, respectively. `redLED` is used to indicate the user is denied entry due to an error in verification or updating their details in the Firebase. `greenLED` is used to indicate the user has been granted permission to enter. `buzzer` is used as an acknowledgement to indicate some card is being read by the RFID reader and is switched ON irrespective of the status of `redLED` or `greenLED`. `rfid` is an MFRC522 constructor that takes the slave select digital pin number (`RFID_SS`) and the reset digital pin number (`RFID_RST`) of the RFID reader as arguments.

```
//Your WIFI details
#define WIFI_SSID "YOUR WIFI SSID"
#define WIFI_PASSWORD "YOUR WIFI PASSWORD"

#define RFID_SS D3 //NodeMCU pin to which SDA/SS pin of RFID is connected
#define RFID_RST D4 //NodeMCU pin to which RST pin of RFID is connected
#define redLED D8 //NodeMCU pin to which Red LED is connected
#define greenLED D9 //NodeMCU pin to which Green LED is connected [RX]
#define buzzer D10 //NodeMCU pin to which Buzzer is connected [TX]

MFRC522 rfid(RFID_SS, RFID_RST); //MFRC522 constructor
```

The variables and functions used in the previous project for obtaining the present date-time using DS3231 RTC module are used in this project as well. `API_KEY` is used to store the Web API key of your Firebase project. Copy the *Web API key* of your Firebase project from the *Project settings > General* (Figure 22.20) and paste it in the place of `Web API Key of your Firebase project`. `DATABASE_URL` stores the Realtime Database URL of your Firebase project. Copy the Realtime Database URL (Figure 22.24) without the *https::///* at the beginning and */* at the end and paste it in the place of `DATABASE URL` of your Realtime Database. In our case, the `DATABASE_URL` is *smart-attendance-rfid-default-rtdb.firebaseio.com*. `USER_EMAIL` and `USER_PASSWORD` store the credentials of the user with which you want to access the Realtime Database of your Firebase project for uploading the data. Paste the user Email and Password (Figure 22.17) in the place of `Authenticated User Email ID` and `Authenticated User Password`, respectively.

```
RTC_DS3231 RTC; //RTC_DS3231 object
DateTime nowDT; //DateTime object
String nowDTStr=""; //Storing the Date-Time in String format

//Firebase Details
#define API_KEY "Web API Key of your Firebase project"
#define DATABASE_URL "DATABASE URL of your Realtime Database"
```

```
#define USER_EMAIL "Authenticated User Email ID"
#define USER_PASSWORD "Authenticated User password"
```

fbdo is a FirebaseData object used to upload data to Realtime Database. auth is a FirebaseAuth object used in authentication of a user using their credentials. config is a FirebaseConfig object used in the initial configuration. You'll get a more clear understanding of these objects after knowing about their usage later in the code.

```
FirebaseData fbdo; //FirebaseData object
FirebaseAuth auth; //FirebaseAuth object
FirebaseConfig config; //FirebaseConfig object
```

newNUID is used to store the NUID of the RFID tag/card. verifiedNUID is an array of Strings used to store the verified RFID tags/cards that are permitted to enter. As an extension to this project, you can also store these verified NUIDs in the Realtime Database instead of as an array in the code because as the number of verified NUIDs increases the size of the code increases as well as the code needs to be re-uploaded whenever a new verified NUID is added. attendance, IN _ OUT, totalInside and error variables are used to upload or update the information in the Realtime Database.

```
String newNUID = ""; //Stores the scanned NUID
//Array of verified NUIDs
String verifiedNUID[] = {"f9e3f197", "4697d83", "764e281", "a26d97f6",
"e39b84c1"};
//Variables used for updating information in Firebase
int attendance;
String IN_OUT;
int totalInside;
bool error = false;
```

NodeMCU's serial is initialized at a baud rate of 115200. The direction of redLED, greenLED and buzzer digital pins is set as OUTPUT as we will write voltage on them for controlling the red LED, green LED and buzzer, respectively. The code for connecting the NodeMCU to WiFi is the same code used in the previous projects. SPI.begin() is used for initialization of SPI communication between NodeMCU and RFID reader. rfid.PCD _ Init() is used for initializing the RFID reader which is also known as PCD (Proximity Coupling Device). The code for initialization and adjusting the time of RTC is the same as discussed in the previous projects.

```
void setup() {
  Serial.begin(115200); //Initialize NodeMCU's Serial
  pinMode(redLED, OUTPUT); //Direction of Red LED
  pinMode(greenLED, OUTPUT); //Direction of Green LED
  pinMode(buzzer, OUTPUT); //Direction of Buzzer

  //Connecting NodeMCU to WiFi
  Serial.print("Connecting to ");
  Serial.println(WIFI_SSID);
  WiFi.begin(WIFI_SSID, WIFI_PASSWORD);
  while(WiFi.status() != WL_CONNECTED) {
    delay(500);
    Serial.print(".");
  }
  Serial.print("\nWiFi connected.\nNodeMCU's Local IP Address: ");
  Serial.println(WiFi.localIP()); //Print the local IP address of NodeMCU

  SPI.begin(); //Initialize SPI communication
```

```
rfid.PCD_Init(); //Initialize RC522 RFID module
RTC.begin(); //Initialize DS3231 RTC module
//Adjusts the time of RTC module when using for the first time or when its
power is disconnected.
if(RTC.lostPower()) {
  //DATE and TIME gets replaced with the date and time of the PC during
compilation.
    RTC.adjust(DateTime(F(__DATE__), F(__TIME__)));
}
```

The dot (.) operator or class member access operator in C++ programming language is used to access the public members of a class object. Public members of a class may contain data members (variables) and/or member functions that can be accessed even outside the class. api _ key and database _ url are the public member variables of the config class object which are assigned with the API _ KEY and DATABASE _ URL respectively using the dot operator. token _ status _ callback is a public member function of the config class object which is assigned with the tokenStatusCallback function using the dot operator. Similarly, the user credentials stored in USER _ EMAIL and USER _ PASSWORD are assigned to auth.user.email and auth.user.password using the dot operator. Firebase. begin() is used to connect the NodeMCU to the Firebase using the configuration and authentication information stored in config and auth. Firebase.reconnectWiFi() is used to reconnect the NodeMCU to WiFi whenever the connection is lost. totalInside stores the total number of people inside the entry. We'll use the one line if-else or ternary operator (: ?) for obtaining the total number of people inside the entry from Realtime Database. You can understand the usage of ternary operator from the following example which obtains the largest number among the two variables.

```
int varA = 10;
int varB = 20;
int largest;

largest = varA > varB ? varA : varB
```

is same as

```
if(varA > varB)
    largest = varA;
else
    largest = varB;
```

Firebase.RTDB.getInt() is used to read the integer value at a particular path in the Realtime Database. It takes two arguments: the first argument is a Firebase data object (fbdo) which is used to point the data in the Realtime Database, and the second argument is the path of the data to be read (/Total _ Inside). If there is no data in the mentioned path, then Firebase.RTDB.getInt() returns false and 0 is assigned to the totalInside. If the data is read successfully, then Firebase. RTDB.getInt() returns true and fbdo.intData() is used to obtain the integer value stored in the mentioned path which is assigned to totalInside. It is important to obtain this data at the beginning because if any person enters or leaves, then the total number of people inside the entry can be updated in the Realtime Database using this value as the reference.

```
config.api_key = API_KEY; //Assign the web API key
config.database_url = DATABASE_URL; //Assign the Realtime Database URL
//Assign the callback function for the long running token generation task
config.token_status_callback = tokenStatusCallback;
//Assign the Authenticated user credentials
auth.user.email = USER_EMAIL;
```

```
    auth.user.password = USER_PASSWORD;

    Firebase.begin(&config, &auth); //Initialize Firebase
    Firebase.reconnectWiFi(true); //Reconnects WiFi if the connection is lost

    totalInside = Firebase.RTDB.getInt(&fbdo, "/Total_Inside") ? fbdo.intData()
    : 0;
}
```

`rfid.PICC _ IsNewCardPresent()` is used to check whether there is any RFID tag/card present in front of the RFID reader. `rfid.PICC _ ReadCardSerial()` is used to check whether the RFID reader is able to read data from the detected RFID tag/card. PICC (Proximity Integrated Circuit Card) is the other name for RFID tag/card. Until an RFID tag/card is detected and its data is read, the code won't execute further.

The double-colon operator (::) or scope resolution operator in C++ programming language is used to distinguish variables, functions or classes of the same name. If you are familiar with C++, then you would have used this operator at least once directly or indirectly. For example, if there is a variable named `var` in the class named `classA` and another variable with the same name `var` in another class named `classB`, then to use `var` in `classA`, `classA::var` is used and for using `var` in `classB`, `classB::var` is used. `MFRC522::PICC _ Type` refers to the `PICC _ Type` class in `MFRC522` class. `piccType` is a variable of `PICC _ Type` class in `MFRC522` class. It stores the RFID tag/card type. If the RFID tag/card is not of Classic MIFARE type, then the code won't execute further.

```
void loop() {
    //Checks whether a PICC is detected and its data is read
    if (!rfid.PICC_IsNewCardPresent() || !rfid.PICC_ReadCardSerial())
        return;

    //Prints the PICC type
    Serial.print("PICC type: ");
    MFRC522::PICC_Type piccType = rfid.PICC_GetType(rfid.uid.sak);
    Serial.println(rfid.PICC_GetTypeName(piccType));

    //Check if the PICC is of Classic MIFARE type
    if(piccType != MFRC522::PICC_TYPE_MIFARE_MINI &&  piccType != MFRC522::PICC_
TYPE_MIFARE_1K && piccType != MFRC522::PICC_TYPE_MIFARE_4K) {
        Serial.println("Your tag is not of MIFARE Classic");
        //Switch ON Red LED & Buzzer for 1 sec
        digitalWrite(redLED, HIGH);
        digitalWrite(buzzer, HIGH);
        delay(1000);
        digitalWrite(redLED, LOW);
        digitalWrite(buzzer, LOW);
        return;
    }
```

The following `for` loop is used to obtain the NUID of an RFID tag/card. `isVerified()` is a custom function written by us to verify whether the obtained NUID is one of the verified NUIDs or not. To see the code inside `isVerified()` function, go to the end of code explanation. If the obtained NUID is not a verified NUID, then the person will be denied entry which will also be indicated using the red LED.

```
    //Obtain the NUID of newly scanned PICC
    for (byte i = 0; i < rfid.uid.size; i++) {
```

```
        newNUID += String(rfid.uid.uidByte[i], HEX); //NUID in HEX (base-16)
        // newNUID += String(rfid.uid.uidByte[i], DEC); //NUID in DEC (base-10)
    }

    if(isVerified(newNUID)) {
```

As we are using a single RFID reader for entry and exit, we need to know previously whether the person is let in or out. Depending on that status, they can be provided entry or exit. This information can be known from the IN _ OUT tag in the Realtime Database.

Firebase.RTDB.getString() is used to read the string value at a particular path in the Realtime Database. It takes the same two arguments as the Firebase.RTDB.getInt(). If you refer to the getInt() and getString() function implementation in Firebase _ ESP _ Client.h library (*https://github.com/mobizt/Firebase-ESP-Client/blob/main/src/README.md*), you'll observe that the second argument (path of the data) is to be given as const char*. As, newNUID is a String adding const char* to it will make the resultant also a String. For example, adding "/" and "/IN _ OUT" which are const char* to newNUID will make the resultant a String. But String cannot be given as the second argument, so c _ str() is used for converting it into a const char* that points to a null-terminated string (note the difference between String and string).

If the data is read successfully, then Firebase.RTDB.getString() returns true and fbdo. stringData() is used to obtain the string value stored in the mentioned path. The obtained string value (either IN or OUT) is assigned to IN _ OUT variable. If there is no data in the mentioned path, then Firebase.RTDB.getString() returns false which is assumed as the person entering the place for the first time. So, OUT is assigned to IN _ OUT variable, and the Attendance of the person in Realtime Database is set to 0 using Firebase.RTDB.setInt(). Unlike getInt(), setInt() accepts three arguments: the first argument is a Firebase data object (fbdo) which is used to point to the data in Realtime Database, the second argument is the path in the Realtime Database where the new data is to be set ("/<NUID>/ Attendance"), and the third argument is the integer data to be set (0). If the integer value is not set at the given path successfully, then the setError() function is executed. It is a custom function written by us to set the error flag (error) to true. To see the code inside setError() function, go to the end of code explanation. fbdo.errorReason() returns a string containing the error description.

```
    //Obtains the string stored in IN_OUT tag
    if(Firebase.RTDB.getString(&fbdo, ("/"+newNUID+"/IN_OUT").c_str())){
      IN_OUT = fbdo.stringData();
    } else {
      IN_OUT = "OUT";
      Firebase.RTDB.setInt(&fbdo, ("/"+newNUID+"/Attendance").c_str(), 0) ?
Serial.println("New Person. Attendance set to 0") : setError(fbdo.
errorReason());
    }
```

If a person is entering the place from outside (OUT), then the following data will be entered or updated in the Realtime Database. If any of the following processes is unsuccessful, then setError() function will be executed to set the error flag (error) to true.

- The attendance of the person stored at /<NUID>/Attendance will be incremented by 1 using getInt() and setInt().
- The position of the person either IN or OUT stored at /<NUID>/IN _ OUT will be updated to IN using setString(). Firebase.RTDB.setString() is used to set string data at a particular path in the Realtime Database. It accepts three arguments: the first two arguments are the same first two arguments as of Firebase.RTDB.setInt(), and the third argument is the String data to be set.

- The time of entry obtained from the getTimeRTC() function is entered at /<NUID>/IN _ TIME using pushString(). getTimeRTC() is a custom function written by us to obtain the current date-time from the DS3231 RTC module. To see the code inside getTimeRTC() function, go to the end of code explanation. Firebase.RTDB.pushString() is similar to setString() function and accepts the same arguments. Unlike setString(), pushString() will push data to the given path in Realtime Database with a random ID.
- The total number of people inside the entry stored at /Total _ Inside will be incremented by 1 using setInt().

```
    if(IN_OUT == "OUT"){
        //If a person enters, their personal attendance, time of entry & total
number of people inside are updated in Firebase
        Serial.print("A person with ID ");
        Serial.print(newNUID);
        Serial.println(" has entered.");
        attendance = Firebase.RTDB.getInt(&fbdo, ("/"+newNUID+"/Attendance").
c_str()) ? fbdo.intData() : setError(fbdo.errorReason());
        Firebase.RTDB.setInt(&fbdo, ("/"+newNUID+"/Attendance").c_str(),
++attendance) ? Serial.println("Attendance Updated") : setError(fbdo.
errorReason());
        Firebase.RTDB.setString(&fbdo, ("/"+newNUID+"/IN_OUT").c_str(), "IN") ?
Serial.println("IN or OUT Updated") : setError(fbdo.errorReason());
        Firebase.RTDB.pushString(&fbdo, ("/"+newNUID+"/IN_TIME").c_str(),
getTimeRTC()) ? Serial.println("Entered Entry time") : setError(fbdo.
errorReason());
        Firebase.RTDB.setInt(&fbdo, "/Total_Inside", ++totalInside) ? Serial.
println("Total Attendance Updated") : setError(fbdo.errorReason());
    }
```

If a person is leaving the place from inside (IN), then the following data will be entered or updated in the Realtime Database. If any of the following processes is unsuccessful, then setError() function will be executed to set the error flag (error) to true.

- The position of the person either IN or OUT stored at /<NUID>/IN _ OUT will be updated to OUT using setString().
- The time of exit obtained from the getTimeRTC() function is entered at /<NUID>/OUT _ TIME using pushString().
- The total number of people inside the entry stored at /Total _ Inside will be decremented by 1 using setInt().

```
else if(IN_OUT == "IN") {
        //If a person exits, time of exit & total number of people inside are
updated in Firebase
        Serial.print("A person with ID ");
        Serial.print(newNUID);
        Serial.println(" has left.");
        Firebase.RTDB.setString(&fbdo, ("/"+newNUID+"/IN_OUT").c_str(), "OUT")
? Serial.println("IN or OUT Updated") : setError(fbdo.errorReason());
        Firebase.RTDB.pushString(&fbdo, ("/"+newNUID+"/OUT_TIME").c_str(),
getTimeRTC()) ? Serial.println("Entered Exit time") : setError(fbdo.
errorReason());
        Firebase.RTDB.setInt(&fbdo, "/Total_Inside", --totalInside) ? Serial.
println("Total Attendance Updated") : setError(fbdo.errorReason());
```

```
      } else
          error = true;

  } else
      error = true;
```

If the error flag (`error`) is true, then the user is denied entry which is indicated using the red LED or else the user is granted permission to enter which is indicated by the green LED. Finally, you can refer to all the custom functions (`isVerified()`, `setError()`, `getTimeRTC()`) at the end of the code explanation.

```
  //If there is any error while reading the card or updating data in Firebase
then Red LED & buzzer is ON
  //or else Green LED & buzzer is ON
  if(error){
    digitalWrite(redLED, HIGH);
    digitalWrite(buzzer, HIGH);
    delay(1000);
    digitalWrite(redLED, LOW);
    digitalWrite(buzzer, LOW);
    error = false;
  } else {
    digitalWrite(greenLED, HIGH);
    digitalWrite(buzzer, HIGH);
    delay(1000);
    digitalWrite(greenLED, LOW);
    digitalWrite(buzzer, LOW);
  }
  newNUID = ""; //Empty the content in newNUID
  Serial.println();
}

//Verifies whether the new NUID is one among the verifiedNUID list
bool isVerified(String newNUID){
  for(int i=0; i < (sizeof(verifiedNUID) / sizeof(verifiedNUID[0])); i++){
    if(newNUID == verifiedNUID[i])
        return true;
  }
  return false;
}

//Sets the Error flag to true
int setError(String errorReason){
  Serial.print("Failure Occured. Reason: ");
  Serial.println(errorReason);
  error = true;

  return 0;
}

//Obtains the present Date-Time from DS3231 RTC module
String getTimeRTC(){
  nowDT = RTC.now();
  char DATE_TIME_FORMAT[] = "DD/MM/YYYY hh:mm:ss";
  nowDTStr = nowDT.toString(DATE_TIME_FORMAT);

  Serial.print("Time: ");
```

```
  Serial.println(nowDTStr);
  return nowDTStr;
}
```

Before uploading the code, change the WIFI _ SSID and WIFI _ PASSWORD with your WiFi credentials and the API _ KEY, DATABASE _ URL, USER _ EMAIL and USER _ PASSWORD with their respective details. Also, don't forget to change the NUIDs in the verifiedNUID array with your list of verified NUIDs. Finally, upload the code to NodeMCU after selecting the appropriate board, COM port and disconnecting the RX and TX connections. After successfully uploading the code, reconnect the RX and TX connections, open the serial monitor and change the baud rate to 115200 to view the data printed on it.

After the NodeMCU gets connected to WiFi and Firebase, try scanning an RFID tag/card to the RFID reader. If you see the red LED blinking, then there is some error in verifying the RFID tag/card or uploading the entry/exit information to the Realtime Database. If you see the green LED blinking, then everything went as expected and the data is successfully uploaded to the Realtime Database. You can open the Realtime Database to see the uploaded data as shown in Figure 22.26.

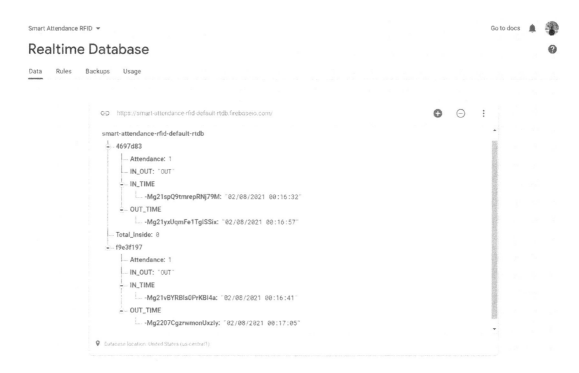

FIGURE 22.26   Entry and exit information of a person entered in the Realtime Database

Many books and technical blogs suggest using the Firebase-Arduino library for uploading data to the Realtime Database. This might be because the books or technical blogs are old or they are unaware of the challenges faced by using the Firebase-Arduino library. The Firebase-Arduino library is one of the oldest Firebase libraries for Arduino IDE, and support for this library is stopped by its creator long ago. There are many flaws in the library which mainly occur due to the updated architecture of Firebase. This library also doesn't offer as many features as compared to the Firebase ESP Client library used in this project.

Before using the Firebase ESP Client library, we have used the Firebase-Arduino library for implementing this project. We have faced a peculiar problem with this library that we have never faced with any of the previously used libraries. The connections and code which worked today won't work after a few days. Initially, we didn't have any idea why is this happening or what the problem is. So, we started to debug the problem by changing the NodeMCU even then the problem persisted with the same result. Next, we thought the error might be because of the Firebase project. So, we have created multiple Firebase projects and tested the code with them but even then nothing seems working. Next, we have checked online to find some blogs or forums explaining similar errors but even there we found nothing useful. We have also changed the WiFi network thinking there might be some firewall issue and checked 10s of times whether the Firebase credentials are entered correctly in the code or not but nothing turned out positive. As a last attempt to sort out the problem, we tested whether the example codes in the Firebase-Arduino library are working or not. To our surprise, even they are not working. So, we have come to the conclusion that the problem is with the library and not with the code or connections. Ideally, there should be no problem with the library as it didn't receive any updates recently. So, nothing should change and the same code which worked a few days back must work now as well. But it didn't work and we also didn't understand what exactly is the problem with the library.

After reading about the library and the functions used in them, we have found nothing wrong with the implementation. Weirdly we found a lot of issues (>100) in the Issues tab. So, we have opened the issues tab hoping to find a similar issue. Luckily, many people were facing the same issue. The replies to one such issue (#506) helped us in resolving the problem.

In case, if you want to use the Firebase-Arduino library for uploading data to the Realtime Database, then you can use the code provided in the resources of this project. Before using the code, you need to install the Firebase-Arduino library and one of its dependencies, ArduinoJson library. You can directly download the Firebase-Arduino library from Github at this link (https://github.com/FirebaseExtended/firebase-arduino). ArduinoJson library can be directly installed from the library manager in Arduino IDE. Search *ArduinoJson* in the search box and install 5.x version (not the latest version) of *ArduinoJson by Benoit Blanchon* among the listed libraries. You can also directly download the 5.x version of the library from Github at this link (https://github.com/bblanchon/ArduinoJson/tree/5.x). It is important to download the 5.x version of ArduinoJson library and not the latest version as the Firebase-Arduino library is built on it. If you have previously installed the latest version of ArduinoJson library (6.x version), then downgrade it to 5.x version or else you'll get a lot of errors while using the Firebase-Arduino library. As explained earlier, there is a problem with the Firebase-Arduino library, so follow the below steps for resolving the issue.

- Open the GRC Fingerprints website (https://www.grc.com/fingerprints.htm) and paste your Database URL (Figure 22.24) without https:// in the Custom Site Fingerprinting section and click *Fingerprint Site*. You'll see the fingerprint of your database URL as shown in Figure 22.27.

FIGURE 22.27   Fingerprint of your database URL

- Next, open the *Arduino > libraries > firebase-arduino > src > FirebaseHttpClient.h* file and go to the end to find a similar command as shown below.

```
static const char kFirebaseFingerprint[] = "03:D6:42:23:03:D1:0C:06:73:F7:E2:
BD:29:47:13:C3:22:71:37:1B";
```

Change the old fingerprint (on the RHS) with the new fingerprint obtained from the GRC fingerprints website and save the file. The new command after changing the fingerprint is shown below.

```
static const char kFirebaseFingerprint[] = "50:89:50:57:90:1F:37:E3:B8:F3:5B:
02:ED:3A:65:6E:6A:34:DB:93";
```

Now, you can use the Firebase-Arduino library for uploading data to the Realtime Database.

You can find all the resources of this project at this Github link **(https://github.com/ anudeep-20/30IoTProjects/tree/main/Project%2022).**

In this project, you have learnt about SPI communication which is one of the widely used interfaces in electronics. You have also learnt about RFID technology and the components used in it. Next, you have learnt to setup a Firebase project and configure the authenticated users for the project. You also learnt to setup a Realtime Database and write security rules for it for protecting the uploaded data. Finally, you have built a smart attendance system using RFID, RTC module and Firebase to track the entry and exit of the people. In the next project, you'll build a system to detect gas leakage or fire and immediately alert the concerned services using an automated call in case of an emergency.

Let's build the next one!

Control Appliances Using Cellular Call or Message

23

Previously, we have controlled appliances using various means such as the webpage, mobile applications, etc. How often we land ourselves in a situation where we are held up in a remote place that is deprived of Internet connectivity! Quite a slender chance and it is even harder to imagine as we need to rely entirely on the cellular network alone in those circumstances. Even in that sort of situations, we can manage as well as control the appliances at far away locations just by means of a cellular call or message. This project will help you to control appliances from anywhere even in remote places with no Internet connectivity at all with a simple cellular call or message.

23.1 MOTIVE OF THE PROJECT

In this project, you'll control an LED using a cellular call or SMS. Just for demonstration purpose, we are controlling an LED, but you can control any appliance using the same procedure.

23.2 GSM SIM800A

Presently, there are many mobile communications standards developed for different generations of cellular networks. One such mobile communication standard is GSM (Global System for Mobile Communications) developed by ETSI (European Telecommunications Standards Institute) for describing the standards of 2G cellular networks. The other standards like UMTS, CDMA and OFDMA are used for subsequent cellular network generations. The architecture and further complicated details of GSM are beyond the scope and not explained in this project.

As 2G cellular network is available in most of the places compared to other generations, we'll use GSM-based modules for controlling appliances. In this project, GSM SIM800A as shown in Figure 23.1 is used for receiving cellular calls and messages. SIM800A cellular chip developed by SIMCOM is located at the centre of the module. The operating temperature of the module is between −40°C and 90°C, and it can be operated with any voltage between 9 V and 12 V. The module won't work as expected with any voltage less than 9 V, and any voltage greater than 12 V can spoil the module. For providing the required voltage to the module, you can use a 12 V DC adapter as shown in Figure 23.2.

DOI: 10.1201/9781003147169-23

FIGURE 23.1 GSM SIM800A

FIGURE 23.2 12 V DC adapter

GSM SIM800A has an onboard SIM (Subscriber Identification Module) cardholder that can be used to insert any 2G SIM. It also has three onboard LEDs, each of which is for a specific purpose.

- STS – It is used to indicate whether the board is powered ON or OFF.
- DC-PWR – It is used to indicate the type of input voltage to the GSM module.
- NTW – It is used to indicate the network connectivity.

Generally, the STS, DC-PWR and NTW LED will be of Green, Red and Blue colour, respectively. But these colours might change depending on the module manufacturer. The NTW LED is very useful in knowing the connectivity of the module to the cellular network through its blink rate. If the NTW LED blinks once per second, then the module is NOT connected to the cellular network and is trying to connect. If the NTW LED blinks three times per second, then the module is connected to the cellular network. To know more about the hardware design and specifications of GSM SIM800A module, refer to its datasheet (https://simcom.ee/documents/SIM800A/SIM800A_Hardware%20Design_V1.02.pdf).

As the size is not a concern in this project, we have used a higher grade module (GSM SIM800A) that can easily connect to the cellular network even inside a closed room. If your project demands a smaller size GSM module that needs to operate on lower voltages (around 5 V), then use the GSM SIM800L module which costs much less and supports almost all the features provided by SIM800A. But the antenna which comes along with the module cannot easily connect to a cellular network compared to GSM SIM800A. To avoid the connectivity problem, generally, the antenna similar to the one used with GSM SIM800A (3dBi GSM antenna) along with a UFL to SMA connector is used as shown in Figure 23.3.

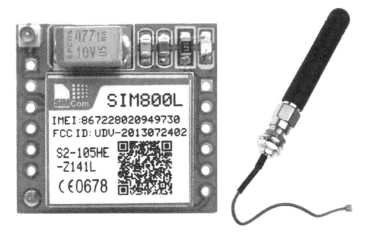

FIGURE 23.3   GSM SIM800L along with antenna

23.3 AT COMMANDS IN GSM

Similar to the HC-05 Bluetooth module, GSM SIM800A also supports AT commands that can be used to control the module. Unlike the HC-05 Bluetooth module, the GSM SIM800A doesn't have any operating modes. So, you need not press any physical button for sending the AT commands. Some of the AT commands and their use are tabulated in Table 23.1. You can find all the AT commands of GSM SIM800A module at this link (https://www.elecrow.com/wiki/images/2/20/SIM800_Series_AT_Command_Manual_V1.09.pdf).

TABLE 23.1   AT commands of GSM SIM800A

AT Command	Function of the AT Command
AT	Testing Command
AT+CGMI	For knowing the name of the GSM manufacturer
AT+CGSN	For knowing the IMEI (International Mobile Equipment Identity) number
AT+CSQ	Provides the signal quality
ATA	Receive an incoming call

For sending the AT commands to the GSM SIM800A, RS232 to USB cable shown in Figure 23.4 can be used. It helps in connecting the GSM SIM800A to the PC. After connecting the cable, your PC might not assign a port to the module and might show an error similar to the ones shown in Section 8.5. The same procedure explained in Section 8.5 can be followed for solving these errors. If a COM port is assigned to the GSM module, then open Putty or Tera Term for sending the AT commands. Unlike for the HC-05 Bluetooth module, no extra configurations need to be done in the Tera Term. But ensure the baud rate is set to 9600.

FIGURE 23.4   RS232 to USB cable

23.4 HARDWARE REQUIRED

TABLE 23.2   Hardware required for the project

HARDWARE	QUANTITY
Arduino Uno	1
GSM SIM800A	1
12 V DC adapter	1
LED	1
Jumper wires	As required

23.5 CONNECTIONS

TABLE 23.3   Connection details

ARDUINO UNO	GSM SIM800A	LED
7	TX	
8	RX	
13		+ve pin (long pin)
GND	GND	−ve pin (short pin)

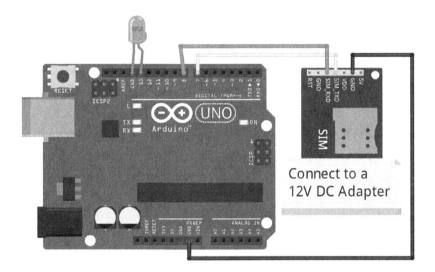

FIGURE 23.5   Connection layout

The list of required components for the project is tabulated in Table 23.2, followed by Table 23.3, which presents the pin-wise connections. Also, the connection diagram is shown in Figure 23.5, which shall help get a visualization of the connections.

23.6 ARDUINO IDE CODE

The following code is for controlling the LED using cellular calls. Instead of default Rx and Tx pins of Arduino Uno (digital pins 0 and 1), digital pins 7 and 8 are configured as Rx and Tx pins, respectively, using the SoftwareSerial library. The LED which needs to be controlled is connected to the digital pin 13 of Arduino Uno. GSM_data is a String type variable used to store the data read from the GSM's serial. SMS_data is another String type variable used to store the data to be sent through SMS (Short Message Service).

```
#include <SoftwareSerial.h> //Library for replicating Rx, Tx

SoftwareSerial GSM_Serial(7, 8); //Rx Pin, Tx Pin

int LED = 13; //Pin to which LED is connected
```

```
String GSM_data = ""; //Stores the data read from GSM module
String SMS_data = ""; //Stores the data to be sent as an SMS
bool LEDstate = false; //Stores the state of LED
```

Both Arduino and GSM serial is initialized at a baud rate of 9600. As we need to control the LED, its direction of operation is set as OUTPUT. Initially, the LED is set to OFF state. AT+CPIN? is used to know whether the SIM card is properly inserted in the GSM or not. If it is inserted properly, then you'll get the response as OK or else ERROR. AT+COPS? is used to know the network operator of the SIM card. If the SIM card is detected, then the network operator will be printed or else 0 or ERROR will be printed. AT+CSQ is used to know the signal strength of the network connection. If the GSM is connected to the network, depending on the connection strength a number will be printed which will help in knowing the signal strength. For knowing the relation between printed number and signal strength (in dBm), refer to the GSM SIM800A AT commands manual (https://www.elecrow.com/wiki/images/2/20/SIM800_Series_AT_Command_Manual_V1.09.pdf).

```
void setup() {
  Serial.begin(9600); //Initialization of Arduino's serial
  GSM_Serial.begin(9600); //Initialization of GSM's serial
  pinMode(LED,OUTPUT); //Define the direction of LED pin
  digitalWrite(LED,LOW); //Initially LED is set to OFF state

  GSM_Serial.println("AT+CPIN?"); //Checks the SIM card presence
  delay(1000);
  GSM_Serial.println("AT+COPS?"); //To know the network operator
  delay(1000);
  GSM_Serial.println("AT+CSQ"); //To know the signal strength
  delay(1000);
}
```

If any new data is available on the GSM's serial, then it is read using readString() function. This data will usually contain trailing or leading whitespace which is removed in place using the trim()function. When a call is made to the GSM, RING will be printed on the GSM's serial. As soon as the call ends, NO CARRIER will be printed on the GSM's serial.

If the GSM gets an incoming call, then the LED will be switched ON, if it is previously OFF or vice versa. As soon as the incoming call ends, LEDstate is toggled and an SMS confirmation is sent to the user. Replace x and y with the country code and phone number to which the acknowledgement SMS has to be sent. For example, if you are living in the UK and want to send the acknowledgement SMS to your phone number (7778889990), then replace xxyyyyyyyyyy with +44778889990. You can also configure to send the acknowledgement SMS to multiple phone numbers with at least 5 second gap between each one of them.

```
void loop() {
  //Checks for any incoming data over GSM serial
  if (GSM_Serial.available()>0) {
    GSM_data = GSM_Serial.readString(); //Reads the data from GSM serial
    GSM_data.trim(); //Removes any trailing or leading white space
    //Print the data read from GSM's serial on the Arduino's serial
    Serial.println(GSM_data);

    //If there is a call then switch ON or OFF the LED
    if(GSM_data == "RING" && LEDstate == false) {
      digitalWrite(LED, HIGH);
      SMS_data = "LED is switched ON";
      Serial.println("LED is switched ON");
    } else if(GSM_data == "RING" && LEDstate == true) {
```

```
      digitalWrite(LED, LOW);
      SMS_data = "LED is switched OFF";
      Serial.println("LED is switched OFF");
    }

    //If call ended then toggle LEDstate & send acknowledgement SMS
    if (GSM_data == "NO CARRIER") {
      LEDstate = !LEDstate;
      sendSMS();
    }
  }
}

//Custom function for sending SMS
void sendSMS(){
  //Replace x with country code and y with mobile number
  GSM_Serial.println("AT+CMGS=\"xxyyyyyyyyyy\"");
  delay(1000);
  GSM_Serial.print(SMS_data); //SMS text to be sent
  delay(1000);
  GSM_Serial.println((char)26); //ASCII code of CTRL+Z. Acts as a Send key.
  delay(1000);
}
```

The following code is for controlling the LED using cellular messages. It is very similar to the previous code but with few modifications. AT+CNMI=2,2,0,0,0 helps in printing any new SMS on GSM's serial as soon as it is received. AT+CMGF=1 is used to set the SMS format to text mode whereas AT+CMGF=0 is used to set the SMS format to PDU (Protocol Data Unit) mode. Text mode is easy for humans to read and interpret whereas PDU mode is easy for computers to interpret. The below example shows a sentence in both text mode and PDU mode.

Text mode: Hello World!!

PDU mode (8-bit): 07914400000000F001000B811000000000F000040D48656C6C6F20576F726C642121

```
#include <SoftwareSerial.h> //Library for replicating Rx, Tx

SoftwareSerial GSM_Serial(7, 8); //Rx Pin, Tx Pin

int LED = 13; //Pin to which LED is connected
String GSM_data = ""; //Stores the data read from GSM module

void setup() {
  Serial.begin(9600); //Initialization of Arduino's serial
  GSM_Serial.begin(9600); //Initialization of GSM's serial
  pinMode(LED,OUTPUT); //Define the direction of LED pin
  digitalWrite(LED,LOW); //Initially LED is set to OFF state

  GSM_Serial.println("AT+CPIN?"); //Checks the SIM card presence
  delay(1000);
  GSM_Serial.println("AT+COPS?"); //To know the network operator
  delay(1000);
  GSM_Serial.println("AT+CMGF=1"); //Sets the message format to text mode
  delay(1000);
  GSM_Serial.println("AT+CNMI=2,2,0,0,0"); //Indicates new SMS
  delay(1000);
}
```

Unlike the previous code, entire data available on GSM's serial won't be read at once but will be read line by line using `readStringUntil('\n')`. Even this data might contain trailing or leading whitespace which will be removed using the `trim()` function. If the GSM receives an SMS, it will be printed on its serial in two lines. The first line consists of metadata which includes the phone number from which the SMS is sent and the date and time at which the SMS is received. The second line consists of the actual message sent from the mobile phone.

```
+CMT: "<COUNTRY CODE - PHONE NUMBER>","","<DATE-TIME>"
<SENT MESSAGE>
```

If the sent message is ON, then the LED will be switched ON. Similarly, if the sent message is OFF, then the LED will be switched OFF. Also, an acknowledgement SMS will be sent after the LED is switched ON or OFF.

```
void loop() {
  //Checks for any incoming data over GSM serial
  if (GSM_Serial.available()>0) {
    //Reads the data from GSM Serial line by line
    GSM_data = GSM_Serial.readStringUntil('\n');
    GSM_data.trim(); //Removes any trailing or leading white space
    //Print the data read from GSM's serial on the Arduino's serial
    Serial.println(GSM_data);

    //Depending on the message, toggle the state of LED
    if(GSM_data == "ON") {
      digitalWrite(LED,HIGH);
      sendSMS("LED is switched ON");
      Serial.println("LED is switched ON");
    } else if(GSM_data == "OFF") {
      digitalWrite(LED,LOW);
      sendSMS("LED is switched OFF");
      Serial.println("LED is swicthed OFF");
    }
  }
}

//Custom function for sending SMS
void sendSMS(){
  GSM_Serial.println("AT+CMGS=\"xxyyyyyyyyyy\""); //Replace x with country
code and y with mobile number
  delay(1000);
  GSM_Serial.print(SMS_data); //SMS text to be sent
  delay(1000);
  GSM_Serial.println((char)26); //ASCII code of CTRL+Z. Acts as a Send key.
  delay(1000);
}
```

Upload both codes one after another to Arduino Uno after selecting the appropriate board and COM port. After successfully uploading the code, open the serial monitor and change the baud rate to 9600. If the connections are perfect and the GSM is connected to the network, you'll see responses for the AT commands in `void setup()`. If you are unable to see the responses, then there might be some problem with the connections, SIM card or the GSM module. If everything works perfect, then make a call or send an SMS to the GSM module and see the LED getting switched ON or OFF.

23.7 TRY IT

Presently, anybody can control your appliances by calling or messaging the GSM. To avoid this, you can add security to it. Modify the code such that only after receiving a particular number of RINGs the appliance can be controlled or modify the code such that the appliance can only be controlled if a password or secret code is sent along with the controlling command. With the knowledge of coding which you have gained until now, try to implement the security feature in both codes on your own.

You can find all the resources of this project at this Github link (https://github.com/anudeep-20/30IoTProjects/tree/main/Project%2023).

In this project, you have learnt about GSM modules and controlling them using AT commands. You have also learnt to build a system that can control an LED using cellular call or SMS. Remember that even though this project deals with controlling a simple electronic component such as LED, you can control any electronic appliance using the same procedure. If it's an electrical appliance, then you can use a relay as a bridge for controlling it. In the next project, you will build a fire alarm system that can alert the concerned emergency services through an automated call, in case of any emergency.

Let's move to the next one!

Fire Alarm Using IoT

24

Fire alarms save lives by providing an early warning, and they play a decisive role in reducing any sort of fire injuries, deaths and damages to things/properties/buildings. Fire detectors can be heat, smoke, carbon monoxide or even radiation sensors, etc. A properly designed fire alarm system gives a timely warning of the threat and allows adequate time for the people to evacuate the premises and also call the authorities before the fire gets out of control. This project will help you to build a fire alarm system that can detect harmful gases and temperature rise within a confined location and can also place an alert through automated call and SMS.

24.1 MOTIVE OF THE PROJECT

In this project, you'll build a fire alarm system that can detect fire, smoke, LPG or rise in room temperature and alert the concerned emergency services by means of an automated call.

24.2 MQ2 GAS SENSOR

Gas sensors are electronic devices that are used to detect different types of gases. They are commonly used to detect harmful or toxic gases that are dangerous to humans. In factories and mines, they are part of alarm systems to detect the leakage of any dangerous gases. In IoT, MQ series sensors are used for detecting various gases. The detectable gases of MQ series sensors shown in Table 24.1 are not only limited to those but they can also detect other gases with low sensitivity.

TABLE 24.1 MQ series sensors and their detectable gases

MQ SENSOR	DETECTABLE GASES
MQ2	Smoke, LPG (liquefied petroleum gas), Carbon monoxide (CO), methane, butane, hydrogen
MQ3	Smoke, ethanol, alcohol
MQ4	Methane, CNG (compressed natural gas)
MQ5	LPG, natural gas
MQ7	CO
MQ8	Hydrogen gas
MQ9	Flammable gases, CO
MQ131	Ozone (O_3)
MQ135	Air quality (ammonia, alcohol, benzene, carbon dioxide, CO, smoke)
MQ136	Hydrogen sulphide gas (H_2S)
MQ137	Ammonia
MQ138	Alcohol, acetone, benzene, hydrogen, formaldehyde gas, toluene, propane
MQ214	Methane, natural gas

DOI: 10.1201/9781003147169-24

Depending on the use case, an appropriate gas sensor needs to be selected. In our case, MQ2 gas sensor is selected as we need to detect smoke and LPG. The capability of MQ2 gas sensor to detect CO (a very harmful gas) is an added advantage. MQ2 is a chemiresistor or MOS (metal oxide semiconductor) type sensor that detects the sensing material based on the change in resistance. The breakout board of MQ2 sensor shown in Figure 24.1 has four pins out of which two are VCC and GND, and the other two are Analog Out (AO) and Digital Out (DO). The analog output voltage from the sensor changes depending on the concentration of gas. The output increases with the increase in the concentration of the gas and decreases with the decrease in the gas concentration. This analog signal is fed to an onboard comparator which converts the analog output to digital output. The sensitivity of digital output can be easily calibrated using the onboard potentiometer.

FIGURE 24.1   MQ2 gas sensor

Some specifications of the MQ2 gas sensor are tabulated in Table 24.2. The concentration of a gas is generally measured in ppm (parts per million). It is the measurement of a particular compound in one million parts of a whole mixture. For example, 500ppm of CO_2 (carbon dioxide) means there are 500 molecules of CO_2 in 1 million molecules of gas. For knowing the complete specifications of MQ2 gas sensor, refer to its datasheet (https://www.mouser.com/datasheet/2/321/605-00008-MQ-2-Datasheet-370464.pdf).

TABLE 24.2   Specifications of MQ2 gas sensor

Operating voltage	5 V±0.1
Load resistance	Adjustable
Heater resistance	33 Ω±5%
Heating consumption	<800 mW
Sensing resistance	3—30 KΩ
Concentration scope (LPG)	200–5,000 ppm

24.3 HARDWARE REQUIRED

TABLE 24.3   Hardware required for the project

HARDWARE	QUANTITY
Arduino Uno	1
GSM SIM800A	1
12 V DC adapter	1
MQ2 Gas sensor	1
LM35 temperature sensor	1
Breadboard	1
Jumper wires	As required

24.4 CONNECTIONS

The list of required components for the project is tabulated in Table 24.3, followed by Table 24.4, which presents the pin-wise connections. Also, the connection diagram shown in Figure 24.2 will help you to get a visualization of the connections.

TABLE 24.4   Connection details

ARDUINO UNO	GSM SIM800A	MQ2 GAS SENSOR	LM35 TEMPERATURE SENSOR
5 V		VCC	+Vs
A0			Out
2		DO	
7	TX		
8	RX		
GND	GND	GND	GND

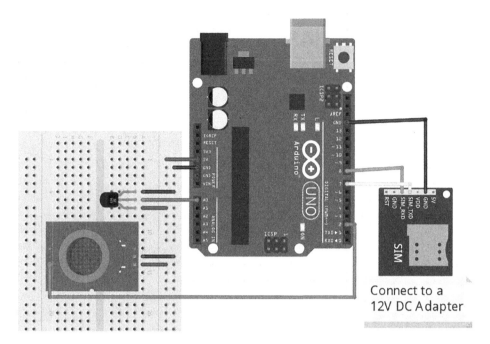

Connect to a
12V DC Adapter

FIGURE 24.2   Connection layout

24.5 ARDUINO IDE CODE

The code used for building a fire alarm is similar to the one used in the previous project but with some modifications. The part of the code used for collecting data from the sensors and sending an alert through call is different from the previous project's code. `SoftwareSerial` library is used to configure digital pins 7 and 8 as Rx and Tx, respectively. In our case, we only need to detect the gas and need not measure its concentration. So, the digital output from MQ2 gas sensor is obtained from the DO pin that is connected to digital pin 2 of Arduino Uno. As the output from LM35 temperature sensor is analog, its out pin is connected to the A0 pin of Arduino Uno. `TEMP _ THRES` is used to store the temperature in °C above which an alert should be triggered.

```
#include <SoftwareSerial.h> //Library for replicating Rx, Tx

SoftwareSerial GSM_Serial(7, 8); //Rx Pin, Tx Pin

#define MQ2 2 //Pin to which MQ2 gas sensor is connected
#define LM35 A0 //Pin to which LM35 is connected
//Temperature (in °C) above which alert should be triggered
#define TEMP_THRES 60
```

VRef is used to store the voltage to which the LM35 temperature sensor's $+V_s$ pin is connected. `resolution` is used to store the resolution with which the `tempRead` must be multiplied to obtain the correct temperature in °C. notGasLeak is used to store the digital readings read from the MQ2 gas sensor. Both Arduino and GSM serial is initialized at a baud rate of 9600. As we are reading voltage signals from MQ2 and LM35, their direction of operation is set as `INPUT`.

```
float VRef = 5.0; //Voltage to which +Vs pin of LM35 is connected
float resolution = (VRef*100)/1023.0; //Calculation of resolution

bool notGasLeak = true; //Status of gas leakage
float tempRead; //Stores the temperature reading

void setup() {
  Serial.begin(9600); //Initialization of Arduino's serial
  GSM_Serial.begin(9600); //Initialization of GSM's serial
  pinMode(MQ2,INPUT); //Define the direction of MQ2 DO pin
  pinMode(LM35,INPUT); //Define the direction of LM35 out pin

  GSM_Serial.println("AT+CPIN?"); //Checks the SIM card presence
  delay(1000);
  GSM_Serial.println("AT+COPS?"); //To know the network operator
  delay(1000);
  GSM_Serial.println("AT+CMGF=1"); //Sets the message format to text mode
  delay(1000);
  GSM_Serial.println("AT+CNMI=2,2,0,0,0"); //Indicates new SMS
  delay(1000);
}

void loop() {
  //Reads the output from DO of MQ2 gas sensor
  notGasLeak = digitalRead(MQ2);
  tempRead = analogRead(LM35); //Reads the analog output from LM35
  tempRead *= resolution; //Conversion of analog output to temperature
```

If the MQ2 gas sensor detects one of its detectable gases mentioned in Table 24.1, then its DO pin will output 0 or else 1. An emergency alert is sent to the stored phone number through call and message if the MQ2 gas sensor detects smoke or LPG or if the room temperature rises above TEMP _ THRES.

```
  if(!notGasLeak) { //If LPG/smoke is detected (MQ2 reading is 0)
    Serial.println("Alert!! Smoke/LPG detected.");
    sendAlert();
  }

  Serial.print("Temperature (in °C): ");
  Serial.println(tempRead);

  if(tempRead > TEMP_THRES) { //If temperature rises above TEMP_THRES
    Serial.println("Alert!! Temperature is rising. Your house might be on
fire.");
    sendAlert();
  }

  delay(200);
}
```

sendAlert() is a custom function used for sending an alert in case of an emergency. This function can be used to send the alert only to one person. But you can modify the code to send the alert call and SMS to multiple phone numbers with at least 5 second gap between each of them. ATDxxyyyyyyyyyy; is the AT command used to call a phone number using the GSM module. Don't forget to replace x and y with the country code and phone number, respectively, before uploading the code.

```
//Custom function for sending an Alert through call and SMS
void sendAlert() {
  //Replace x with country code and y with mobile number
  GSM_Serial.println("AT+CMGS=\"xxyyyyyyyyyy\"");
  delay(1000);
  GSM_Serial.print("Emergency!! Your house might be under risk of LPG leakage
or fire."); //SMS text to be sent
  delay(1000);
  GSM_Serial.println((char)26); //ASCII code of CTRL+Z. Acts as a Send key.
  delay(5000);
  //Replace x with country code and y with mobile number
  GSM_Serial.println("ATDxxyyyyyyyyyy;");
  delay(1000);
}
```

Finally, upload the code to Arduino Uno after selecting the appropriate board and COM port. After successfully uploading the code, open the serial monitor and change the baud rate to 9600. If the connections are perfect and the GSM is connected to the network, then you'll see responses for the AT commands in void setup(). If you are unable to see the responses, then there might be some problem with the connections, SIM card or the GSM module. If everything works perfect, then blow smoke onto the MQ2 gas sensor and wait for at least 5 seconds to get an emergency SMS and call. Similarly, increase the LM35 temperature above 60°C using a fire source and wait for at least 5 seconds to get a similar emergency SMS and call.

You can find all the resources of this project at this Github link (https://github.com/ anudeep-20/30IoTProjects/tree/main/Project%2024).

In this project, you have learnt about MQ series sensors and the gases they can detect. You have also learnt to build a fire alarm system that can detect harmful gases and temperature rise and send an alert through automated call and SMS. In the next project, you'll build a milk quality testing device that can be used to assess the quality of the milk at home without the necessity of visiting a dairy products quality testing laboratory.

Let's build the next project!

Build a Milk Quality Testing Device

25

Adulteration in milk could have an adverse effect on health of people of all ages where the contaminants like urea, starch, glucose, formalin along with detergent, etc. are used as adulterants to increase the thickness and viscosity of the milk as well as to preserve it for a longer period. These adulterations decrease the nutritive value of milk and can pose serious health hazards leading to fatal diseases. Existing common detection techniques are not always convenient, and implementing easy detection methods at the consumer level can aid a long way to help endorse the product quality. This project will help you to build a milk testing device that ensures a quick turnaround time and also aids in quality estimation.

25.1 MOTIVE OF THE PROJECT

In this project, you'll build a milk quality testing device from scratch that can test the quality of drinkable cow milk or buffalo milk without the need to visit a food testing laboratory. It uses a pH sensor, a lactometer and a temperature sensor for accurate measurements.

Disclaimer: The results obtained from the milk quality testing device built in this project should not be considered for any analysis or experimentation purpose. Even though the pH sensor used in this project is laboratory or industry grade, the procedure for obtaining accurate results is tedious and you may commit mistakes in between. The quality results returned by the device should only be used to get an estimation of the milk quality but not the literal milk quality.

25.2 pH SENSOR

In chemistry, pH (potential of hydrogen or power of hydrogen) is a logarithmic scale used to specify the acidity or basicity of an aqueous solution. A solution that has the solvent as water is known as an aqueous solution. The pH scale is inversely proportional to the molar concentration of hydrogen ions (H^+) in the solution:

$$pH = -\log_{10}\left[H^+\right] \tag{25.1}$$

The pH of acidic solutions (higher concentration of H^+ ions) will always be less than the pH of basic solutions (lower concentration of H^+ ions). At 25°C, the pH of acidic solutions will be less than 7, the pH of basic solutions will be greater than 7 and the pH of neutral solutions (like pure water) will be equal to 7. Usually, the pH of an aqueous solution will be between 0 to 14, but rarely it may go below 0 for very strong acids or go above 14 for very strong bases. The pH of aqueous solutions is temperature dependant, and it can vary with an increase or decrease in temperature.

There are many types of pH sensors available in the market, but the pH probe used with them will be the same. Figures 25.1 and 25.2 are two of the commonly used pH sensors. The design of these pH sensors

DOI: 10.1201/9781003147169-25

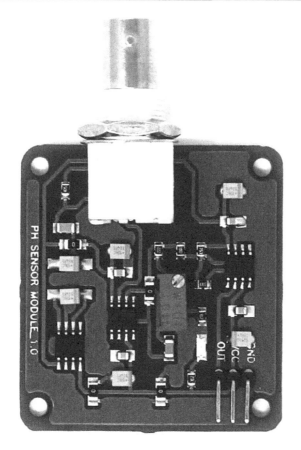

FIGURE 25.1   pH sensor module (Type 1)

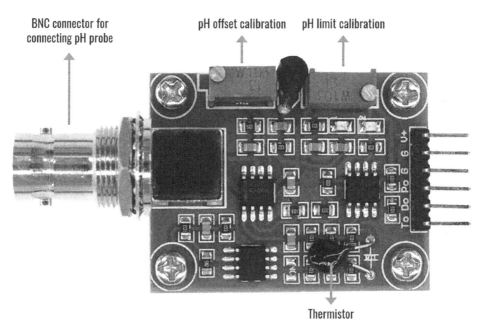

FIGURE 25.2   pH sensor module (Type 2)

might change from time to time or place to place but their working remains the same. Figure 25.3 shows the pH probe used along with the pH sensor. The internal functioning of pH probe is a little complex, and it is not explained in this project. If you want to know about the detailed working of the pH probe, then refer to this link (https://www.ysi.com/ysi-blog/water-blogged-blog/2019/02/anatomy-of-ph-electrodes). In this project, we'll use the pH sensor shown in Figure 25.2 along with the pH probe. So, if pH sensor is mentioned anywhere in the project, then it means we are referring to the sensor shown in Figure 25.2. Some of the specifications of pH sensor are tabulated in Table 25.1. The usage of the pins and potentiometers on the pH sensor are tabulated in Table 25.2.

FIGURE 25.3   pH probe

TABLE 25.1   Specifications of pH sensor

Heating voltage	5V±0.2
Working current	5–10 mA
Detectable pH range	0–14
Detectable temperature range	0–60°C
Response time	≤5 seconds
Stability time	≤1 minute
Power consumption	≤0.5 W
Working humidity	95%

At 25°C, the pH of unadulterated milk will be between 6.5 and 6.9 (acidic). This value might change depending on various factors like the source from which milk is obtained, the number of days elapsed since procurement from the source (freshness of the milk), etc. In the research paper published by Kanwal, R., et al., they have analyzed various factors of milk depending on the source from which it is taken (https://scialert.net/fulltext/?doi=ajps.2004.300.305). The analysis of variations in milk pH depending on the source from which it is taken is tabulated in Table 25.3.

TABLE 25.2   Pinout of the pH sensor

V+	5 V DC (input)
G	GND for pH sensor
G	GND for pH probe
PO	pH analog output
DO	3.3 V output when a pH limit is reached
TO	Temperature output
BNC (Bayonet Neill-Concelman) connector	Used to connect the pH probe
Offset potentiometer (near the BNC connector)	Used to calibrate the pH analog reading
Limit setting potentiometer	Used to set the pH limit at which DO pin should trigger

TABLE 25.3   Variations in the pH of milk depending on the source

SOURCE OF MILK	RANGE OF PH
Cow	6.63–6.68
Buffalo	6.60–6.90
Goat	6.34–6.68
Sheep	6.40–6.80

From the analysis, you can observe that cow milk has the highest mean pH and sheep milk has the least mean pH. In this project, we'll only consider pH variations in cow milk and buffalo milk as we are building a device to test those. The variations in milk pH over time are analysed by Marouf, A., et al., in their research paper (https://www.arcjournals.org/pdfs/ijarps/v4-i12/1.pdf). They have taken six cow milk samples directly from the farm and exposed them to a laser with six different power outputs. Each milk sample is considered separately, and its change in pH is analyzed over time. The results of their analysis are reproduced in Figure 25.4.

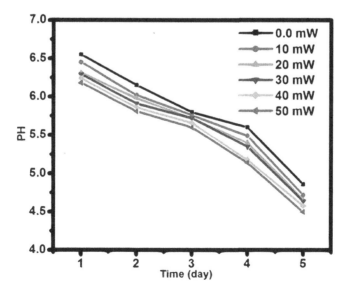

FIGURE 25.4   Variations in pH over time

Even though it shows six different milk samples, our focus is only on the milk sample which was not exposed to laser (0.0 mW, square or top-most line). You can observe from the analysis that milk becomes more and more acidic with time. This is because the lactic acid bacteria in the milk converts lactose sugar into lactic acid. So, pH is an important factor in the determination of the quality and freshness of milk.

The pH of an aqueous solution also depends on temperature. It has no effect or minimal effect between 10°C and 30°C, but the pH changes about 0.06 as the temperature increases to 50°C. In most use cases, the temperature of milk will be below 30°C. So, the temperature correction for pH is not required. Even if it is >30°C, the error will be at most 0.06 until 50°C.

25.3 CALIBRATION OF pH SENSOR

The important part of pH sensor usage is calibration. It is so important that without performing this step the pH sensor is almost useless. Just like any other analog sensor, the pH sensor will also output voltage through PO pin that will represent the pH value of the aqueous solution. Ideally, the output voltage should be linearly distributed between the pH scale, i.e, 0 V should represent pH 0 and 5 V should represent pH 14. But by default, the pH sensor is set at 0 V for pH 7 which means the output voltage will go negative when measuring the pH of acidic solutions. So, the pH sensor needs to be calibrated such that it outputs 2.5 V for pH 7.

To calibrate the pH sensor, first do either of the following,

- Connect the pH probe to the BNC connector and place it in a neutral solution whose pH is 7 (like pure water).
- Short the BNC connector using a wire. It can be done by wrapping one side of an electrically conducting wire around the outside of the BNC connector and push the other end into the hole of the BNC connector. The short circuit is equivalent to a neutral pH reading of 7.

Next, use a multimeter to measure the voltage between PO and G pins of the pH sensor. Adjust the offset potentiometer until it measures 2.5 V. If you don't have a multimeter, then connect the Arduino Uno and pH sensor as shown in Figure 25.5 and by referring to Table 25.4.

Upload the following code to Arduino Uno after selecting the appropriate board and COM port. Next, open the serial monitor and rotate the offset potentiometer until you see 2.5 V reading.

```
#define pHpin A0
float pHvoltage;

void setup() {
  Serial.begin(9600);
}

void loop() {
  pHvoltage = analogRead(pHpin) * (5.0 / 1023.0);
  Serial.print(pHvoltage);
  Serial.println(" V");
  delay(500);
}
```

The calibration of the pH sensor need not be done again even if the pH probe is changed. After calibrating the pH sensor, remove the short circuit of the BNC connector and connect the pH probe for calibrating it. The readings of the pH probe change frequently and need to be calibrated for maintaining consistency. Don't worry! This calibration can be done using code and doesn't require any hardware modifications.

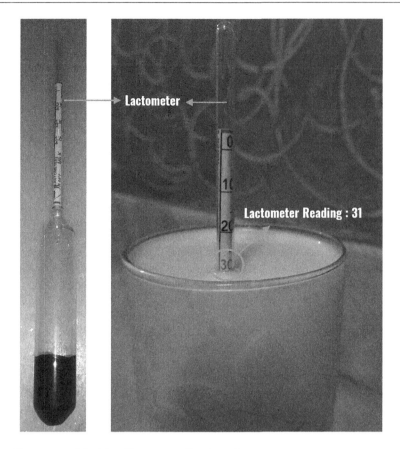

FIGURE 25.5   pH sensor and Arduino Uno connection layout

TABLE 25.4   pH sensor and Arduino Uno connection details

ARDUINO UNO	pH SENSOR
5 V	V+
GND	G, G
A0	PO

After the pH sensor calibration, place the pH probe inside any known pH aqueous solution. You might see a difference between the actual pH and the measured pH of that solution. For calibrating the pH probe, this difference (known as offset) must be added to all the pH readings measured using it. In our Arduino IDE code, we have automated this pH probe calibration so that the code need not be changed whenever calibration is required.

25.4 LACTOMETER

Lactometer is an instrument used for testing the purity of milk which works on the principle of specific gravity (Archimede's principle). It measures the density of milk relative to water. If the specific gravity of the milk being tested is within the approved range, then the milk is considered pure or else it is considered adulterated. Previously, for testing the quality of milk only, a lactometer is used but due to advancement in technology, milk is being adulterated such that its lactometer reading (LR) is well within the approved range. To identify these types of adulterations, we'll consider other parameters also for accurate assessment of the milk quality.

At 20°C, the density of unadulterated milk will be from 1.026 gram/mL to 1.032 g/mL (gram/millilitre), and its lactometer reading will be from 26 to 32, whereas the density of water will be 1.000 g/ml and its lactometer reading will be 0. If more water is added to the milk, then its lactometer reading will go below 26, or if any solids are added to the milk, then its lactometer reading will go above 32. For measuring the lactometer reading of a sample of milk, stir the milk sample and gently place the lactometer in it. Wait for some time until the lactometer is free from oscillations. The reading just above the surface of milk is its lactometer reading. Presently, there are no digital lactometers available in the market for directly interfacing with a microcontroller. So, we'll use a normal lactometer as shown in Figure 25.6 and enter its reading to the device using an HC-05 Bluetooth module.

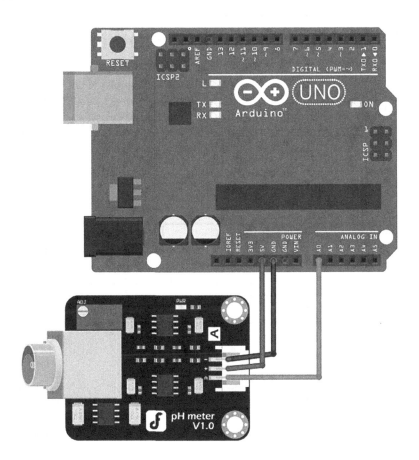

FIGURE 25.6 Lactometer

The density of milk is temperature dependent. It means the lactometer reading changes depending on the temperature of milk. So, the observed lactometer reading must be corrected depending on the temperature of milk. If the temperature of milk is above 20°C, then 0.2 must be added to the observed lactometer reading for every 1°C change. Similarly, if the temperature of milk is below 20°C, then 0.2 must be subtracted from the observed lactometer reading for every 1°C change. The final lactometer reading after correcting it with the temperature difference is known as Corrected Lactometer Reading (CLR).

Lactometer reading changes depending on the source from which it is taken as well. The research paper by Marouf, A., et al., also provides the analysis of change in lactometer reading depending on the source from which it is taken (https://www.arcjournals.org/pdfs/ijarps/v4-i12/1.pdf). The results of the analysis mentioned in the research paper are tabulated in Table 25.5. From the analysis, you can observe that the mean lactometer reading of cow milk is higher than that of others. It means unadulterated cow milk has more fat content than milk from other sources. In this project, we'll only consider the lactometer reading variations in cow milk and buffalo milk and leave aside the rest.

TABLE 25.5 Variations in the lactometer reading of milk depending on the source

SOURCE OF MILK	RANGE OF LACTOMETER READING	MEAN LACTOMETER READING
Cow	28–34	30.00
Buffalo	26–29	27.65
Goat	27–30	28.65
Sheep	27–29	28.05

25.5 1-WIRE COMMUNICATION PROTOCOL

The 1-wire communication protocol is similar in concept to the I²C protocol but with slow data transmission rates and longer range. It is a low-cost serial signalling protocol designed by Dallas Semiconductor Corp. (now acquired by Maxim Integrated). A typical 1-wire network consists of a single master connected to one or more slaves through the 1-wire bus as shown in Figure 25.7. To accommodate the use of only one wire, the 1-wire communication protocol combines the clock, data and power into a single signal.

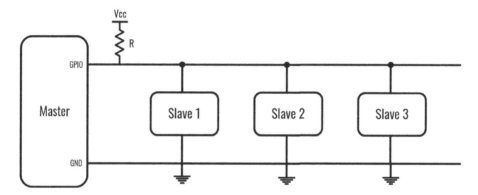

FIGURE 25.7 1-Wire network

When no information is being transferred, the data line will be in HIGH state. As the devices on the 1-wire bus don't have a dedicated power line, they use an internal capacitor that is charged while the line is idle (HIGH state) to derive the required power (parasite power). The minimum operating voltage of these devices is 2.8 V. If the supply current to these devices is not sufficient, then a pull-up resistor is required. Generally, the pull-up resistance value is chosen between 1 kΩ and 4.7 kΩ which will set the current range between 5 mA and 1.06 mA.

The 1-wire communication protocol is bidirectional and half-duplex, and the data transmission is bit sequential with the LSB transmitted first. Each of the maxim's 1-wire devices comes with a factory-lasered unique and unalterable 64-bit ID stored in the ROM (Read-Only Memory) section of the device. It is a combination of 8-bit CRC (Cyclic Redundancy Check), 48-bit Serial number and 8-bit family code as shown in Figure 25.8. Using a search ROM algorithm, the master can get the unique IDs of all the slave devices on the 1-wire bus and need not remember them.

CRC (8-bit)	Serial number (48-bit)	Family code (8-bit)

FIGURE 25.8 64-bit device ID

In a 1-wire communication protocol, all the communications are initiated by the master to which the slaves respond. Direct communication between slaves is not allowed and need to be done only through the master. For starting a transmission, the master will pull the data line to LOW for at least 480 µs. This will reset all the slave devices on the 1-wire bus. If any slave device is present on the 1-wire bus, it'll show its presence by pulling the data line to LOW for at least 60 µs (presence pulse) after the master releases it. To transmit a binary number 1, the master sends a small LOW pulse ranging between 1µs and 15µs. Similarly, to transmit a binary number 0, the master sends a 60 µs LOW pulse. Every slave device has a monostable multivibrator that acts as its internal timer. It starts at the falling edge of the LOW pulse. It is easily affected by the analog tolerances so the pulse for transmitting a binary number 0 should be exactly 60 µs and the pulse for transmitting a binary number 1 should not exceed 15 µs.

For receiving each bit from the slave device, the master sends a 1–15 µs LOW pulse. If the transmitting slave device wants to send a binary number 1, then it does nothing. If it wants to send a binary number 0, then it pulls the data line to LOW for 60 µs. Usually, the data is sent or received in groups of 8-bits. For every data transfer on the 1-wire bus, a three-phase sequence (reset, device selection and device function) is used as shown in Figure 25.9. The supported speeds in a 1-wire bus are 15.4 Kbps in standard mode and 125 Kbps in overdrive mode.

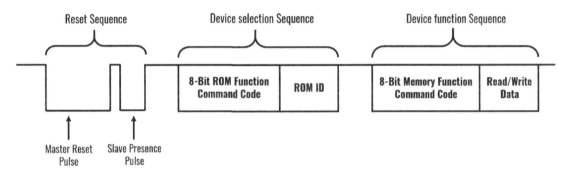

FIGURE 25.9   1-Wire data transfer sequence

25.6 DS18B20 TEMPERATURE SENSOR

Similar to the LM35 temperature sensor, DS18B20 is also a temperature sensor developed by Dallas Semiconductor Corp. (now acquired by Maxim Integrated). It uses the 1-wire communication protocol for sending or receiving data from the microcontroller. It comes in two form factors as shown in Figure 25.10. The form factor of the sensor shown on the left side is similar to the LM35 temperature sensor. This can be used at all the places wherever LM35 temperature sensor is employed. The second type of form factor shown on the right side is a waterproof version of the DS18B20 temperature sensor. The sensor can be used to measure the temperature of liquids without the fear of spoiling them. This is the reason why we have chosen the DS18B20 temperature sensor besides the commonly used LM35 temperature sensor in this project.

After seeing the DS18B20 temperature sensor, you might be getting a doubt. As it is based on the 1-wire communication protocol, it should be having only two lines but why is it having three lines? This is a valid doubt which everyone might get. DS18B20 temperature sensor uses an additional V_{DD} line for uninterrupted power supply. But it can also be used with only two lines (DQ and GND) in the Parasitic Power mode. Some of its specifications are tabulated in Table 25.6. To know the complete functionality and specifications of the DS18B20 temperature sensor, refer to its datasheet (https://datasheets.maximintegrated.com/en/ds/DS18B20.pdf).

FIGURE 25.10   Form factors of DS18B20 temperature sensor

TABLE 25.6   Specifications of DS18B20 temperature sensor

Supply voltage	3–5.5 V
Active current	≤1.5mA (Typically, 1 mA)
Temperature range	−55–125°C
Accuracy (−10–+85°C)	±0.5°C
Accuracy (−30–+100°C)	±1°C
Accuracy (−55–+125°C)	±2°C
Standby power	0 W
Resolution	9–12 bits (default 12 bits)
Conversion time	≤750 ms
Power consumption	≤0.5 W
Working humidity	95%

25.7 HARDWARE REQUIRED

TABLE 25.7   Hardware required for the project

HARDWARE	QUANTITY
Arduino Uno	1
pH sensor	1
pH probe	1
DS18B20 Temperature sensor	1
SSD1306 OLED module	1
HC-05 Bluetooth module	1
Latching switch (non-momentary)	1
LEDs	3 (1×Red, 1×Yellow, 1×Green)
Resistor (4.7 kΩ)	1
Breadboard	1
Jumper wires	As required

25.8 CONNECTIONS

The list of required components for the project is tabulated in Table 25.7, followed by Table 25.8, which presents the pin-wise connections. Also, the connection diagram shown in Figure 25.11 will help you to get a visualization of the connections.

TABLE 25.8 Connection details

ARDUINO UNO	PH SENSOR	DS18B20 TEMPERATURE SENSOR	SSD1306 OLED MODULE	HC-05 BLUETOOTH MODULE	LATCHING SWITCH	RED LED	YELLOW LED	GREEN LED	RESISTOR (4.7 KΩ)
5V	V+	VDD	VCC		Pin 1				Pin 1
2		DQ		VCC	Pin 2				Pin 2
3				TxD					
5				RxD					
8						+ve pin (long pin)			
9							+ve pin (long pin)		
10								+ve pin (long pin)	
A0	PO								
A4			SDA						
A5			SCL						
GND	G, G	GND	GND	GND		−ve pin (short pin)	−ve pin (short pin)	−ve pin (short pin)	

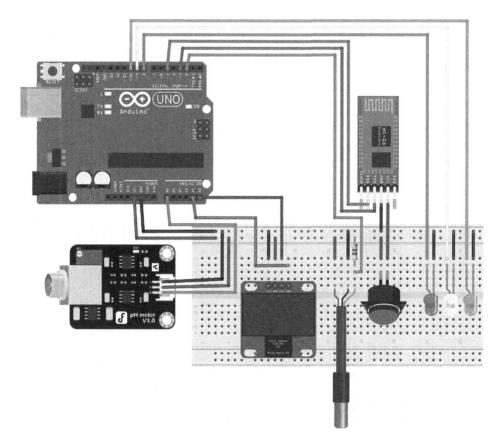

FIGURE 25.11 Connection layout

25.9 ARDUINO IDE CODE

`OneWire.h` library is used for interfacing 1-wire slave devices to the master. For installing it, open the library manager in Arduino IDE and search for *OneWire*. Install the latest version of *OneWire by Paul Stoffregen* among the listed libraries. You can also directly download the library from Github at this link (https://github.com/PaulStoffregen/OneWire). `DallasTemperature.h` library is used to obtain temperature readings from the 1-wire-based temperature sensors. For installing it, open the library manager in Arduino IDE and search for *DallasTemperature*. Install the latest version of *DallasTemperature by Miles Burton* among the listed libraries. You can also directly download the library from Github at this link (https://github.com/milesburton/Arduino-Temperature-Control-Library).

```
#include <OneWire.h> //Library for 1-Wire communication protocol
#include <DallasTemperature.h> //Library for Temperature sensor
#include <SoftwareSerial.h> //Library for replicating Rx & Tx
#include <Adafruit_SSD1306.h> //Library for SSD1306 OLED module
```

The following preprocessors store the digital pins to which the DS18B20 temperature sensor, HC-05 Bluetooth module, LEDs and pH sensor are connected. The analog values from the pH sensor will slightly fluctuate. So, to get a more stable output, we'll average some analog values. The number of analog values to be averaged is stored in the nSampleAvg. The reference voltage of Arduino Uno is stored in ref-Voltage. It is used for converting analog values to its equivalent voltage. calRefpH stores the pH value of the solution used for calibrating the pH probe. The code used for displaying content on the SSD1306 OLED display using Adafruit _ SSD1306.h library is the same as the previous projects. Instead of default Rx and Tx pins of Arduino Uno, BT _ Tx and BT _ Rx are configured as Rx and Tx pins, respectively, using the SoftwareSerial library.

```
//Initialising the required pins
#define tempPin 2
#define BT_Tx 3
#define BT_Rx 5
#define redLED 8
#define yellowLED 9
#define greenLED 10
#define pHpin A0
//Constants used for calculating the pH
#define nSampleAvg 10
#define refVoltage 5.0
#define calRefpH 7.0

#define SCREEN_WIDTH 128 //OLED screen width (in pixels)
#define SCREEN_HEIGHT 64 //OLED screen height (in pixels)
#define OLED_ADDRESS 0x3C //OLED I2C Address

//Used for displaying content on the OLED screen
Adafruit_SSD1306 display(SCREEN_WIDTH, SCREEN_HEIGHT, &Wire, -1);
SoftwareSerial BTserial(BT_Tx,BT_Rx); //Rx Pin, Tx Pin
```

oneWire is a OneWire constructor that takes the digital pin to which the temperature sensor's DQ line is connected (tempPin) as the only argument. tempSensor is a DallasTemperature constructor that takes the address where oneWire is stored as the only argument. All the following declared variables are used for either storing or calculating pH, lactometer reading, CLR and final quality grade.

```
//Used for obtaining data from DS18B20 Temperature sensor
OneWire oneWire(tempPin);
DallasTemperature tempSensor(&oneWire);

//Variables used for calculating pH
float cummAValue = 0;
float calVoltage = 2.5;
float calpHvalue;
float pHvoltage;
float pHvalue;
float calibration;
unsigned long start = millis();
float tempValueC = 0; //Stores the temperature in °C

float LR = 0.0; //Stores the lactometer reading
float CLR = 0.0; //Stores the Corrected lactometer reading
String quality = "NA"; //Stores the final milk quality grade
```

All the Serial statements in the code are commented out for two reasons. First of all, printing on the serial monitor is not necessary as all the required information is continuously printed on the OLED screen. The second reason being, if Serial is enabled, you might see some glitches on the OLED screen.

So, uncomment the serial statements only if you want to debug the code. Bluetooth's serial (`BTserial`) is initialized at a baud rate of 9600. `tempSensor.begin()` is used to initiate the communication between the DS18B20 temperature sensor and Arduino Uno. As we are controlling LEDs, their direction of operation is set as `OUTPUT`. Similarly, as we are reading the analog values from the pH sensor, its direction of operation is set as `INPUT`.

```
void setup() {
  //Serial.begin(9600);
  BTserial.begin(9600); //Initialization of Bluetooth's serial
  tempSensor.begin(); //Initialization of DS18B20 Temperature sensor
  //Defining the direction of GPIO pins
  pinMode(redLED, OUTPUT);
  pinMode(yellowLED, OUTPUT);
  pinMode(greenLED, OUTPUT);
  pinMode(pHpin, INPUT);

  display.begin(SSD1306_SWITCHCAPVCC, OLED_ADDRESS); //Initialize OLED module

  //Initial display configuration
  display.setTextSize(1);
  display.setTextColor(SSD1306_WHITE);
  display.clearDisplay();

  //pH probe calibration
  display.setCursor(2,20);
  display.println("Place the pH probe in water and wait for 1 minute.");
  display.display();
  delay(3000);
```

The following part of the code is for pH probe calibration. It is required so that the entire code need not be re-uploaded whenever the pH probe calibration needs to be changed. Before starting the code execution, the pH probe must be placed in a neutral solution (like pure water) whose pH is exactly 7. If you don't have a neutral solution, then place it in a known pH solution and change the `calRefpH` preprocessor value with its pH value.

It has to be remembered that it takes at least 1 minute for the fluctuations in the pH values to be minimized after placing the pH probe inside a liquid. During that time, the analog values from the pH sensor are averaged and converted to voltage. This voltage is converted to pH by multiplying 2.8 to it. The calibration value is calculated by subtracting the obtained pH value (`calpHvalue`) from the actual pH value (`calRefpH`). Completion of pH probe calibration will be indicated by the three LEDs. They'll all glow at a time for three seconds.

The reason for multiplying 2.8 for converting voltage to its equivalent pH is simple. The analog output or voltage from the pH sensor is directly proportional to the pH of liquid, i.e, they both are linearly related. If you remember, the pH sensor is calibrated such that at 0 V the pH is 0, at 2.5 V the pH is 7 and at 5 V the pH is 14. Using this information, we can construct a line equation with voltage (V) and pH as the parameters. As you might have studied in your high school, the line equation using two known points is

$$y - y_1 = \frac{y_2 - y_1}{x_2 - x_1} * (x - x_1) \tag{25.2}$$

where

    x and y are the variable parameters (V and pH)

    x_1, y_1 is the first point (0,0)

    x_2, y_2 is the second point (2.5, 7)

By substituting the known point values in Equation 25.2, we get the following relation between V and pH:

$$pH - 0 = \frac{7-0}{2.5-0} * (V - 0) \tag{25.3}$$

$$pH = V * \frac{7}{2.5} \tag{25.4}$$

So, the voltage must be multiplied by $\frac{7}{2.5}$ or 2.8 for converting into its equivalent pH value. The graphical representation of the relation between the voltage from the pH sensor and the pH of the liquid is shown in Figure 25.12.

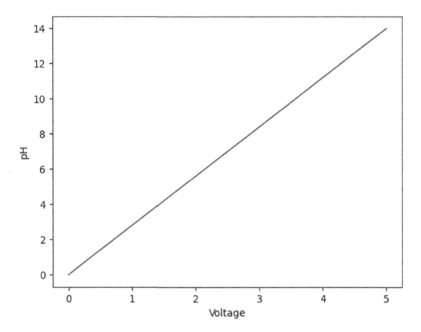

FIGURE 25.12   Relation between voltage and pH of the liquid

```
while(millis() - start <= 60000){
  cummAValue = 0;
  for (int i = 0; i < nSampleAvg; i++)
  {
    cummAValue += analogRead(pHpin);
    delay(300);
  }
  calVoltage = (cummAValue/nSampleAvg) * (refVoltage/1023.0);
}
calpHvalue = calVoltage * 2.8;
calibration = calRefpH - calpHvalue;

//Indicating the completion of pH probe calibration using LEDs
digitalWrite(redLED, HIGH);
digitalWrite(yellowLED, HIGH);
digitalWrite(greenLED, HIGH);
delay(3000);

display.clearDisplay();
```

```
  digitalWrite(redLED, LOW);
  digitalWrite(yellowLED, LOW);
  digitalWrite(greenLED, LOW);
  delay(1000);
}
```

After the calibration of pH probe, you can measure the pH of any aqueous solution. Now, place the pH probe inside an aqueous solution whose pH has to be measured. In our case, it is milk. The analog values from the pH sensor are averaged and converted to voltage. This voltage is converted to pH by multiplying with 2.8. The final pH value of the liquid is obtained by adding the calibration value to it.

Even though the pH value of liquid might not be correct initially, wait for at least a minute to obtain the correct pH value. You can eliminate the use of two variables (`calpHvalue` and `calibration`) by using the following equation:

```
pHvalue = pHvoltage * 2.8 + calibration
pHvalue = pHvoltage * 2.8 + calRefpH - calpHvalue
pHvalue = pHvoltage * 2.8 + calRefpH - calVoltage * 2.8
pHvalue = calRefpH + ((pHvoltage - calVoltage) * 2.8)
```

If you want to use the modified equation, then comment the `pHvalue = (pHvoltage * 2.8) + calibration;` equation and uncomment `pHvalue = calRefpH + ((pHvoltage - calVoltage)*2.8);` equation.

```
void loop() {
  //Obtaining the pH value
  cummAValue = 0;
  for (int i = 0; i < nSampleAvg; i++) {
    cummAValue += analogRead(pHpin);
    delay(300);
  }
  pHvoltage = (cummAValue/nSampleAvg) * (refVoltage/1023.0);
  pHvalue = (pHvoltage * 2.8) + calibration;
  //pHvalue = calRefpH + ((pHvoltage - calVoltage)*2.8);
  /*Serial.print("pH= ");
  Serial.print(pHvalue);*/
```

After the pH value of the liquid stabilizes, insert the DS18B20 temperature sensor into the liquid. `requestTemperatures()` gets the temperature reading in °C from all the temperature sensors on the 1-wire bus. This information will be stored in the `tempSensor` object in the form of an array. As we have only one temperature sensor on the 1-wire bus, its information will be stored at the 0th index. `getTempCByIndex(0)` gets the temperature of liquid in °C.

```
  //Reading the temperature in °C
  tempSensor.requestTemperatures();
  tempValueC = tempSensor.getTempCByIndex(0);
  /*Serial.print(" Temperature: ");
  Serial.print(tempValueC);
  Serial.print(" °C");*/
```

Next, place the lactometer inside the milk whose quality has to be tested. After recording the lactometer reading, switch ON the HC-05 Bluetooth module using the latching switch. Pair your mobile phone to the HC-05 Bluetooth module and enter the lactometer reading using the Bluetooth terminal mobile application as explained in Section 4.7. Do not enter any text along with the lactometer reading.

The lactometer reading of the milk will be read from the `BTserial` and then it is converted to float. `CLR` is calculated by adding the temperature correction factor. It is nothing but the difference between the actual temperature of the milk and 20°C multiplied by 0.2.

```
//Receiving LR from Bluetooth and calculating the CLR
if(BTserial.available() > 0){
  LR = BTserial.readString().toFloat();
  if(tempValueC > 0)
    CLR = LR + (tempValueC - 20)*0.2;
  /*Serial.print(" LR: ");
  Serial.print(LR);
  Serial.print(" CLR: ");
  Serial.print(CLR);*/
}
```

If CLR is calculated, then the final grade of milk is determined using Table 25.9. These metrics are based on the literature survey done by us. The quality metrics of milk differs from region to region. In some places, the lactometer reading of milk might be above 32 due to the high-fat content in it. It doesn't mean the milk is adulterated. If the quality metrics mentioned in Table 25.9 doesn't give proper results for the milk available in your region, then take some milk samples from your region and test them. Depending on the results, update Table 25.9 and the Arduino IDE code accordingly.

TABLE 25.9 Quality metrics for determination of milk quality

pH RANGE	CLR RANGE	FINAL QUALITY	USAGE	LED
6.5–6.8	26–32	GOOD	Everyone (including infants)	GREEN
6.3–6.5 and 6.8–7.0	24–26 and 32–34	AVERAGE	Advisable only for adults	GREEN
6.0–6.3 and 7.0–7.3	21–24 and 34–37	BAD	Cannot be used for drinking but can be used for cooking	YELLOW
≤6.0 and ≥7.3	≤21 and ≥37	VERY BAD	Cannot be used for any purpose	RED

Even though the BAD quality milk is not advisable for drinking, it is used in limited quantities for cooking. For example, broken milk (which comes under the BAD quality) is used to make kalakand (a form of cheesecake) sweets in India. Broken milk is formed due to the overgrowth of bacteria that causes the milk to undergo change in taste, smell and texture. Drinking it directly may harm your health, but cooking the same is not harmful. Finally, all the calculated parameters are displayed on the OLED screen.

```
//Assessing the final quality of milk using pH and CLR
if(CLR)
  if((pHvalue <= 6.0 or pHvalue >= 7.3) or (CLR <= 21 or CLR >= 37)){
    quality = "VERY BAD";
    digitalWrite(redLED, HIGH);
    delay(2000);
    digitalWrite(redLED, LOW);
  } else if((pHvalue <= 6.3 or pHvalue >= 7.0) or (CLR <= 24 or CLR >= 34)){
    quality = "BAD";
    digitalWrite(yellowLED, HIGH);
    delay(2000);
    digitalWrite(yellowLED, LOW);
  } else if((pHvalue <= 6.5 or pHvalue >= 6.8) or (CLR <= 26 or CLR >= 32)){
    quality = "AVERAGE";
```

```
      digitalWrite(greenLED, HIGH);
      delay(2000);
      digitalWrite(greenLED, LOW);
   } else if((pHvalue > 6.5 and pHvalue < 6.8) and (CLR > 26 and CLR < 32)){
      quality = "GOOD";
      digitalWrite(greenLED, HIGH);
      delay(2000);
      digitalWrite(greenLED, LOW);
   }
/*Serial.print(" Quality: ");
Serial.println(quality);*/

//Printing all the required details on the OLED screen
display.clearDisplay();
display.setCursor(0,0);
display.print("pH: ");
display.setCursor(20,0);
display.print(pHvalue);
display.setCursor(0,15);
display.print("Temperature: ");
display.setCursor(80,15);
display.print(tempValueC);
display.setCursor(0,30);
display.print("LR: ");
display.setCursor(30,30);
display.print(LR);
display.setCursor(60,30);
display.print("CLR: ");
display.setCursor(90,30);
display.print(CLR);
display.setCursor(30,45);
display.print(quality);
display.display();

   delay(3000);
}
```

After uploading the code to Arduino Uno, follow the below steps for determining the quality of milk:

- Place the pH probe inside a neutral solution (like pure water) and then power the Arduino Uno board. If it is already powered, then click the reset button on the Arduino Uno board. Don't remove the pH probe for at least 1 minute. Completion of 1 minute will be indicated by the three LEDs. The pH probe can also be placed inside a known pH solution, but its value must be changed in calRefpH before uploading the code.
- Now, place the pH probe inside the milk which you want to test. Wait for another 1 minute for the pH values to stabilize. This value will be printed on the OLED screen.
- After the pH value of milk is stabilized, place the lactometer and DS18B20 temperature sensor inside the milk. Wait for at least 30 seconds until the temperature reading gets stabilized, and the lactometer is free of oscillation.
- Next, switch ON the HC-05 Bluetooth module using the latching switch and connect your mobile phone to enter the lactometer reading. Depending on the temperature, the lactometer reading is corrected and the CLR is printed on the OLED screen.
- Depending on the pH value and CLR, the overall quality grade of milk is determined and it is printed on the OLED screen. Each quality grade is associated with a colour LED. Depending on the milk quality grade, a corresponding LED will blink.

You can find all the resources of this project at this Github link (https://github.com/ anudeep-20/30IoTProjects/tree/main/Project%2025).

In this project, you have learnt about the working of pH sensor and its calibration. You have also learnt about the lactometer and correcting its reading based on the temperature of milk. Finally, you have learnt about the 1-wire protocol and the usage of a sensor based on it. Using all these, you have built a milk quality testing device that can test the quality of milk at home. By this project, you have covered most of the important concepts, boards and sensors used in IoT. The next step is to minimize the size of the projects you are building. In the next project, you'll learn about ATtiny85 which can be used to reduce the size of your projects by at least 5 times.

Let's move to the next project!

Miniaturize Your IoT Projects Using ATtiny 26

Innovative IoT implementations are usually complex projects with the complexity being multidimensional. One of the main complexity in these projects is the requirement of smaller and portable electronic components that are faster, smarter and more power efficient. The end user market is also being attracted towards miniaturized devices as it offers the same functionality but in a thin, small and light form factor. This project will help you to shrink the size of your IoT projects without compromising their functionality by making use of ATtiny85 microcontroller.

26.1 MOTIVE OF THE PROJECT

In this project, you'll learn to shrink the size of your IoT projects using an ATtiny85 microcontroller. Most of the functionalities that can be done by Arduino Uno can be done by ATtiny85 but with a much smaller form factor. To give a perspective, ATtiny85 is 50 times smaller than Arduino Uno.

26.2 ATTINY85 MICROCONTROLLER

ATtiny microcontrollers are a subfamily of 8-bit AVR microcontrollers developed by ATMEL (now acquired by Microchip Technology). As the name suggests, ATtiny microcontrollers are smaller in size compared to other microcontrollers. Due to their smaller size, they have fewer GPIO pins and reduced memory size and features.

ATtiny85 shown in Figure 26.1 is a low-power consuming 8-bit AVR RISC-based microcontroller. It has 8 pins among which 5 can be used as GPIO pins. Among them, all the 5 pins can be used as digital pins, 4 can be used as analog pins and 3 can be used as PWM pins. Despite its smaller size, it supports all the commonly used protocols. i.e., UART, SPI and I²C. Unlike the ATmega328P microcontroller used in Arduino Uno, it can be used without an external oscillator. Some of its features are tabulated in Table 26.1. To know about its complete features, refer to its datasheet (https://ww1.microchip.com/downloads/en/DeviceDoc/Atmel-2586-AVR-8-bit-Microcontroller-ATtiny25-ATtiny45-ATtiny85_Datasheet.pdf). The pinout of ATtiny85 microcontroller is shown in Figure 26.2. Each pin of ATtiny85 can be used for multiple purposes. For example, the 7th pin of ATtiny85 can be used as Digital Pin 2 or Analog Pin 4 or SCK in SPI or SCL in I²C protocol. Details of the functionalities of each pin are tabulated in Table 26.2. If your project requires more than 5 GPIO pins, then use ATtiny84 which has the same architecture and features but has more GPIO pins (11).

DOI: 10.1201/9781003147169-26

FIGURE 26.1   ATtiny85 microcontroller

TABLE 26.1   Specifications of ATtiny85 microcontroller

Operating voltage	1.8–5.5 V
Program memory size	8 KB
Number of comparators	1
CPU speed (MIPS/DMIPS)	20
Data EEPROM	512 bytes
ADC channels	4
Max ADC resolution (bits)	10
Max 8 Bit Digital timers	2
Temperature range	−40–85°C

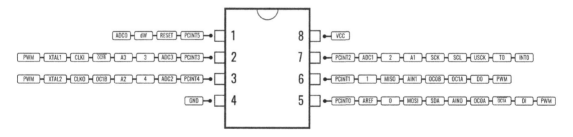

FIGURE 26.2   ATtiny85 pinout

TABLE 26.2   ATtiny85 pinout details

ATTINY85 PIN NUMBER	FUNCTION	DETAILS
1	PCINT5	Pin Change Interrupt 0, Source 5
	RESET	Reset Pin
	dW	debug Wire I/O
	ADC0	ADC Input Channel 0
2	PCINT3	Pin Change Interrupt 0, Source 3

(Continued)

TABLE 26.2 (*Continued*)   ATtiny85 pinout details

ATTINY85 PIN NUMBER	FUNCTION	DETAILS
	ADC3	ADC Input Channel 3
	3	Digital Pin 3
	A3	Analog Pin 3
		Complementary Timer/Counter1 Compare Match B Output
	CLKI	External Clock Input
	XTAL1	Crystal Oscillator Pin 1
	PWM	Pulse Width Modulation Pin
3	PCINT4	Pin Change Interrupt 0, Source 4
	ADC2	ADC Input Channel 2
	4	Digital Pin 4
	A2	Analog Pin 2
	OC1B	Timer/Counter1 Compare Match B Output
	CLKO	System Clock Output
	XTAL2	Crystal Oscillator Pin 2
	PWM	Pulse Width Modulation Pin
4	GND	Connected to Ground
5	PCINT0	Pin Change Interrupt 0, Source 0
	AREF	External Analog Reference
	0	Digital Pin 0
	MOSI	Master Output Slave Input (SPI)
	SDA	Serial Data (I2C or Two-Wire Mode)
	AIN0	Analog Comparator, Positive Input
	OC0A	Timer/Counter0 Compare Match A Output
		Complementary Timer/Counter1 Compare Match A Output
	DI	Data Input (Three-Wire Mode)
	PWM	Pulse Width Modulation Pin
6	PCINT1	Pin Change Interrupt 0, Source 1
	1	Digital Pin 1
	MISO	Master Input Slave Output (SPI)
	AIN1	Analog Comparator, Negative Input
	OC0B	Timer/Counter0 Compare Match B output
	OC1A	Timer/Counter1 Compare Match A Output
	DO	Data Output (Three-Wire Mode)
	PWM	Pulse Width Modulation Pin
7	PCINT2	Pin Change Interrupt 0, Source 2
	ADC1	ADC Input Channel 1
	2	Digital Pin 2
	A1	Analog Pin 1
	SCK	Serial Clock (SPI)
	SCL	Serial Clock (I2C or Two-Wire Mode)
	USCK	Universal Serial Interface Clock (Three-Wire Mode)
	T0	Timer/Counter0 Clock Source
	INT0	External Interrupt 0 Input
8	VCC	Connected to Input Voltage

26.3 INITIAL SETUP OF ATtiny85

Unlike Arduino Uno, ATtiny85 doesn't have any USB interface for directly uploading code from a PC. So, we need an external programmer like the one shown in Figure 26.3 for programming it. Even though programming a microcontroller with a dedicated programmer is easy, purchasing extra components for a particular use case is indeed a waste of money and resources. Arduino Uno also has an inbuilt programmer which is used to program its onboard ATmega328P microcontroller. To make use of resources effectively, we'll use the inbuilt programmer in Arduino Uno to program the ATtiny85.

FIGURE 26.3 USB AVR programmer for ATMEL microcontrollers

There are few steps that need to be done for programming ATtiny microcontrollers using Arduino IDE and Arduino Uno board. Open *File* > *Preferences* (Ctrl+Comma) in Arduino IDE and click the small window box beside *Additional Boards Manager URLs*. A pop-up window will open as shown in Figure 26.4, paste this link (https://raw.githubusercontent.com/damellis/attiny/ide-1.6.x-boards-manager/package_damellis_attiny_index.json) in it and click *OK*. Open the *Boards Manager...* in *Tools* > *Board* as shown in Figure 26.5 and search for *attiny*. From the listed boards, install the latest version of *attiny by David A. Mellis* as shown in Figure 26.6. After installing it, again go to *Tools* > *Board*. Now, you can see a new set of boards named *ATtiny Microcontrollers* as shown in Figure 26.7. The Arduino IDE setup for programming ATtiny microcontrollers needs to be done only once.

The next step is to make the Arduino Uno board as an ISP programmer so that it can program external microcontrollers. It can be done by uploading the *ArduinoISP* code to Arduino Uno. The *ArduinoISP* code can be found at *File* > *Examples* > *11.ArduinoISP* > *ArduinoISP* in Arduino IDE. Upload the *ArduinoISP* code to Arduino Uno after selecting the appropriate board and COM port. This step has to be done before uploading the code to the ATtiny microcontroller.

After successfully uploading the *ArduinoISP* code, connect the Arduino Uno and ATtiny85 as shown in Figure 26.8 and Table 26.3. Before uploading the code to ATtiny85, select the *Board* as *ATtiny Microcontrollers* > *ATtiny25/45/85*. Select the *Processor* as *ATtiny85*, *Clock* as *Internal 8MHz* and the COM port

FIGURE 26.4 Additional boards manager URLs

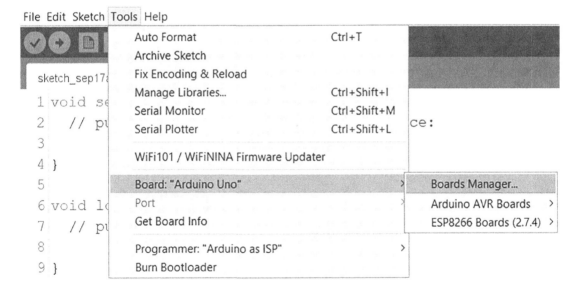

FIGURE 26.5 Open boards manager

assigned to your Arduino Uno as shown in Figure 26.9. Finally, select the programmer as *Arduino as ISP* and click Burn Bootloader as shown in Figure 26.10. If burning the bootloader is successful, then your connections and setup are perfect and you can upload any code to the ATtiny85 microcontroller. If you get an error, then jump to Section 26.7 for resolving it.

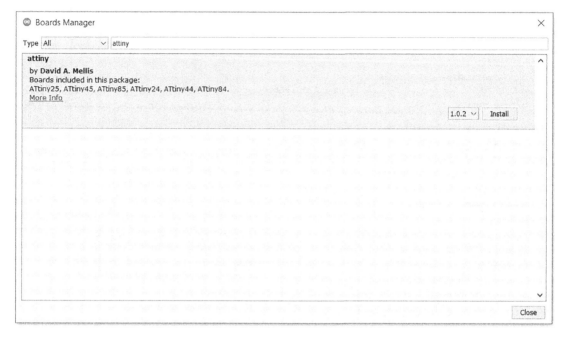

FIGURE 26.6   ATtiny boards

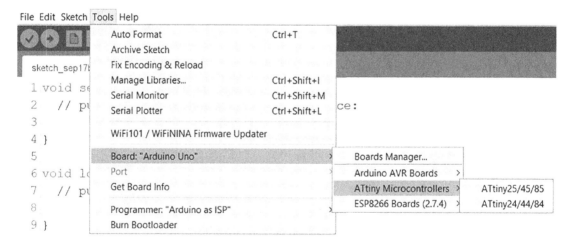

FIGURE 26.7   ATtiny microcontrollers

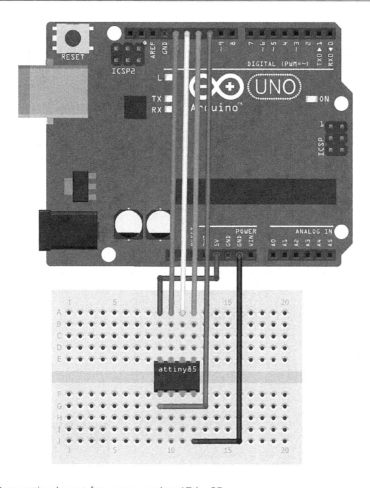

FIGURE 26.8 Connection layout for programming ATtiny85

TABLE 26.3 Connection details for programming ATtiny85

ARDUINO UNO	ATTINY85
VCC	Pin 8 (VCC)
10	Pin 1 (RESET)
11	Pin 5 (MOSI)
12	Pin 6 (MISO)
13	Pin 7 (SCK)
GND	Pin 4 (GND)

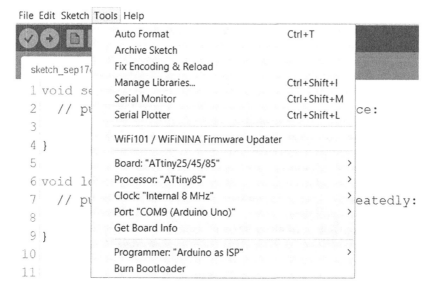

FIGURE 26.9 Selecting the board, processor, clock and port

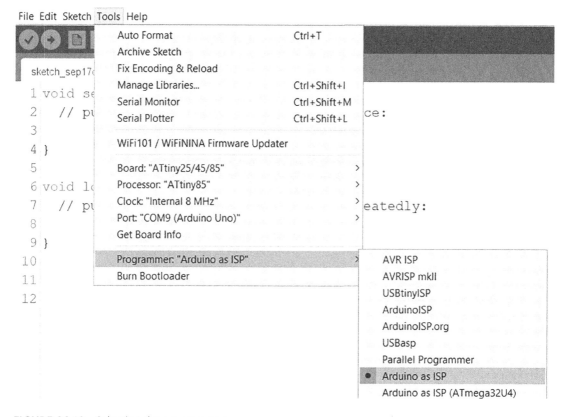

FIGURE 26.10 Selecting the programmer

26.4 UPLOADING CODE TO ATtiny85

As an example, we'll upload the LED Blink code to ATtiny85. You can also upload any code using the same setup explained previously. Keeping the connections as shown in Figure 26.8, upload the following code after selecting the appropriate board and COM port.

```
#define LED 4

void setup() {
  //Initialize digital pin LED_BUILTIN as an output
  pinMode(LED, OUTPUT);
}
void loop() {
  //Turn the LED on (HIGH is the voltage level)
  digitalWrite(LED, HIGH);
  delay(1000); //Wait for a second
  //Turn the LED off by making the voltage LOW
  digitalWrite(LED, LOW);
  delay(1000); //Wait for a second
}
```

Remember you should not include any `Serial` statements in the code as ATtiny85 doesn't have a default serial. If included, the compiler will throw an error.

26.5 HARDWARE REQUIRED

TABLE 26.4 Hardware required for the project

HARDWARE	QUANTITY
ATtiny85 microcontroller	1
LED	1
9 V battery	1
LM7805 voltage regulator	1
0.33 μF ceramic capacitor	1
0.1 μF ceramic capacitor	1
Resistor (1 kΩ)	1
Breadboard	1
Jumper wires	As required

26.6 CONNECTIONS

After successfully uploading the code to ATtiny85, remove its connections with Arduino Uno and connect the components listed in Table 26.4 as shown in Figure 26.11. You can also refer to Table 26.5 which presents the pin-wise connection details. After powering the system, you can see the LED blinking indicating that your setup, connections and code are all perfect.

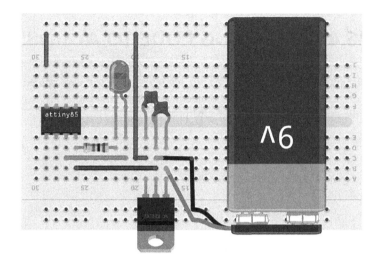

FIGURE 26.11 Connection layout

TABLE 26.5 Connection details

ATTINY85 MICROCONTROLLER	LED	RESISTOR (1 KΩ)	LM7805 VOLTAGE REGULATOR	0.33 μF CERAMIC CAPACITOR	0.1 μF CERAMIC CAPACITOR	9 V BATTERY
			Input	Pin 1		+ve pin
Pin 8 (VCC)			Output		Pin 1	
Pin 3 (Digital Pin 4)	+ve pin (long pin)					
	−ve pin (short pin)	Pin 1				
Pin 4 (GND)		Pin 2	GND	Pin 2	Pin 2	−ve pin

26.7 COMMON ERRORS

As the size of microcontroller decreases, handling it becomes more and more difficult. Even in the case of ATtiny85 microcontroller, you might face errors during its initial setup or code upload. Some of those errors and their solutions are explained below.

The most common error you might see while uploading the code is *Yikes! Invalid device signature.* It is given in Figure 26.12. The possible reasons for the error are:

- The ATtiny85 you are using might be damaged or might not be working. Change the ATtiny85 with a working one and try again.
- ATtiny85 might not be connected properly to Arduino Uno. Double-check or triple-check the connections.
- You might have selected the wrong board or wrong clock speed. Change them accordingly.

Another error you might encounter is *avrdude: stk500_recv() programmer not responding* as shown in Figure 26.13 or the Arduino IDE will be stuck at *Uploading...* (full level bar) for a long time (>1 minute) as shown in Figure 26.14. The possible reasons for the error are as follows:

```
An error occurred while uploading the sketch
Sketch uses 280 bytes (3%) of program storage space. Maximum is 8192 bytes.
Global variables use 9 bytes (1%) of dynamic memory, leaving 503 bytes for local variables. Maximum is 512 bytes.
An error occurred while uploading the sketch
avrdude: Yikes!  Invalid device signature.
        Double check connections and try again, or use -F to override
        this check.
```

FIGURE 26.12 Error while uploading the sketch

- This error occurs mainly when you don't upload the ArduinoISP code to Arduino Uno before uploading the code to ATtiny85. So, upload the ArduinoISP code to Arduino Uno and try again.
- ATtiny85 might not be connected properly to Arduino Uno. Double-check or triple-check the connections.
- The ATtiny85 you are using might be damaged or might not be working. Change the ATtiny85 with a working one and try again.
- You might have selected the wrong board or wrong clock speed. Change them accordingly.

```
Problem uploading to board.  See http://www.arduino.cc/en/Guide/Troubleshooting#upload for suggestions
Sketch uses 280 bytes (3%) of program storage space. Maximum is 8192 bytes.
Global variables use 9 bytes (1%) of dynamic memory, leaving 503 bytes for local variables. Maximum is 512 bytes.
avrdude: stk500_recv(): programmer is not responding
avrdude: stk500_getsync() attempt 1 of 10: not in sync: resp=0x03
avrdude: stk500_recv(): programmer is not responding
avrdude: stk500_getsync() attempt 2 of 10: not in sync: resp=0x03
avrdude: stk500_recv(): programmer is not responding
avrdude: stk500_getsync() attempt 3 of 10: not in sync: resp=0x03
avrdude: stk500_recv(): programmer is not responding
avrdude: stk500_getsync() attempt 4 of 10: not in sync: resp=0x03
Problem uploading to board.  See http://www.arduino.cc/en/Guide/Troubleshooting#upload for suggestions.
avrdude: stk500_recv(): programmer is not responding
avrdude: stk500_getsync() attempt 5 of 10: not in sync: resp=0x03
avrdude: stk500_recv(): programmer is not responding
avrdude: stk500_getsync() attempt 6 of 10: not in sync: resp=0x03
avrdude: stk500_recv(): programmer is not responding
avrdude: stk500_getsync() attempt 7 of 10: not in sync: resp=0x03
avrdude: stk500_recv(): programmer is not responding
avrdude: stk500_getsync() attempt 8 of 10: not in sync: resp=0x03
avrdude: stk500_recv(): programmer is not responding
avrdude: stk500_getsync() attempt 9 of 10: not in sync: resp=0x03
avrdude: stk500_recv(): programmer is not responding
avrdude: stk500_getsync() attempt 10 of 10: not in sync: resp=0x03
```

FIGURE 26.13 Programmer not responding

```
Uploading
Sketch uses 280 bytes (3%) of program storage space. Maximum is 8192 bytes.
Global variables use 9 bytes (1%) of dynamic memory, leaving 503 bytes for local variables. Maximum is 512 bytes.
```

FIGURE 26.14 Arduino IDE stuck at uploading

You can find all the resources of this project at this Github link (https://github.com/anudeep-20/30IoTProjects/tree/main/Project%2026).

In this project, you have learnt about the ATtiny85 microcontroller and programming it using Arduino Uno and Arduino IDE. You have also learnt about the common errors encountered while using the ATtiny85 microcontroller and methods to resolve them. Finally, you have uploaded your first code (LED Blink) to the ATtiny85 microcontroller without any errors. In the next project, you'll build a wrist wearable pedometer from scratch using ATtiny85.

Let's build the next one!

Wearable Pedometer Using ATtiny85

27

A pedometer is a portable electronic device that will count the number of steps walked by a person and provide an estimate of the distance walked. It is generally worn on the person's wrist or waistband. The readings from a pedometer generally serve as a motivational factor for physical activity by providing continuous feedback for self-monitoring one's status visually and attractively. They may aid the individuals to achieve step recommendations for health benefits looking from a general public health perspective. Usage of pedometers to endorse increased physical activity through walking has gained popularity as they tend to be inexpensive and easy to use. Researches support the usefulness of pedometers usage to increase physical activity and decrease body mass index and blood pressure. This project will help you to build a pedometer and track the progress to stay plugged into your health and fitness goals and at the same time inspire engagement in additional activities.

27.1 MOTIVE OF THE PROJECT

In this project, you'll build a wrist-wearable pedometer from scratch using an ATtiny85 microcontroller, MPU6050 and OLED module. It can count the number of steps walked by the person wearing it.

27.2 WORKING OF PEDOMETER

A pedometer counts the number of steps walked by a person by detecting their motion. Getting the pedometer-only devices from the market is a little difficult as the pedometer technology is integrated into most of our daily used devices like smart bands and smartwatches. Even smartphones can detect the number of steps we have walked, but their accuracy is poor compared to the wrist-wearable devices as it uses the internal accelerometer and gyroscope, and it is not worn on the hand.

The wrist-wearable pedometers calculate the number of steps walked by a person by analyzing the movement of their hand. If you observe carefully, when a person walks, their hand moves in an oscillatory motion similar to a simple pendulum as shown in Figure 27.1. This motion is analyzed by the pedometer, and the number of steps walked by the person is calculated.

The accuracy of a pedometer varies widely from device to device. It depends on a lot of factors like the hardware used in the pedometer, the algorithm used to calculate the number of steps walked, the placement or orientation of the device, etc. The hardware of a pedometer mainly consists of an accelerometer, gyroscope sensor and a microcontroller used for processing the signals from these two sensors. The software is written mainly based on an algorithm developed to calculate the number of steps walked. The placement and orientation of the device are very important for accurate calculation of steps walked. It has to be placed as advised by the manufacturer or else it might not yield the desired results. Even the most accurate pedometers developed by the topmost companies will have an error of ±5%. The pedometer

DOI: 10.1201/9781003147169-27

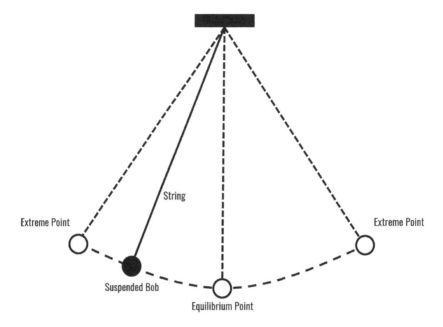

FIGURE 27.1   Simple pendulum motion

developed in this project will not give the utmost accuracy but gives a decent accuracy sufficient to track our daily activity. We will also suggest ways to improve the algorithm for improving accuracy.

27.3 WORKING OF 3-AXIS ACCELEROMETER

An accelerometer is an electronic device used to measure the acceleration of an object in one or more axes. Acceleration is nothing but the rate of change in velocity. Accelerometers have a wide range of applications. They are used in defence applications, aircraft, smartphones, UAVs and many more.

For understanding the working of accelerometers, you need to understand the physics behind the acceleration measurement using a spring-mass system. Let us consider a load of mass, m attached to a wall by a spring of spring coefficient, K. If we apply a force, F, to the load, then it'll accelerate with an acceleration of a and displaces a distance of d in the direction of force as shown in Figure 27.2.

The force acted on the load by the spring, F_s, can be written as follows:

$$F_s = K * d \tag{27.1}$$

According to Newton's second law of motion, the rate of change of momentum of an object is directly proportional to the force applied and occurs in the same direction as the applied force. For objects with constant mass, the law can be restated as follows:

$$F = m * a \tag{27.2}$$

where
   m is the mass of the object;
   F is the force applied to the object;
   a is the acceleration in the direction of force.

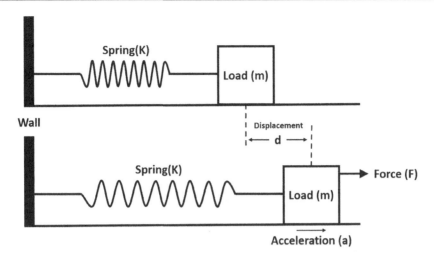

FIGURE 27.2   Spring-mass system

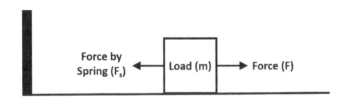

FIGURE 27.3   Force acting on the object

In our case, the object with constant mass is the load attached to the spring.

According to Newton's third law, forces always occur in pairs and they are equal in magnitude but opposite in direction. So, the force of spring, F_s, is equal in magnitude to the force, F, applied to the load as shown in Figure 27.3.

$$F_s = F \tag{27.3}$$

Upon substitution we get,

$$K * d = m * a \tag{27.4}$$

$$a = d * \frac{K}{m} \tag{27.5}$$

In the above equation, the mass of the load, m, and the spring coefficient, K, are both constants. So, the acceleration, a, is directly proportional to the displacement, d. To find the acceleration of the load, we need to find the displacement. It can be found using either resistive method, inductive method or capacitive method. We'll look into the commonly used one, capacitive method.

Capacitance, C, between two plates of Area, A, and separated by distance, d, can be defined as

$$C = \varepsilon * \frac{A}{d} \tag{27.6}$$

where ε is the permittivity of the dielectric between the two plates.

The distance between the plates is inversely proportional to its capacitance. So, the capacitance decreases as the distance between the plates increases and vice-versa. Electronically, calculation of capacitance is easy. To calculate the change in capacitance due to acceleration of an object, we will make use of a 3D capacitor cube with a load of mass, m, attached to the wall by a spring of spring coefficient, K, on all three sides as shown in Figure 27.4. All the walls are capacitive, and the distance between them is d_1 when no acceleration is acting upon it. Even though the image shows the spring system only on a 2D plane, in reality, it is a 3D cube with springs attached to the load both into the page ($\otimes$) as well as from the page ($\odot$) directions.

If the 3D capacitor cube is attached to an object which is accelerating at an acceleration of a in X-direction, then the load will compress in the same direction as shown in Figure 27.5 resulting in a change

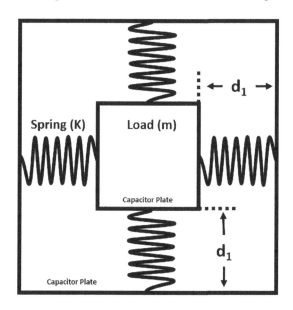

FIGURE 27.4   3D capacitor cube

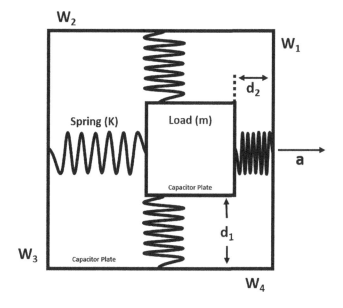

FIGURE 27.5   Accelerating 3D capacitor cube

in capacitance. The capacitance between the wall W_1 and the load increases as the distance between them decreases. Similarly, the capacitance between the wall W_3 and load decreases as the distance between them increases.

$$C_{W1L} = \varepsilon * \frac{A}{d_2} \tag{27.7}$$

$$C_{W3L} = \varepsilon * \frac{A}{2 * d_1 - d_2} \tag{27.8}$$

where
   C_{W1L} is the capacitance between the wall W_1 and load;
   C_{W3L} is the capacitance between the wall W_3 and load;
   d_2 is the new distance between the wall W_1 and load.

By re-arranging the equations, the new distance between the wall W_1 and the load can be written as

$$d_2 = \varepsilon * \frac{A}{C_{W1L}} \tag{27.9}$$

So, the final displacement of load, Δd, can be written as

$$d_1 - d_2 = d_1 - \varepsilon * \frac{A}{C_{W1L}} \tag{27.10}$$

$$\Delta d = d_1 - \varepsilon * \frac{A}{C_{W1L}} \tag{27.11}$$

By substituting Equation 27.11 in Equation 27.5, we get

$$a = \Delta d * \frac{K}{m} \tag{27.12}$$

$$a = \left(d_1 - \varepsilon * \frac{A}{C_{W1L}} \right) * \frac{K}{m} \tag{27.13}$$

By substituting all the known values into Equation 27.13, we can get the acceleration of the object. But attaching such a huge 3D capacitor box to an object for measuring its acceleration is not practical as it increases the overall size and weight of the object. For measuring the acceleration using the same concept but with a smaller size, MEMS (Micro Electro Mechanical Systems) is used. This system contains both electronic and mechanical components fabricated at a scale of micrometre (μm or 10^{-6}m). So, they can be easily placed inside an accelerometer IC.

MEMS accelerometer consists of an external fixed assembly and an internal movable assembly attached to the fixed assembly by springs as shown in Figure 27.6. The movable assembly has plates attached to it which form a capacitor with the fixed assembly. If the MEMS accelerometer is attached to an accelerating object, then the plates on the movable assembly displace in the direction of acceleration as shown in Figure 27.7. By measuring the new capacitance, we can find the displacement of the movable assembly upon which we can calculate the acceleration of the object. By placing the same system on all three axes, we can find the acceleration of the object on all three axes.

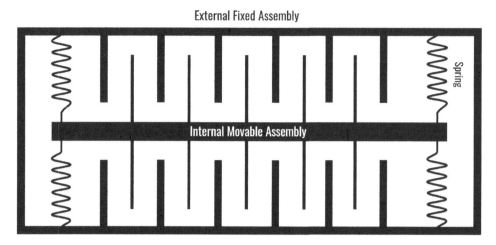

FIGURE 27.6 MEMS accelerometer

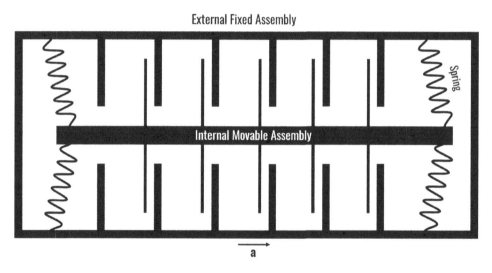

FIGURE 27.7 Accelerating MEMS accelerometer

27.4 WORKING OF GYROSCOPE

A gyroscope is a device used to measure or maintain the orientation and angular velocity of an object. Similar to accelerometers, gyroscopes also have a wide range of applications. They are used in smartphones, gyrocompasses, inertial navigation systems, stability assistance in autonomous vehicles and many more.

For understanding the internal working of a gyroscope, you need to know the Coriolis effect that helps us in measuring the rotation or angular velocity of an object. Consider a body of mass, m, moving in X-direction with a linear velocity, v. If an angular velocity, ω, acts on the body in Z-direction, then the body experiences a Coriolis force, F_c, in the Y-direction as shown in Figure 27.8.

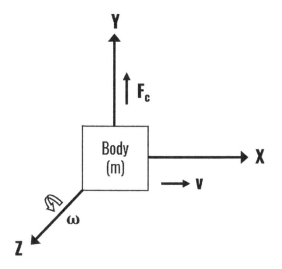

FIGURE 27.8 Coriolis force

Coriolis force is a vector quantity whose equation can be written as

$$\vec{F_c} = -2m * \vec{\omega} \times \vec{v} \qquad\qquad (27.14)$$

The direction of Coriolis force can be easily found using the right-hand rule as shown in Figure 27.9. It is seen perpendicular to both the linear velocities in X-direction and angular velocity in Z-direction.

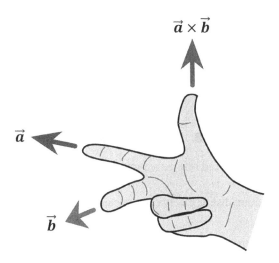

FIGURE 27.9 Right-hand rule

    To determine the angular velocity using the Coriolis effect, we'll use a two-mass system attached to capacitor plates by springs that continuously move in the opposite direction with the same linear velocity as shown in Figures 27.10 and 27.11. If an angular velocity, ω, is applied in the ⊙ Z-direction, then the Coriolis force will act in opposite direction on the two masses as shown in Figure 27.12. This is because the direction of linear velocity of both masses is different. Due to Coriolis force, the masses will move either in the downward or upward direction depending on their linear velocity direction. This causes the capacitance between the plates to increase or decrease respectively.

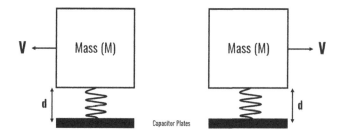

FIGURE 27.10   Two-mass system moving away from each other

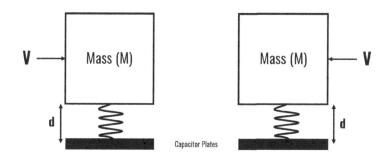

FIGURE 27.11   Two-mass system moving towards each other

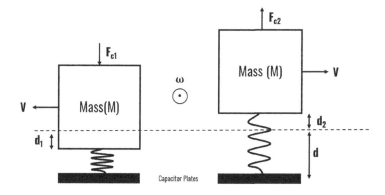

FIGURE 27.12   Coriolis force experienced by the two-mass system

By substituting the separation distance between the capacitor plates in Equation 27.6, we get

$$C_1 = \varepsilon * \frac{A}{d - d_1} \tag{27.15}$$

$$C_2 = \varepsilon * \frac{A}{d + d_2} \tag{27.16}$$

where
  C_1 is the capacitance between the first mass and the capacitor plate beneath it;
  C_2 is the capacitance between the second mass and the capacitor plate beneath it;
  d is the distance between the mass and the capacitor plate beneath it when there is no Coriolis force;
  d_1 is the displacement of the first mass;
  d_2 is the displacement of the second mass.

As the displacement of two masses is in opposite direction, the resultant displacement Δd will be

$$\Delta d = d_1 + d_2 \tag{27.17}$$

$$\Delta d = \varepsilon * A * \left(\frac{1}{C_2} - \frac{1}{C_1} \right) \tag{27.18}$$

By substituting Δd in Equation 27.11, we'll get the acceleration of the two-mass system.

$$a = \varepsilon * A * \left(\frac{1}{C_2} - \frac{1}{C_1} \right) * \frac{K}{2m} \tag{27.19}$$

where K is the spring coefficient;
m is the mass of a single body whereas $2m$ is the combined mass of the system.

Substitute the acceleration, a, in Equation 27.2 and equate it to Coriolis force.

$$2m * \varepsilon * A * \left(\frac{1}{C_2} - \frac{1}{C_1} \right) * \frac{K}{2m} = -2m * \vec{\omega} \times \vec{v} \tag{27.20}$$

$$\vec{\omega} \times \vec{v} = \frac{K}{2m} * \varepsilon * A * \left(\frac{1}{C_1} - \frac{1}{C_2} \right) \tag{27.21}$$

By substituting all the known quantities in Equation 27.21, the angular velocity, $\vec{\omega}$, applied on the two-mass system can be obtained. We can calculate the angular velocity of a rotating object by attaching the two-mass system to it. But it might increase the object's overall size and weight. To calculate the angular velocity of an object using the same concept but with a smaller size, a MEMS gyroscope is used.

MEMS gyroscope consists of four proof masses (M_1, M_2, M_3, and M_4) as shown in Figure 27.13 that continuously oscillate inward and outward simultaneously in a horizontal plane so that they react to the

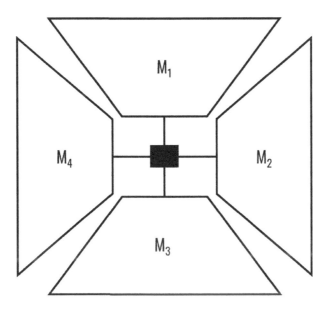

FIGURE 27.13   MEMS gyroscope

Coriolis effect. When the MEMS gyroscope is subjected to angular velocity, the Coriolis force acting on the proof masses will change its direction of oscillation from horizontal to vertical.

Depending on the axis of angular velocity, there are three modes in the MEMS gyroscope namely roll mode, pitch mode and yaw mode. If the angular velocity is applied along the X-axis, then the proof masses M_1 and M_3 will move up and down out of the plane due to the Coriolis effect which causes the roll angle to change. If the angular velocity is applied along the Y-axis, then the proof masses M_2 and M_4 will move up and down out of the plane due to the Coriolis effect which causes the pitch angle to change. If the angular velocity is applied along the Z-axis, then all the proof masses will move in the same horizontal plane but in the opposite direction causing the yaw angle to change.

The continuous movement of proof masses will cause a capacitance change which will be used to calculate the angular velocity of the object on which the MEMS gyroscope is placed.

27.5 MPU6050 SENSOR

MPU6050 sensor is a combination of MEMS 3-Axis Accelerometer, MEMS gyroscope and a Digital Motion Processor in a single IC. It has 6 DOF (Degrees of Freedom), three for accelerometer and three for gyroscope. DOF is nothing but the number of output values that are free to vary. MPU6050 IC works on 3.3 V, but the MPU6050 breakout board as shown in Figure 27.14 can have 5 V input. It converts the 5 V input to 3.3 V as an input to the IC using a voltage regulator. When the MPU6050 is in sleep state or idle, it consumes only 5 μA of current. But it consumes around 3.6 mA of current while measuring the linear acceleration or angular velocity.

MPU6050 also has an embedded temperature sensor that can measure the temperature from −40°C to 85°C with an accuracy of ±1°C. It is mainly used as an offset for calibration of accelerometer or gyroscope due to the temperature changes but not to measure the environmental temperature.

FIGURE 27.14   MPU6050 breakout board

MPU6050 can measure the linear acceleration in four programmable full-scale ranges of ±2g, ±4g, ±8g and ±16g using the accelerometer. g is the acceleration due to gravity whose value is 9.80665 m/s². MPU6050 can also measure the angular velocity in four programmable full-scale ranges of ±250°/s, ±500°/s, ±1000°/s and ±2000°/s using the gyroscope. Remember that "smaller the range, the more sensitive" will be the output values. Depending on the project, you need to adjust the range of both the accelerometer and gyroscope. By default, the range of accelerometer is ±2g and the range of gyroscope is ±250°/s.

To avoid I²C address conflicts with other I²C devices on the same bus, MPU6050 provides two different I²C addresses. ADO pin is used to choose the I²C address according to our use case. If the ADO pin is pulled HIGH by connecting it to 3.3 V, then the MPU6050 uses a 0x69 I²C address. If the ADO pin is left unconnected, then the MPU6050 uses its default I²C address, 0x68. It is because the ADO pin is internally connected to a 4.7 KΩ pull-down resistor. Sometimes when the ADO pin is left unconnected, the MPU6050 might not transfer the data as expected and returns 0. So, to avoid this we recommend connecting the ADO pin to GND for using 0x68 I²C address.

The MPU6050 sensor has many more features which can be used according to your use case. If you want to know about its complete features and specifications, refer to its datasheet (https://invensense.tdk.com/wp-content/uploads/2015/02/MPU-6000-Datasheet1.pdf). You can also know about its complete register map at this link (https://invensense.tdk.com/wp-content/uploads/2015/02/MPU-6000-Register-Map1.pdf).

27.6 ALGORITHM

As explained before, when a person is walking, they consciously or unconsciously move their hand in an oscillatory motion. The angular movement due to oscillation of hand can be detected using a gyroscope. So, we'll only use the output values from gyroscope in MPU6050. To visualize the changes in gyroscope values while a person is walking, we'll use a breadboard that can transmit the gyroscope values. This breadboard needs to be worn by a walking person using a rubber band or tape.

Initially, upload the following code to the ATtiny85 using Arduino Uno as explained in the previous project. The code will be explained later in the project. After uploading the code, connect the MPU6050 and HC-05 Bluetooth module as shown in Figure 27.15. For easier understanding, the connection details are tabulated in Table 27.1.

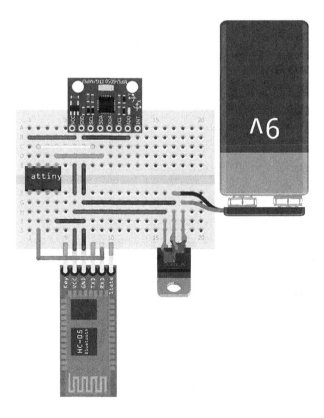

FIGURE 27.15   Connection layout for gyroscope values

TABLE 27.1 Connection details for gyroscope values

ATTINY85 MICROCONTROLLER	MPU6050 SENSOR	HC-05 BLUETOOTH MODULE	LM7805 VOLTAGE REGULATOR	0.33 µF CERAMIC CAPACITOR	0.1 µF CERAMIC CAPACITOR	9 V BATTERY
			Input	Pin 1		+ve pin
Pin 8 (VCC)	VCC	VCC	Output		Pin 1	
Pin 5	SDA					
Pin 7	SCL					
Pin 2		RXD				
Pin 4 (GND)	GND, ADO	GND	GND	Pin 2	Pin 2	−ve pin

```
#include <TinyWireM.h>
#include <SoftwareSerial.h>
SoftwareSerial BTserial(1 ,3);

#define I2CAddr 0x68
#define FSrange 0x08
#define SSF 65.5

float gyroX, gyroY, gyroZ;

void setup() {
  BTserial.begin(9600);
  TinyWireM.begin();
  TinyWireM.beginTransmission(I2CAddr);
  TinyWireM.write(0x6B);
  TinyWireM.write(0x00);
  TinyWireM.endTransmission(true);
  TinyWireM.beginTransmission(I2CAddr);
  TinyWireM.write(0x1B);
  TinyWireM.write(FSrange);
  TinyWireM.endTransmission(true);
}

void loop() {
  TinyWireM.beginTransmission(I2CAddr);
  TinyWireM.write(0x43);
  TinyWireM.endTransmission(false);
  TinyWireM.requestFrom(I2CAddr,6);
  gyroX = (TinyWireM.read()<<8 | TinyWireM.read()) / SSF;
  gyroY = (TinyWireM.read()<<8 | TinyWireM.read()) / SSF;
  gyroZ = (TinyWireM.read()<<8 | TinyWireM.read()) / SSF;

  BTserial.print(gyroX);
  BTserial.print(",");
  BTserial.print(gyroY);
  BTserial.print(",");
  BTserial.println(gyroZ);
  delay(100);
}
```

Connect the HC-05 Bluetooth module to your PC as explained in Section 12.1.4 and open the serial plotter in Arduino IDE. Make sure the baud rate of the serial plotter is set to 9600. Also, don't forget to select the appropriate COM port assigned by your PC to the HC-05 Bluetooth module before opening the serial plotter. Now, wear or attach the breadboard to one of your wrists and walk for 30 seconds and then run for 30 seconds. Don't move too far away from your PC or else the Bluetooth connection might disconnect.

If you have performed everything as explained, then you'll see a graph as shown in Figure 27.16. If you observe the graph carefully, the gyroscope values in X-direction and Y-direction are not dependent on the person's hand movement, but the Z-direction gyroscope values change according to their hand movement while walking. The maximum and minimum values might change from person to person as each person's walking style is different from another. But, on average, the gyroscope values in Z-direction will be ±180°/s if the person is walking, and it will be ±300°/s if the person is running. A person is considered to walk or run a step when the gyroscope's Z-direction value crosses the maximum and minimum once. So, we need to write the code such that it counts a step when the gyroscope Z-direction value crosses the minimum and maximum consecutively.

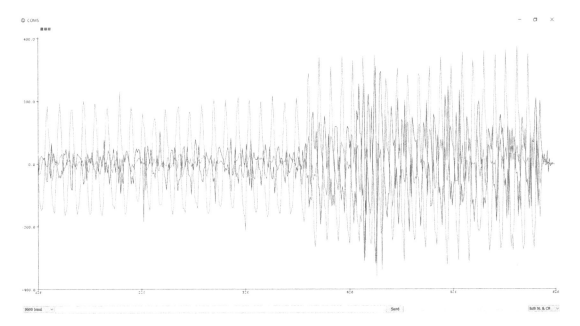

FIGURE 27.16   Change in gyroscope values while a person is walking and running

27.7 ARDUINO IDE CODE

As discussed earlier, ATtiny85 has very less program memory space compared to Arduino Uno or NodeMCU. So, it cannot store and run large programs. The `Adafruit _ SSD1306.h` library and `Wire.h` library consume a lot of program memory and cannot be used with ATtiny85. So, similar libraries with reduced functionality that consume less program space are used for ATtiny microcontrollers. `TinyOzOLED.h` is used for displaying content on OLED. You can directly download it from Github (https://github.com/SensorsIot/TinyOzOled). `TinyWireM.h` is used for I²C functionalities in ATtiny microcontrollers. For installing it, open the library manager in Arduino IDE and search for *TinyWireM*. Install the latest version of *TinyWireM by Adafruit* among the listed libraries. You can also directly download the library from Github at this link (https://github.com/adafruit/TinyWireM).

For counting a step while the person is walking or running, the Z-direction gyroscope value must cross −180° and +180° consecutively. This reason for it is explained in the previous section. THRESMIN and THRESMAX store the threshold for minimum and maximum Z-direction gyroscope values for counting a step respectively. You can also change these values according to your conditions. I2CAddr stores the I²C address of MPU6050 sensor. If the ADO pin is connected to 3.3 V, then the I²C address is 0x69. If the ADO pin is left unconnected or connected to GND, then the I²C address is 0x68. In our case, we have chosen the I²C address as 0x68 by connecting the ADO pin to GND.

The full-scale range of gyroscope in MPU6050 must be chosen according to the use case. Choosing a higher full-scale range will compromise the accuracy. We have observed that the maximum and minimum rate of change of angle while a person is running is ±400°/s. Among the available full-scale ranges for gyroscope, we have chosen ±500°/s. The values obtained from the gyroscope are not the true angular velocity values. They must be divided by a sensitivity scale factor for conversion. The value of the sensitivity scale factor can be obtained from the datasheet of MPU6050.

To set the full-scale range of gyroscope, a register in MPU6050 must be modified. For example, if the value of the register is set to 0x00, then the full-scale range will be ±250°/s. The possible MPU6050 register values and their respective sensitivity scale factor are tabulated in Table 27.2.

TABLE 27.2 Details for configuring full-scale range in gyroscope

FS_SEL	GYROSCOPE VALUES	MPU6050 REGISTER VALUE	SENSITIVITY SCALE FACTOR
0	±250°/s	0x00	131.0
1	±500°/s	0x08	65.5
2	±1000°/s	0x10	32.8
3	±2000°/s	0x18	16.4

```
#include <TinyOzOLED.h> //OLED library for ATtiny
#include <TinyWireM.h> //I2C library for ATtiny

#define THRESMIN -180 //Z-direction gyroscope minimum for step count
#define THRESMAX 180 //Z-direction gyroscope maximum for step count
#define I2CAddr 0x68 //I2C address of MPU6050
#define FSrange 0x08 //Full scale range of gyroscope
#define SSF 65.5 //Sensitivity Scale factor
```

gyroX, gyroY and gyroZ store all the three gyroscope values from the MPU6050 sensor. minZ and maxZ store the current minimum and current maximum Z-direction gyroscope values, respectively. pos and neg are two flags used to check whether the Z-direction gyroscope value has crossed the THRESMAX and THRESMIN, respectively, or not. steps keeps track of the number of steps walked by the person. charArrSteps is used to convert the steps into a character array for displaying it on the OLED.

```
//Variables used for calculating steps
float gyroX, gyroY, gyroZ;
float minZ, maxZ;
bool pos = false, neg = false;
int steps = 0;
char charArrSteps[40];
```

OzOLED.init() is used to initialize the I²C communication between the ATtiny and OLED. OzOled.clearDisplay() clears all the data on the OLED screen. OzOled.setNormalDisplay() sets the display to normal mode. OzOled.sendCommand(0xA1) sets the orientation of OLED screen to horizontal. OzOled.sendCommand(0xC8) sets the orientation of OLED screen to vertical. For printing text on the OLED screen using OzOLED library, it must be converted to a character array. The OzOled.printString() is used for printing a string on the OLED screen. It accepts four arguments: the first argument is the character array to be printed, the second and third arguments are the position at which the character array must be printed on the OLED screen and the fourth argument is the length of the character array.

```
void setup() {
  //Initial OLED setup
  OzOled.init();
  OzOled.clearDisplay();
  OzOled.setNormalDisplay();
  OzOled.sendCommand(0xA1);
  OzOled.sendCommand(0xC8);

  ("Steps: " + String(steps)).toCharArray(charArrSteps, 40);
  OzOled.printString(charArrSteps, 1, 2, sizeof(charArrSteps));
```

TinyWireM.begin() is used to initialize the I²C communication bus. TinyWireM.beginTransmission(I2CAddr) is used to initialize the transmission between the ATtiny and MPU6050. TinyWireM.write(0x6B) accesses the power management register of MPU6050. TinyWireM.write(0x00) sets the sleep register to 0 for preventing the MPU6050 going into sleep mode. TinyWireM.endTransmission() ends the data transmission to a slave device on an I²C bus. It accepts a single argument. If it is true, then the master releases the I²C bus so that other slave devices on the same I²C bus can transmit data to the master. If it is false, then the master doesn't release the I²C bus. By default, it is true. TinyWireM.write(0x1B) is used to access the register where the full-scale range of gyroscope can be modified. TinyWireM.write(FSrange) sets the full-scale range as stored in the FSrange preprocessor.

```
  //Initial I2C and MPU6050 setup
  TinyWireM.begin();
  TinyWireM.beginTransmission(I2CAddr);
  TinyWireM.write(0x6B);
  TinyWireM.write(0x00);
  TinyWireM.endTransmission(true);
  TinyWireM.beginTransmission(I2CAddr);
  TinyWireM.write(0x1B);
  TinyWireM.write(FSrange);
  TinyWireM.endTransmission(true);
}
```

TinyWireM.write(0x43) is used to access the register where the gyroscope values are stored. TinyWireM.requestFrom(I2CAddr,6) requests 6 bytes of gyroscope data from the MPU6050. Gyroscope values for each direction will be stored in 2 bytes. TinyWireM.read()<<8 | TinyWireM.read() is used to read two bytes of data from the MPU6050 in one step. This value is divided by the sensitivity scale factor stored in the SSF preprocessor.

```
void loop() {
  //Read the gyro values from MPU6050
  TinyWireM.beginTransmission(I2CAddr);
  TinyWireM.write(0x43);
  TinyWireM.endTransmission(false);
  TinyWireM.requestFrom(I2CAddr,6);
  gyroX = (TinyWireM.read()<<8 | TinyWireM.read()) / SSF;
  gyroY = (TinyWireM.read()<<8 | TinyWireM.read()) / SSF;
  gyroZ = (TinyWireM.read()<<8 | TinyWireM.read()) / SSF;
```

If the read Z-direction gyro value is greater than the current maxZ, then the new maxZ is the read Z-direction gyro value. Similarly, if the read Z-direction gyro value is less than the current minZ, then the new minZ is the read Z-direction gyro value. If the current maxZ crosses the THRESMAX or the current minZ crosses the THRESMIN, then the pos flag or neg flag is set to true respectively. If both the flags are true, then the `steps` variable is incremented by 1. The same information is printed on the OLED screen as well. After detection of a step, all the supporting variables `minZ`, `maxZ`, `pos` and `neg` are reset.

```
//Calculate the Minimum Value & Maximum value of the hand movement
if(gyroZ > maxZ){
  maxZ = gyroZ;
}
if(gyroZ < minZ){
  minZ = gyroZ;
}

if(maxZ > THRESMAX){
  pos = true;
}
if(minZ < THRESMIN){
  neg = true;
}
```

```
//After detection of a step, update the parameters and print the number of
steps on OLED
  if (pos && neg){
    steps++;
    ("Steps: " + String(steps)).toCharArray(charArrSteps, 40);
    OzOled.printString(charArrSteps, 1, 2,sizeof(charArrSteps));
    pos = false;
    neg = false;
    minZ = 0;
    maxZ = 0;
  }
}
```

Finally, upload the code to ATtiny85 as explained in the previous project.

27.8 HARDWARE REQUIRED

TABLE 27.3 Hardware required for the project

HARDWARE	QUANTITY
ATtiny85 microcontroller	1
MPU6050 sensor	1
SSD1306 OLED module	1
9 V Battery	1
LM7805 voltage regulator	1
0.33 µF Ceramic capacitor	1
0.1 µF Ceramic capacitor	1
Breadboard	1
Jumper wires	As required

27.9 CONNECTIONS

After successfully uploading the code to ATtiny85, remove its connections with Arduino Uno and connect the components as in Table 27.3 and by referring to Figure 27.17. You can also refer to Table 27.4 which presents the pin-wise connection details. After powering the system, wear it to your wrist using a rubber band or attach it using an adhesive tape and walk for some distance. You can see the number of steps being updated on the OLED screen in real-time.

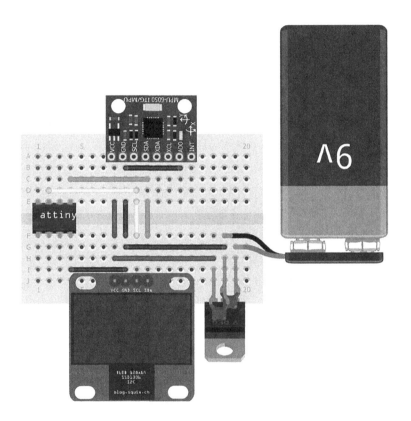

FIGURE 27.17   Connection layout

TABLE 27.4   Connection details

ATTINY85 MICROCONTROLLER	MPU6050 SENSOR	SSD1306 OLED MODULE	LM7805 VOLTAGE REGULATOR	0.33 µF CERAMIC CAPACITOR	0.1 µF CERAMIC CAPACITOR	9 V BATTERY
			Input	Pin 1		+ve pin
Pin 8 (VCC)	VCC	VCC	Output		Pin 1	
Pin 5	SDA	SDA				
Pin 7	SCL	SCL				
Pin 4 (GND)	GND, ADO	GND	GND	Pin 2	Pin 2	−ve pin

You can find all the resources of this project at this Github link (https://github.com/ anudeep-20/30IoTProjects/tree/main/Project%2027).

In this project, you have learnt about the internal working of accelerometer, gyroscope, MEMS technology and MPU6050 sensor. Using this knowledge, you have built a pedometer from scratch that can count the number of steps walked by a person wearing it. In the next project, you'll build a solution for an industrial problem that can predict the danger area around a heavy lift crane using the same sensors you used in the previous projects.

Let's build the next project!

Danger Area Prediction System

28

Cranes are the commonly used machinery for lifting heavy weight objects in many places like industries, ports, construction sites, etc. Due to the weight of these objects, the crane might go into a state of imbalance or the wire holding them might cut off. This may cause the objects being lifted to fall on the ground staff or far away people resulting in serious injuries or death. Many such incidents are being reported irrespective of the measures taken to prevent them. These might be due to the negligence of ground staff or the crane operator or due to overloading the crane or the crane might not be maintained properly. Whatever be the reason, reducing the number of injuries and fatalities due to the sudden or unnoticed fall of heavy objects is an industrial problem that needs proper attention. This project will help you to build a solution for the same using IoT.

28.1 MOTIVE OF THE PROJECT

In this project, you'll build a prototype that can predict and indicate the danger zone inside which the hanging weight from the crane might fall. This information will help the workers to avoid using the danger area until the heavy weight is safely placed, thereby avoiding any possible injuries or damages.

28.2 ALGORITHM

The reason for the imbalance of crane or increase in the tension of lifting wire is due to the wobble or oscillation of heavy weights lifted by the crane. If the heavy weight is separated from the crane during oscillation, then its projectile depends on its position of the oscillation. The velocity of heavy weight will be minimum when it is at either of the extreme position, and it will be the maximum when it is at the equilibrium position.

The free-body diagram of the heavy weight separated at the extreme position is shown in Figure 28.1. As it has no velocity ($u = 0$), the only force acting on it is the gravitational force (F_g) in the downward direction. So, the heavy weight will vertically fall down if it got cut at either of the extreme positions.

The free-body diagram of the heavy weight separated at the equilibrium position is shown in Figure 28.2. As its initial velocity (u) is non-zero, it travels in a projectile as shown in Figure 28.3. H is the height at which the heavy weight is separated, and ΔX is the horizontal displacement of the heavy weight.

From the projectile, we can interpret the initial velocity and acceleration of the heavy weight in X and Y directions. The coordinate space is defined such that the right side and upward direction are taken as positive.

$$u_X = u \tag{28.1}$$

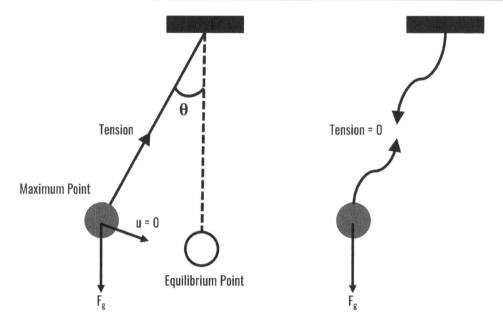

FIGURE 28.1   Free-body diagram of heavy weight at its extreme position

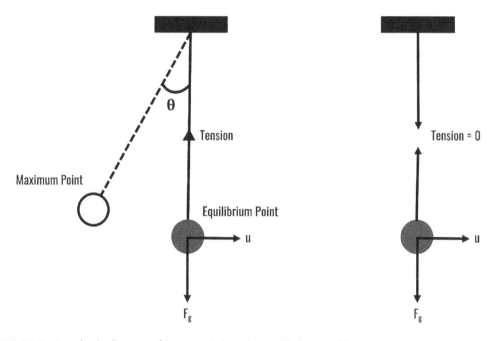

FIGURE 28.2   Free-body diagram of heavy weight at its equilibrium position

$$u_Y = 0 \tag{28.2}$$

$$a_X = 0 \tag{28.3}$$

$$a_Y = -g \tag{28.4}$$

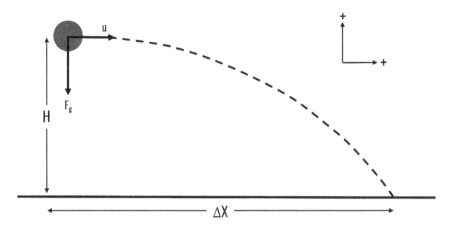

FIGURE 28.3   Projectile of heavy weight

where
  u_X and u_Y are the initial velocity in X-direction and Y-direction, respectively;
  a_X and a_Y are the acceleration in X-direction and Y-direction, respectively;
  g is the acceleration due to gravity.

Using these quantities and kinematic equations, we can find the horizontal displacement of the projectile (ΔX). Equations 28.5, 28.6 and 28.7 are the kinematic equations in X-direction, while Equations 28.8, 28.9 and 28.10 are the kinematic equations in Y-direction.

$$v_X = u_X + a_X * t \tag{28.5}$$

$$\Delta X = u_X * t + \frac{1}{2} * a_X * t^2 \tag{28.6}$$

$$v_X^2 = u_X^2 + 2a_X * \Delta X \tag{28.7}$$

$$v_Y = u_Y + a_Y * t \tag{28.8}$$

$$\Delta Y = u_Y * t + \frac{1}{2} * a_Y * t^2 \tag{28.9}$$

$$v_Y^2 = u_Y^2 + 2a_Y * \Delta Y \tag{28.10}$$

where
  v_X and v_Y are the final velocity in X-direction and Y-direction, respectively;
  ΔY is the vertical displacement of the projectile;
  t is the total time of the projectile.

For finding the total time of projectile, substitute Equation 28.2 and Equation 28.4 in Equation 28.9.

$$0 - H = 0 * t + \frac{1}{2} * (-g) * t^2 \tag{28.11}$$

$$t = \sqrt{\frac{2H}{g}} \tag{28.12}$$

By substituting Equation 28.1 and Equation 28.12 in Equation 28.6, we get the horizontal displacement.

$$\Delta X = u * \left(\sqrt{\frac{2H}{g}} \right) + \frac{1}{2} * 0 * \left(\sqrt{\frac{2H}{g}} \right)^2 \tag{28.13}$$

$$\Delta X = u * \left(\sqrt{\frac{2H}{g}} \right) \tag{28.14}$$

From Equation 28.14, you can observe that the heavy weight's horizontal displacement depends on the height at which it is separated (H) and its initial horizontal velocity (u). The hanging height of the heavy weight can be found using an ultrasonic sensor, but finding its initial linear velocity is difficult. Previously used sensors like MPU6050 can only provide linear acceleration and angular velocity.

There are many other factors like air resistance that are neglected during the calculation of horizontal displacement. For smaller objects, the air resistance can be neglected as it interacts with a smaller number of air molecules, but in industries, heavy objects with a large surface area are lifted using cranes due to which the air resistance will be an important factor to be considered. Also, the ultrasonic sensor cannot be placed anywhere on the heavy weight for finding its hanging height. It has to be placed at the heavy weight's centre of gravity for obtaining the correct horizontal displacement.

As we cannot obtain all the parameters required for the calculation of ΔX, we'll use another method for predicting the danger area. Similar to the previous project, we'll observe the oscillation of hanging heavy weight using an MPU6050 sensor. Depending on the intensity of oscillation, the area of danger zone can be classified into multiple levels. As we cannot observe the oscillation of heavy weight on a real crane, we'll build a prototype as shown in Figure 28.4 that mimics the heavy weight lifting of cranes. It consists of two parts: transmitter and receiver. The transmitter should be placed near or on the hanging heavy weight so that the heavy weight oscillation can be recorded, and the receiver should be placed at the topmost position of the crane so that the danger area indication laser can be projected onto the ground.

The communication between the transmitter and receiver can be wired or wireless. Even though any of them can be used, we have chosen wireless communication over wired because it is difficult to have

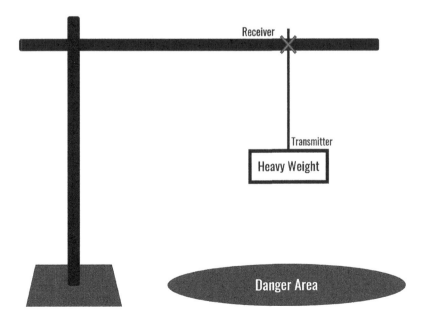

FIGURE 28.4 Prototype of danger area prediction system

wires to the receiver from an oscillating object. For the wireless communication between the transmitter and receiver, we'll use two HC-05 Bluetooth modules configured as Master and Slave. Refer to Section 11.4 or Section 11.5 for Master-Slave configuration of HC-05 Bluetooth modules.

The transmitter circuit must be as small as possible so that it can be easily placed on an oscillating weight. So, we'll use an ATtiny85 microcontroller connected to an MPU6050 sensor and one of the Master-Slave configured HC-05 Bluetooth modules. Upload the same code used in the previous project for obtaining the gyroscope values to ATtiny85 and connect the transmitter circuit as shown in Figure 27.15. The receiver circuit can be of any size as it is placed at a non-oscillating part of the crane. So, we have used an Arduino Uno for convenience and to visualize the transmitted gyroscope data. Connect the receiver circuit as shown in Figure 11.12 and upload the following code to Arduino Uno.

```
String rcvdStr;
int commaIndex, secCommaIndex;
float gyroX, gyroY, gyroZ;

void setup() {
  Serial.begin(9600);
}

void loop() {
  if (Serial.available() > 0) {
    rcvdStr = Serial.readStringUntil('\n');
    commaIndex = rcvdStr.indexOf(',');
    secCommaIndex = rcvdStr.indexOf(',', commaIndex + 1);

    gyroX = rcvdStr.substring(0, commaIndex).toFloat();
    gyroY = rcvdStr.substring(commaIndex + 1, secCommaIndex).toFloat();
    gyroZ = rcvdStr.substring(secCommaIndex + 1).toFloat();

    Serial.print(gyroX);
    Serial.print(",");
    Serial.print(gyroY);
    Serial.print(",");
    Serial.println(gyroZ);
  }
}
```

After both the HC-05 Bluetooth modules get paired, open the serial plotter. If the heavy weight object oscillates, then you'll see a graph similar to the one shown in Figure 28.5. The oscillation of heavy weight will be fierce at the beginning, but it will gradually reduce to settle at its equilibrium position. The oscillation of heavy weight will usually be in the *XY* plane. Unless it rotates along itself while oscillating, you won't see gyroscope readings in Z-direction.

Depending on the sum of magnitude of maximum angular velocity in positive and negative directions of one oscillation cycle, the danger area is divided into four categories: critical, severe, moderate and safe. For ease of explanation, we'll term the categorization parameter as OSCILLATIONSUM. All the oscillations with OSCILLATIONSUM greater than 300°/s are categorized in the critical danger zone. The area of danger zone under this category will be the highest. The oscillations with an OSCILLATIONSUM greater than 150°/s but less than 300°/s are categorized in the severe danger zone. The oscillations with OSCILLATIONSUM greater than 75°/s but less than 150°/s are categorized in the moderate danger zone. The danger area under this category will be the least. All the remaining oscillations with OSCILLATIONSUM less than 75°/s are categorized in the safe zone. There will be no danger area under this category. The OSCILLATIONSUM values used for categorizing the danger zone are not universal, and they can be changed according to your conditions.

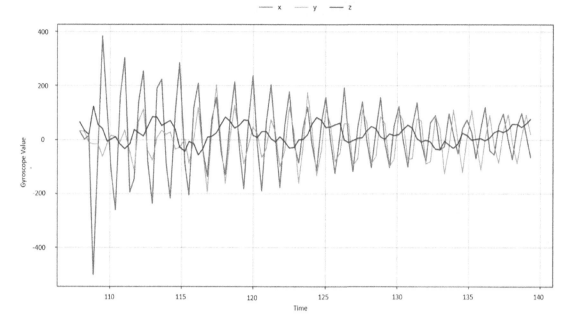

FIGURE 28.5   Gyroscope values from the transmitter

28.3 HARDWARE REQUIRED

TABLE 28.1   Hardware required for transmitter

HARDWARE	QUANTITY
ATtiny85 microcontroller	1
Servo motor	1
HC-05 Bluetooth module	1
9 V Battery	1
LM7805 voltage regulator	1
0.33 µF Ceramic capacitor	1
0.1 µF Ceramic capacitor	1
Breadboard	1
Jumper wires	As required

TABLE 28.2   Hardware required for receiver

HARDWARE	QUANTITY
Arduino Uno	1
Servo motor	1
HC-05 Bluetooth module	1
LEDs	3 (1×Red, 1×Yellow, 1×Green)
Laser module	1
Breadboard	1
Jumper wires	As required

28.4 CONNECTIONS

The hardware (Table 28.1), connections and code of the transmitter are the same, and no need to re-connect or re-upload the code unless you encounter an error. However, the hardware (Table 28.2), connections and code of the receiver need to be changed. Connect the receiver circuit as shown in Figure 28.6, and the pin-wise connections can be referred from Table 28.3.

The arrangement of transmitter and receiver modules on the prototype is shown in Figure 28.7. The transmitter module as shown in Figure 28.8 is placed inside a box and hung to a stand made of iron. Remember to balance the weight inside the box or else it might bend to one side and the oscillations won't be proper. You can add some extra weight equivalent to the 9 V battery weight at the other end. The arrangement of servo motor and laser module in the receiver is shown in Figure 28.9. The laser module must be tied to the rotating part of the servo motor.

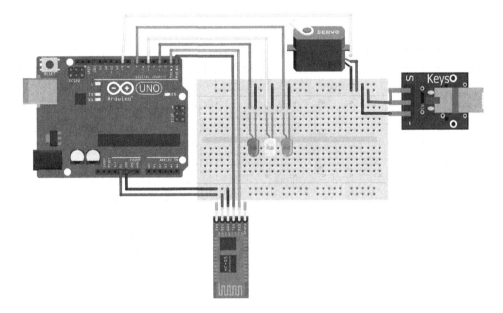

FIGURE 28.6   Connection layout of receiver

TABLE 28.3   Connection details of receiver

ARDUINO UNO	SERVO MOTOR	HC-05 BLUETOOTH MODULE	LASER MODULE	RED LED	YELLOW LED	GREEN LED
5 V	VCC (Red)	VCC	+ve pin			
0 (RXD)		TXD				
1 (TXD)		RXD				
3				+ve pin (long pin)		
5					+ve pin (long pin)	
7						+ve pin (long pin)
9	OUT (Yellow)					
GND	GND (Black)	GND	−ve pin	−ve pin (short pin)	−ve pin (short pin)	−ve pin (short pin)

FIGURE 28.7   Arrangement of transmitter and receiver

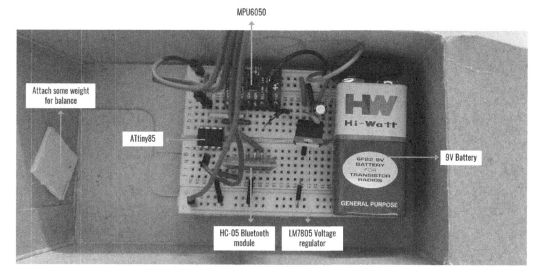

FIGURE 28.8   Connections of the transmitter module

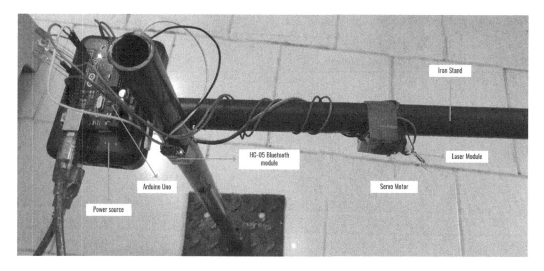

FIGURE 28.9   Connections of the receiver module

28.5 ARDUINO IDE CODE

For easily controlling the servo motor, we'll use Servo.h library. RED, YELLOW, GREEN and SERVO store the Arduino Uno's digital pins to which red LED, yellow LED, green LED and OUT pin of servo motor are connected respectively. CRITICALTHRES, SEVERETHRES and MODERATETHRES store the OSCILLATIONSUM thresholds for critical, severe and moderate categories respectively. ARRAYLEN stores the length of arrays which store the gyroscope values. initPos stores the initial angle of servo motor. Initially, the servo motor should be at an angle such that the laser attached to it must vertically face downward. Rot1, Rot2 and Rot3 store the angle of servo motor to be rotated from its initial position for critical, severe and moderate categories respectively.

myServo is a Servo class variable used for easy access of Servo.h library functions. rcvdStr stores the data received from the transmitter. commaIndex and secCommaIndex are helper variables used for splitting the rcvdStr with comma (,) as a separator. gyroX, gyroY and gyroZ are used to store the gyroscope values received from the transmitter. minValue and maxValue are used to store the minimum and maximum gyroscope values in an oscillation cycle of the heavy weight. oscl stores the OSCILLATIONSUM of the heavy weight which is nothing but the sum of magnitude of minValue and maxValue. i and j are iterators for arrays. isP and isN are two flags to indicate the sign of gyroscope values. xarrP and xarrN store the positive and negative gyroscope values, respectively.

```
#include <Servo.h> //Library for servo motor

//Digital pins to which LEDs & servo motor is connected
#define RED 3
#define YELLOW 5
#define GREEN 7
#define SERVO 9

//Helper constant values
#define CRITICALTHRES 300
#define SEVERETHRES 150
#define MODERATETHRES 75
#define ARRAYLEN 15

#define initPos 180
#define Rot1 40
#define Rot2 25
#define Rot3 15

Servo myServo; //Servo class object

//Variables used in the determination of danger area
String rcvdStr;
int commaIndex, secCommaIndex;
float gyroX, gyroY, gyroZ, minValue, maxValue, oscl;
int i = 0, j = 0;
bool isP = true, isN = true;
float xarrP[ARRAYLEN] = {0.0};
float xarrN[ARRAYLEN] = {0.0};
```

Arduino Uno's serial is initialized at 9600 baud rate. Remember to configure both the Master-Slave HC-05 Bluetooth modules also at 9600 baud rate. As we are writing voltage to the LEDs, its direction of operation is set to OUTPUT. myServo is linked to SERVO and the servo motor's initial angle is set to initPos.

```
void setup() {
  Serial.begin(9600); //Initialization of Arduino's Serial
  //Defining the direction of operation of LEDs
  pinMode(RED, OUTPUT);
  pinMode(YELLOW, OUTPUT);
  pinMode(GREEN, OUTPUT);
  myServo.attach(SERVO); //Linking to the Arduino Uno's PWM pin
  myServo.write(initPos); //Initial position of servo motor
}
```

If data is available on the Arduino Uno's `Serial`, then each line is read at a time. `indexOf` is an inbuilt function of Arduino IDE used to get the index of a particular character in a string. It accepts two arguments; the first one is the character to search and the second one is the index from which the search should begin. If the second argument is not provided then the search will start from the beginning of string. `substring()` is another inbuilt function of Arduino IDE used to get a substring of a string. It takes two arguments; the first one is the start index and the second one is the exclusive end index. It means the index of the string mentioned in the second argument won't be included in the substring. If the second argument is not provided, then the substring continues to the end of the string. `toFloat()` is used to convert a string to float.

```
void loop() {
  if (Serial.available() > 0) {
  //Split the received data from transmitter with comma as separator
    rcvdStr = Serial.readStringUntil('\n');
    commaIndex = rcvdStr.indexOf(',');
    secCommaIndex = rcvdStr.indexOf(',', commaIndex + 1);

    gyroX = rcvdStr.substring(0, commaIndex).toFloat();
    gyroY = rcvdStr.substring(commaIndex + 1, secCommaIndex).toFloat();
    gyroZ = rcvdStr.substring(secCommaIndex + 1).toFloat();
```

Even though both *X*-direction and *Y*-direction gyroscope values vary during the oscillations of heavy object weight, only the *X*-direction gyroscope values are used to predict the danger area. We want you to include the *Y*-direction gyroscope values in the present algorithm for predicting the danger area and observe the difference. If the `gyroX` values are negative, then they are stored in `xarrN` or else they are stored in `xarrP`. If the sign of `gyroX` values changes from negative to positive, then the minimum gyroscope value in the oscillation cycle is found. Similarly, if the sign of `gyroX` values changes from positive to negative, then the maximum gyroscope value in the oscillation cycle is found. The OSCILLATIONSUM is found by adding the magnitude of minimum and maximum gyroscope values in an oscillation cycle.

```
    //Store the positive and negative X-direction gyroscope values in their
respective arrays
    if(gyroX >= 0) {
      xarrP[i] = gyroX;
      i += 1;
    }
    if(gyroX < 0) {
      xarrN[i] = gyroX;
      i += 1;
    }
    //Finding the maximum and minimum gyroscope value in X-direction
    if (gyroX >= 0 and isN) {
      minValue = xarrN[0];
      for (j = 0; j < ARRAYLEN; j++) {
```

```
      if (xarrN[j] < minValue) {
        minValue = xarrN[j];
      }
    }
    Serial.print("minValue: ");
    Serial.println(minValue);
    isP = true;
    isN = false;
    i = 0;
  }

  if (gyroX < 0 and isP) {
    maxValue = xarrP[0];
    for (j = 0; j < ARRAYLEN; j++) {
      Serial.print(xarrP[j]);
      if (xarrP[j] > maxValue) {
        maxValue = xarrP[j];
      }
      Serial.print(',');
    }
    Serial.println("");
    Serial.print("maxValue: ");
    Serial.println(maxValue);
    isP = false;
    isN = true;
    i = 0;
  }

  //Calculate the OSCILLATIONSUM value
  oscl = maxValue + abs(minValue);
```

Depending on the OSCILLATIONSUM value, the servo motor is rotated to indicate the danger zone. The laser module attached to the servo motor can only project a dot onto the ground whose distance from the heavy weight's equilibrium position can be considered as the radius of the danger area. You can also use multiple lasers to indicate the circumference of the danger area. But the power consumption increases as the number of lasers increases. Also, a LED is used to visually indicate the danger area category. Red LED is used to indicate critical category, yellow LED is used to indicate the severe category and green LED is used to indicate the moderate category. For the safe category, no LED is switched ON.

```
  //Rotate the servo motor according to the OSCILLATIONSUM
  //Also, ON the respective category LED
  if (oscl > CRITICALTHRES) {
    myServo.write(initPos - Rot1);
    digitalWrite(RED, HIGH);
    digitalWrite(YELLOW, LOW);
    digitalWrite(GREEN, LOW);
  } else if (oscl > SEVERETHRES) {
    myServo.write(initPos - Rot2);
    digitalWrite(RED, LOW);
    digitalWrite(YELLOW, HIGH);
    digitalWrite(GREEN, LOW);
  } else if (oscl > MODERATETHRES) {
    myServo.write(initPos - Rot3);
    digitalWrite(RED, LOW);
    digitalWrite(YELLOW, LOW);
    digitalWrite(GREEN, HIGH);
```

```
    } else {
      myServo.write(initPos);
      digitalWrite(RED, LOW);
      digitalWrite(YELLOW, LOW);
      digitalWrite(GREEN, LOW);
    }

    //Reset the index of the arrays
    if (i >= ARRAYLEN)
      i = 0;
  }
}
```

Finally, upload the code to receiver Arduino Uno after selecting the appropriate board and COM port. Wait until both the HC-05 Bluetooth modules get paired and oscillate the heavy weight. You can observe that the servo motor will rotate according to the intensity of oscillations, and the laser module will project the laser onto the ground indicating the danger area. Also, a colour LED will glow according to the intensity of oscillations.

You can find all the resources of this project at this Github link (https://github.com/ anudeep-20/30IoTProjects/tree/main/Project%2028).

In this project, you have built a prototype for an industrial problem to reduce the number of fatalities and injuries due to mishaps in lifting heavy weight objects by cranes. Until now you have only learnt about microcontrollers and evaluation boards, but in the next project, you'll learn about Raspberry Pi which is a commonly used mini-computer in IoT.

Let's move to the next project!

Introduction and Setup of Raspberry Pi

29

Raspberry Pi is a credit card-sized computer that can deliver most of the features and capabilities of a normal PC, that too with low power consumption. It is prevalently used in applications for real-time image/video processing, IoT as well as in robotics. Python is the primary programming language of Raspberry Pi. It is used to control the Raspberry Pi's peripherals and its onboard GPIO pins. The launch of Raspberry Pi 4 augurs compute-intensive embedded applications and is three times faster, possessing more memory capacity than its predecessors. This project will help you to better understand the Raspberry Pi using an exercise to control an LED.

29.1 MOTIVE OF THE PROJECT

In this project, you'll learn to setup a new Raspberry Pi and control an LED using it. Unlike the previous projects, Arduino IDE won't be used to control the LED, but python programming language will be used.

29.2 RASPBERRY Pi

Raspberry Pi (often called Rpi) is a computationally powerful mini-computer developed by the Raspberry Pi Foundation along with Broadcom. It has almost all the useful peripherals like USB, Ethernet, WiFi, HDMI, Audio Jack, memory card interface, external display connector, external camera connector and GPIO pins for controlling external hardware. There are many versions of Raspberry Pi ranging from Raspberry Pi Zero to Raspberry Pi 4. The peripherals and specifications of Raspberry Pi differ a lot from model to model. So, purchase a Raspberry Pi according to your use case.

All the projects in this book use Raspberry Pi 4B (4GB RAM), which is shown in Figure 29.1. But, you can also use any Raspberry Pi model equivalent to Raspberry Pi 2 or above. Figure 29.2 details the pin configurations of Raspberry Pi 4B. Some of the specifications of Raspberry Pi 4B are tabulated in Table 29.1. To know more about Raspberry Pi 4B, refer to its datasheet (https://datasheets.raspberrypi.org/rpi4/raspberry-pi-4-datasheet.pdf).

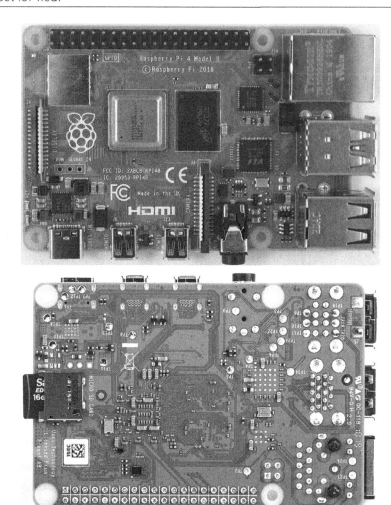

FIGURE 29.1   Raspberry Pi 4B

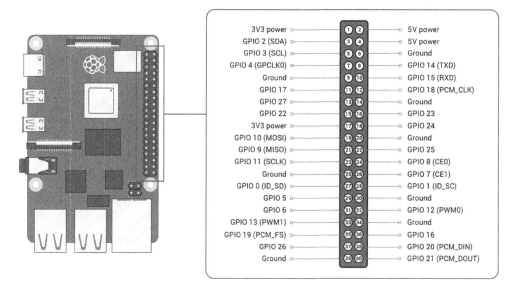

FIGURE 29.2   Raspberry Pi 4B pin configuration

TABLE 29.1   Specifications of Raspberry Pi 4B

SoC	Broadcom BCM2711
CPU	4×Cortex-A72 1.5 GHz
GPU	Broadcom VideoCore VI @ 500 MHz
RAM	1GB, 2GB, 4GB or 8GB
USB ports	4 (4×USB 2.0 ports, 2×USB 3.0 ports)
HDMI	2×Micro-HDMI
Audio input/output	3.5 mm phone jack
Ethernet	10/100/1000 Mbit/s
WiFi	dual-band 2.4/5 GHz
Bluetooth	5.0
Low-level peripherals	17×GPIO, 4×UART, 4×SPI and 4×I2C connectors
Power source	5 V through USB C or GPIO header
Power rating	600 mA (3 W) average when idle, 1.25 A (6.25 W) maximum under stress, recommended 3 A (15 W) power supply

29.3 RASPBERRY Pi SETUP USING DEDICATED DISPLAY

Raspberry Pi is not a plug-and-play device. Similar to a PC, you need to do some basic setup before using it. If you have a dedicated display, keyboard and mouse, then setting up the Raspberry Pi is very easy. Remember, you cannot consider your laptop screen as a dedicated display. Table 29.2 shows all the required hardware for setting up your Raspberry Pi with a dedicated display.

TABLE 29.2   Hardware required for Raspberry Pi setup using a dedicated display

HARDWARE	QUANTITY
Raspberry Pi 4B	1
5 V DC adapter	1
USB A to USB C cable	1
Micro SD card (at least 8GB)	1
SD card reader	1
Monitor*	1
HDMI/VGA to mini-HDMI cable (display connector)*	1
7 inch Raspberry Pi display**	1
FFC (flat flexible cable)**	1
Keyboard	1
Mouse	1

Micro SD card acts as a hard disk for the Raspberry Pi which stores the OS and all our files. To have a smooth experience, a 16GB micro SD card is preferred but you can work with an 8GB micro SD card as well. You can use either a monitor or a 7-inch Raspberry Pi display but not both. If you are using a monitor, then you require an HDMI/VGA to mini-HDMI cable or else an FFC for connecting to the Raspberry Pi.

Initially, insert the micro SD card into your SD card reader and connect it to a PC for formatting it. Next, you need to download an OS and flash it to the formatted micro SD card. You can do all these steps using a single software, Raspberry Pi Imager developed by the Raspberry Pi organization. It can be downloaded from this link (https://www.raspberrypi.org/software/). Choose the appropriate software variant and click download. As we are using a PC with Windows OS, we'll download Raspberry Pi Imager for Windows as shown in Figure 29.3.

Install Raspberry Pi OS using Raspberry Pi Imager

Raspberry Pi Imager is the quick and easy way to install Raspberry Pi OS and other operating systems to a microSD card, ready to use with your Raspberry Pi. Watch our 45-second video to learn how to install an operating system using Raspberry Pi Imager.

Download and install Raspberry Pi Imager to a computer with an SD card reader. Put the SD card you'll use with your Raspberry Pi into the reader and run Raspberry Pi Imager.

Download for Windows

Download for macOS

Download for Ubuntu for x86

To install on **Raspberry Pi OS**, type
`sudo apt install rpi-imager`
in a Terminal window.

FIGURE 29.3   Download of Raspberry Pi Imager

Installing the Raspberry Pi Imager software is very easy and straightforward. After successful installation, run the software and you'll see a screen as shown in Figure 29.4. Insert the micro SD card into the SD card reader and connect it to your PC. To format the micro SD card, click on *CHOOSE OS* and select *Erase* as shown in Figure 29.5. Next, click on *CHOOSE STORAGE* and select your SD card from the list of storage options as shown in Figure 29.6. Finally, click on write to start the format. If the formatting is successful, then you'll see a pop-up window as shown in Figure 29.7.

The next step is to flash OS to the formatted micro SD card. Raspberry Pi supports a lot of operating systems like Raspberry Pi OS, Ubuntu Core, RISC OS Pi, Windows IoT Core and many more. But we'll use the official and most commonly used OS, the Raspberry Pi OS. Click *CHOOSE OS* again and select the Raspberry Pi OS according to your memory card size. If you are using a micro SD card of size 16GB, then select the first option, *Raspberry Pi OS (32-bit)* as shown in Figure 29.8. If you are using a micro SD card of size 8GB or less, then click on the second option, *Raspberry Pi OS (other)* and select *Raspberry Pi OS Lite (32-bit)*. If your micro SD card size is 32GB or more, then select *Raspberry Pi OS Full (32-bit)* as shown in Figure 29.9. As our micro SD card size is 128GB, we'll install the *Raspberry Pi OS Full (32-bit)*. You can also flash other Raspberry Pi–supported OS according to your use case by selecting the other options. If you have an already downloaded .img file of the OS, then you can directly flash it by selecting the last option, *Use custom*. Next, select the storage for flashing the OS and click *Write*.

If the flashing of OS is successful, then you'll see a pop-up window as shown in Figure 29.10. Remove the micro SD card from the SD card reader and insert it into the Raspberry Pi. You can find the SD card slot at the back of Raspberry Pi. Always remember to power OFF the Raspberry Pi before inserting or removing the micro SD card. If not, the Raspberry Pi might get damaged or corrupted. Finally, connect the display, keyboard and mouse to the Raspberry Pi and power it. After the Raspberry Pi gets booted, you can see the home screen of Raspberry Pi OS as shown in Figure 29.11. You can either configure the basic Raspberry Pi settings by clicking *Next* or do it later by clicking *Cancel*. By clicking *Next*, you can set the *Country*, *Language*, *Timezone* and keyboard type of the Raspberry Pi as shown in Figure 29.12. After setting the details, again click *Next* to change the Raspberry Pi's password as shown in Figure 29.13. After changing the password, click *Next*. If you see any black border around the screen, then check the box and click *Next* as shown in Figure 29.14. Connect to a WiFi network among the available networks and click *Next* as shown

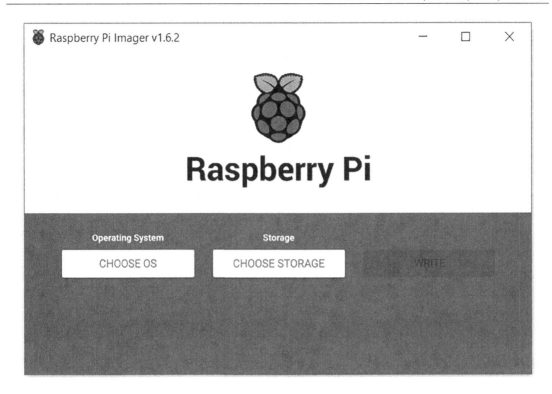

FIGURE 29.4   Raspberry Pi Imager v1.6.2

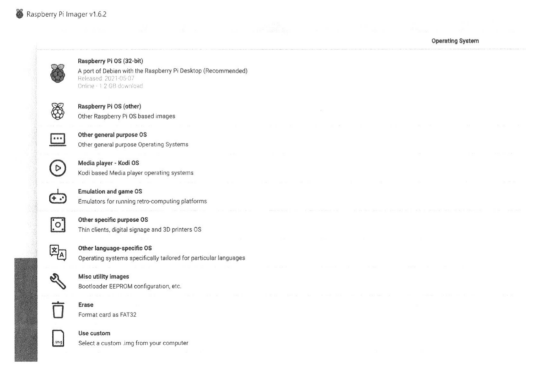

FIGURE 29.5   Selection of erase option

FIGURE 29.6   Selection of storage

FIGURE 29.7   Format successful message

in Figure 29.15. If you have a stable WiFi connection, then update your Raspberry Pi by clicking *Next* as shown in Figure 29.16. It will take some time to download and install the updates depending on your WiFi speed. After the update is completed, you'll see a pop-up window as shown in Figure 29.17. Click *OK*. If the setup is completed without any errors, then you'll see the Setup Complete window as shown in Figure 29.18.

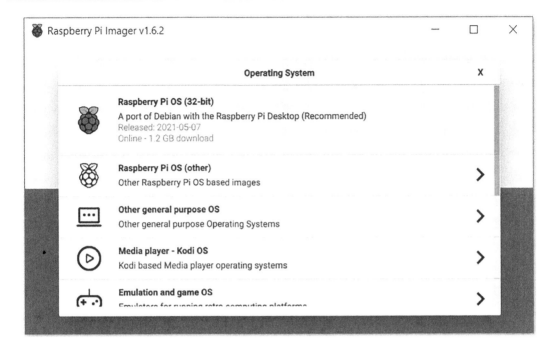

FIGURE 29.8   Selection of OS

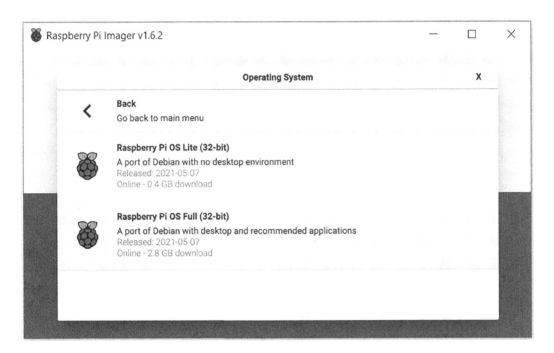

FIGURE 29.9   Selection of OS

In case, if you intend to use the display temporarily (only for the setup) or if you want to control the Raspberry Pi remotely, then VNC can be used. To enable the VNC server, connect the Raspberry Pi to WiFi and open the Raspberry Pi Software Configuration Tool. You can open it either from *Applications Menu > Preferences > Raspberry Pi Configuration* as shown in Figure 29.19 or by entering the command

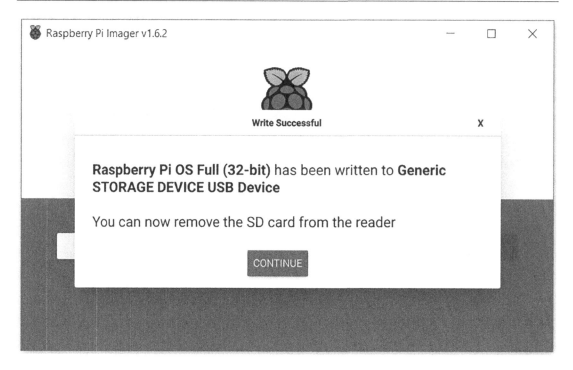

FIGURE 29.10   Completion of OS flash

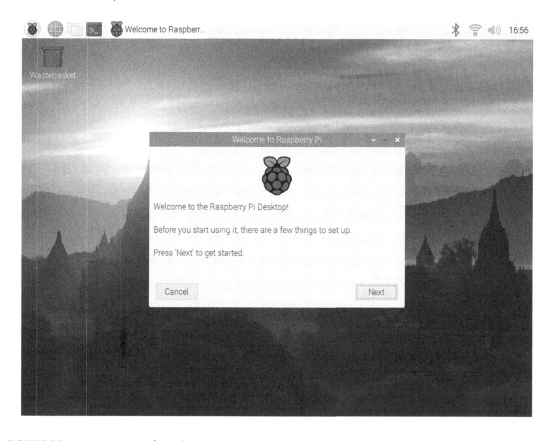

FIGURE 29.11   Homepage of Raspberry Pi

FIGURE 29.12   Set country, language and time zone

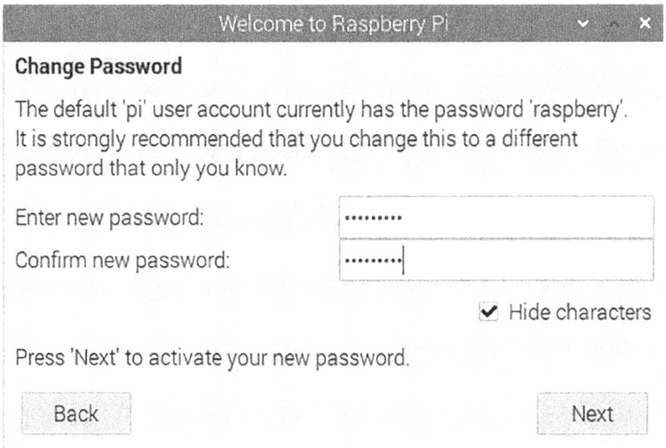

FIGURE 29.13   Change password

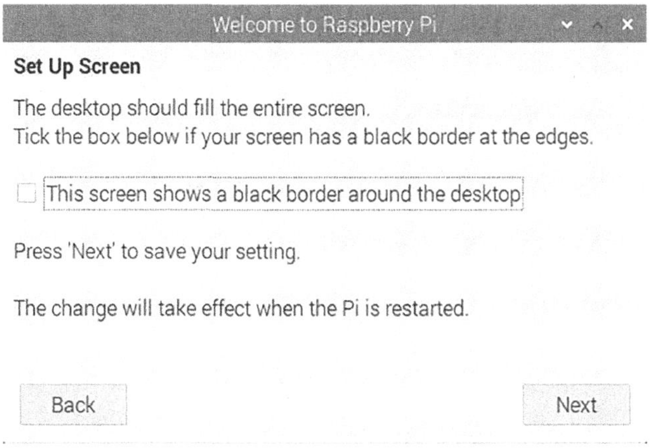

FIGURE 29.14   Screen setup

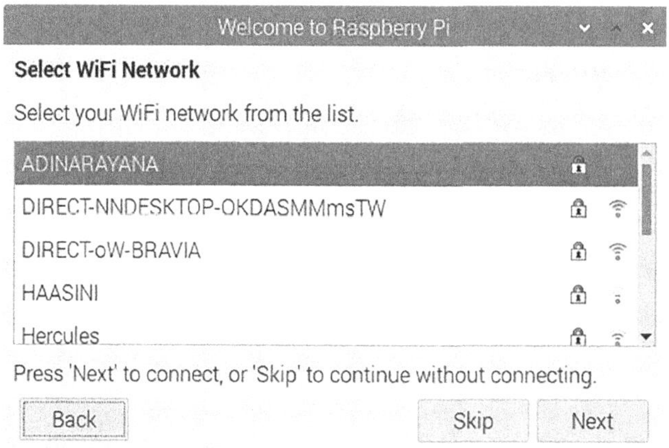

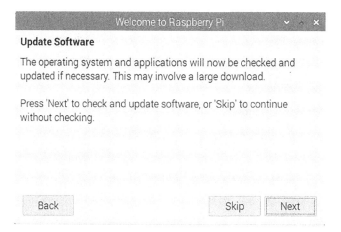

FIGURE 29.15   Connect to WiFi

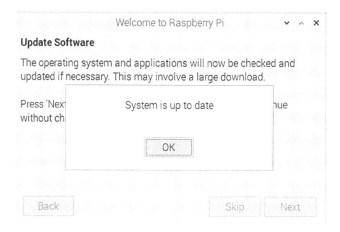

FIGURE 29.16   Update Raspberry Pi software

FIGURE 29.17   Software up to date

FIGURE 29.18 Setup complete

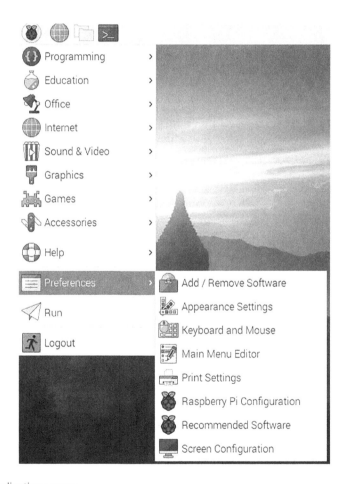

FIGURE 29.19 Applications menu

`sudo raspi-config` in the terminal. If you have opened it from Applications Menu, then you'll see a window as shown in Figure 29.20. Go to the *Interfaces* tab, enable both *SSH* and *VNC*, and click OK as shown in Figure 29.21.

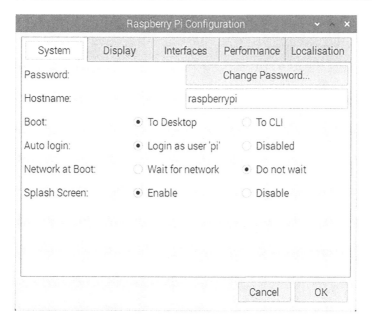

FIGURE 29.20 Raspberry Pi configuration

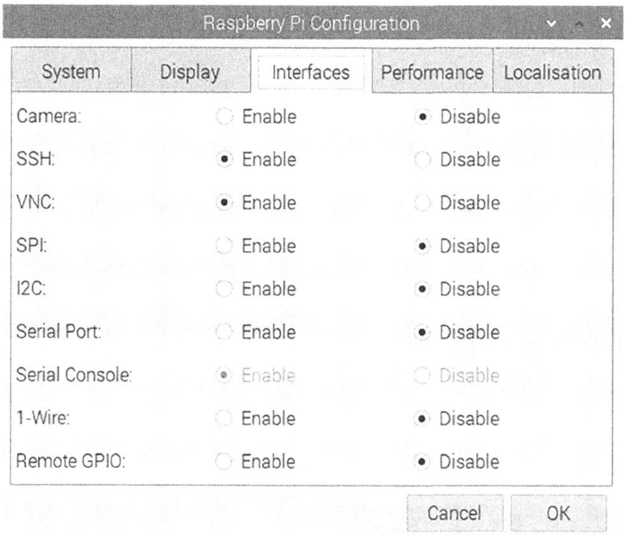

FIGURE 29.21 Interfaces tab of Raspberry Pi configuration

If the VNC server is enabled, then you'll see a VNC icon at the top-right corner. Click it and note down the IP address of Raspberry Pi as shown in Figure 29.22. Next, download and install the appropriate VNC viewer (https://www.realvnc.com/en/connect/download/viewer/) variant on the PC from which the Raspberry Pi needs to be accessed. After installing the VNC viewer, open it and enter the IP address of Raspberry Pi in the Address box at the top as shown in Figure 29.23. If you are accessing the Raspberry Pi using VNC for the first time, then you'll see an Identity Check window as shown in Figure 29.24. As you know and trust your Raspberry Pi device, click *Continue*. Enter the Username and password of your Raspberry Pi and click OK as shown in Figure 29.25. If you haven't changed the username and password

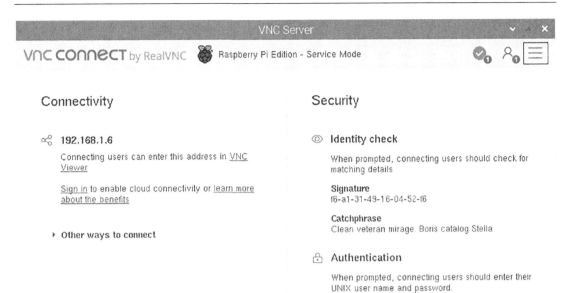

FIGURE 29.22   VNC server

FIGURE 29.23   Address box in VNC viewer

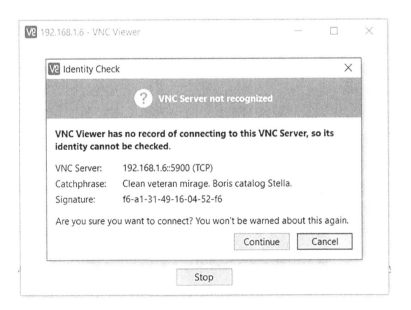

FIGURE 29.24   Identity check

of your Raspberry Pi or if you are logging in for the first time, then by default the username is *pi* and the password is *raspberry*. After logging in, you can see the home page of Raspberry Pi and control it remotely from your PC without the need for a dedicated display.

FIGURE 29.25   Authentication

On logging in to the Raspberry Pi, if you happen to see *Cannot currently show the desktop* as shown in Figure 29.26, then again boot the Raspberry Pi using a dedicated monitor. Open the *Screen Configuration* window from *Applications Menu* > Preferences. Choose a resolution other than the default one from *Configure* > *Screens* > *HDMI-1* > *Resolution*. As the PC from which we'll remotely control the Raspberry Pi is 1920×1080 resolution, we have chosen *1920×1080* as shown in Figure 29.27. For the changes to take place, again go to *Configure* and click *Apply*. Now, you can access the Raspberry Pi from the VNC viewer without any error.

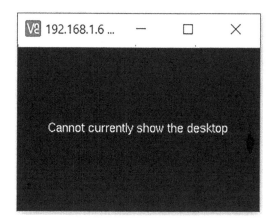

FIGURE 29.26   VNC viewer error

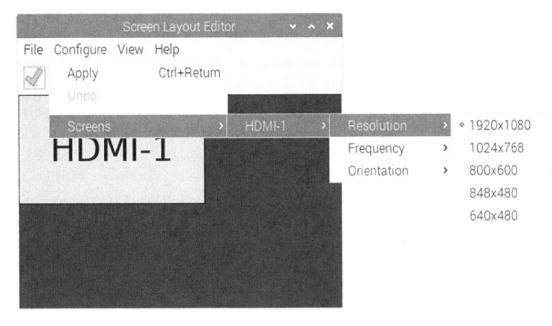

FIGURE 29.27   Screen layout editor

29.4 SETUP OF HEADLESS RASPBERRY Pi

If you don't have a dedicated display, keyboard and mouse for setting up the Raspberry Pi, then there is another method. This involves a little complex process compared to the previous method, but the advantage is that you can use your PC monitor for the initial configuration. The hardware and software required for headless setup of Raspberry Pi are tabulated in Tables 29.3 and 29.4.

TABLE 29.3   Hardware required for the setup of headless Raspberry Pi

HARDWARE	QUANTITY
Raspberry Pi 4B	1
5 V DC adapter	1
USB A to USB C cable	1
Micro SD card (at least 8GB)	1
SD card reader	1

TABLE 29.4   Software required for the setup of headless Raspberry Pi

SOFTWARE
Raspberry Pi Imager
Angry IP scanner
Tera Term or Putty
VNC viewer

The process until flashing the Raspberry Pi OS to the micro SD card is the same as the previous method. If the flashing is successful, you'll see a boot drive in the File Explorer or Finder as shown in Figure 29.28. This drive contains all the flashed OS files. Open it and create two new files named *SSH* and *wpa_supplicant.conf* as shown in Figure 29.29. Open the *wpa_supplicant.conf* file and paste the following content in it. Change <Country Code> with your country's 2 letter ISO 1366-1 code, <Name of your WiFi> and <Password for your WiFi> with the name and password of your WiFi, respectively. You can refer to this link (https://en.wikipedia.org/wiki/ISO_3166-1; last edited June 12, 2022) to find your country's ISO 1366-1 code. *SSH* (Secure Shell) is an empty file with no extension.

```
ctrl_interface=DIR=/var/run/wpa_supplicant GROUP=netdev
country=<Country Code>
update_config=1

network={
 ssid="<Name of your WiFi>"
 psk="<Password for your WiFi>"
}
```

FIGURE 29.28   boot drive

This PC > boot (E:)

Name	Date modified	Type	Size
SSH	05-10-2021 16:29	File	0 KB
wpa_supplicant.conf	05-10-2021 16:29	CONF File	1 KB

FIGURE 29.29   SSH and wpa_supplicant.conf files

Now you can eject the SD card reader and insert the micro SD card into the Raspberry Pi. Power the Raspberry Pi and wait until its onboard green LED is stabilized. Next, download and install the appropriate Angry IP scanner (https://angryip.org/download/) variant on your PC. You can use any other IP scanner as well. Connect your PC to the same WiFi network mentioned in the *wpa_supplicant.conf* and open the Angry IP scanner. Click on *Start* as shown in Figure 29.30 to scan the IP addresses mentioned

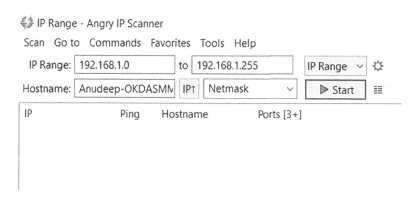

FIGURE 29.30   Angry IP scanner

in the *IP Range*. After the scan gets completed, you'll see the scan statistics as shown in Figure 29.31. All the alive hosts will be either in Green or Blue colour, whereas the remaining hosts will be in Red colour. If the Raspberry Pi is connected to the WiFi network mentioned in *wpa_supplicant.conf*, then one of the alive hosts will have *raspberrypi.local* Hostname. The IP address corresponding to that alive host is the IP address of your Raspberry Pi. In our case, it is 192.168.1.6.

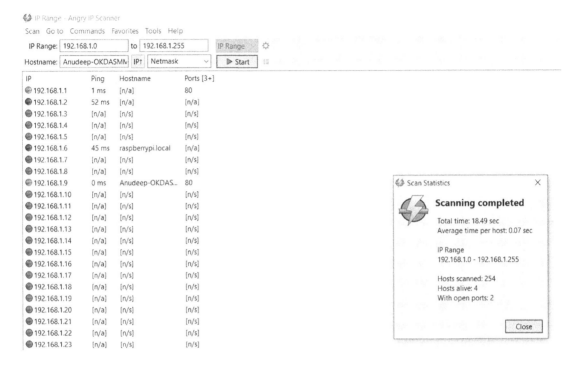

FIGURE 29.31   Scan statistics

To access the Raspberry Pi through SSH, you can use either Tera Term or Putty. Open Tera Term and select *TCP/IP*. Next, select the service as *SSH*, enter the Raspberry Pi's IP address in the *Host* column and click *OK* as shown in Figure 29.32. You'll see a security warning as shown in Figure 29.33. As your

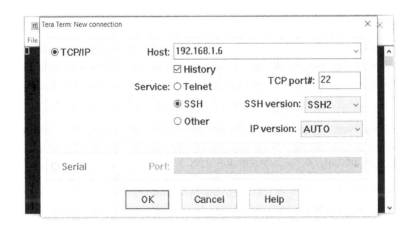

FIGURE 29.32   Tera term: new connection

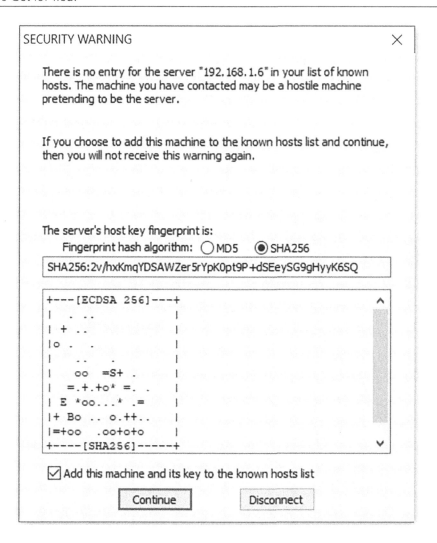

FIGURE 29.33 Security warning

Raspberry Pi is a known host, you can click *Continue*. Next, you'll see an SSH Authentication window as shown in Figure 29.34. If you haven't changed the *User name* and *passphrase* of your Raspberry Pi or if you are logging in for the first time, then by default the User name is *pi* and the passphrase is *raspberry*. Enter the User name and passphrase and click *OK*. If the Authentication is successful, then you'll see the Raspberry Pi terminal as shown in Figure 29.35.

Using the terminal, you can control the whole Raspberry Pi and can also do all the things that can be done from a GUI. But controlling from a terminal is very difficult and time consuming because every operation requires a command. So, we'll enable the VNC server for controlling the Raspberry Pi through GUI. Open the *Raspberry Pi Software Configuration Tool* shown in Figure 29.36 by entering the command `sudo raspi-config`. Using the down arrow, go to *Interface Options* and click Enter which will navigate to another screen as shown in Figure 29.37. Select *VNC* using the down arrow and click Enter. Using Tab on your keyboard, select *Yes* and click Enter as shown in Figure 29.38. If the VNC server is enabled, you'll see an acknowledgement message as shown in Figure 29.39. Similarly, enable *SSH* from the *Interface Options*.

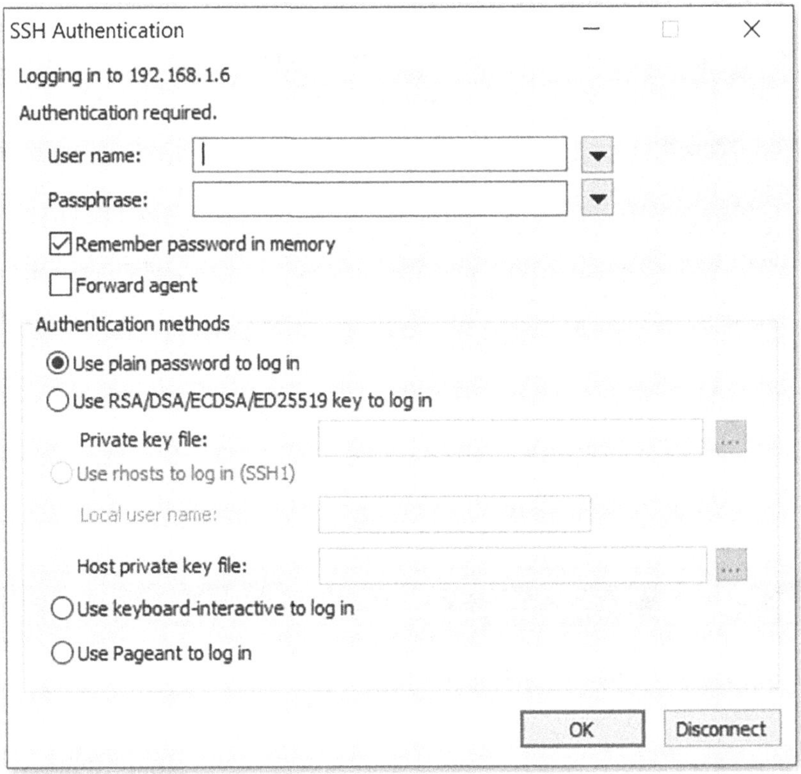

FIGURE 29.34   SSH authentication

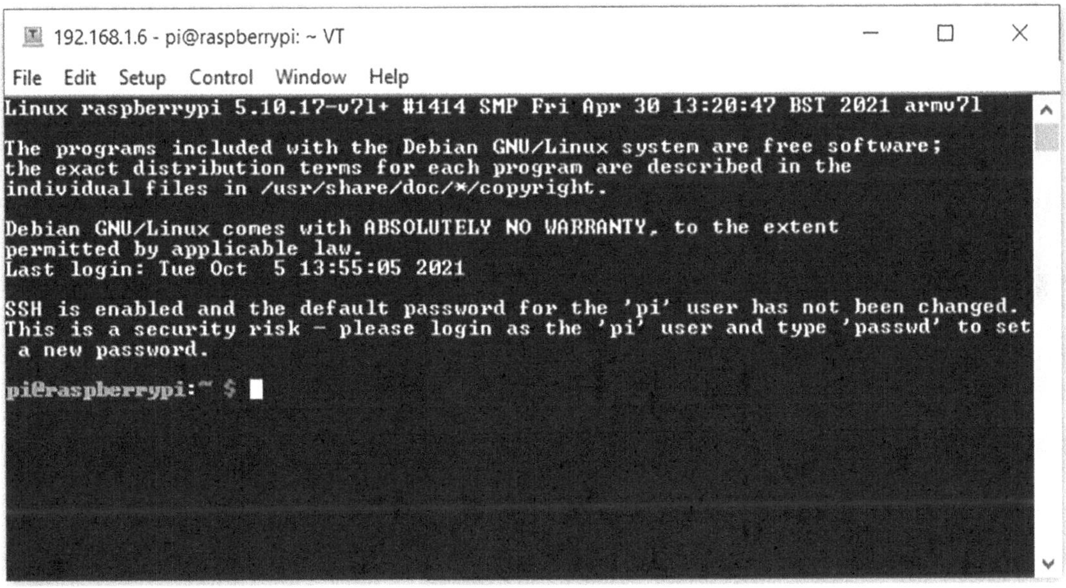

FIGURE 29.35   Raspberry Pi terminal

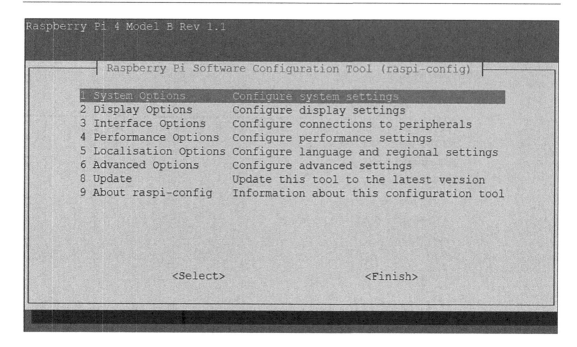

FIGURE 29.36   Raspberry Pi software configuration tool

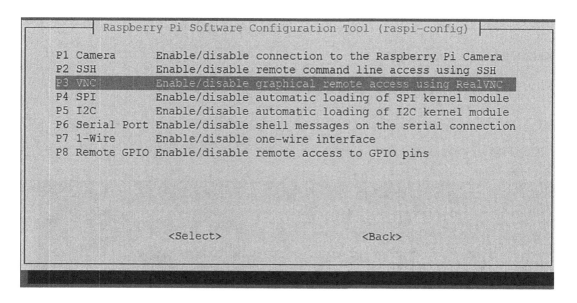

FIGURE 29.37   Interface options

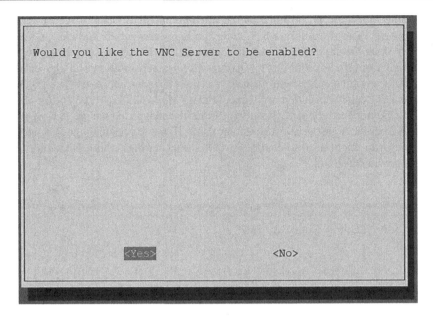

FIGURE 29.38 Enable VNC server

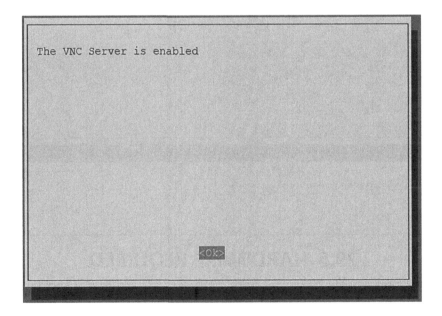

FIGURE 29.39 VNC server enabled acknowledgement

The remaining procedure is the same as the previous method. Open VNC viewer, enter the IP address of Raspberry Pi in the Address box, enter the login credentials and click *OK*. If you notice *Cannot currently show the desktop* as shown in Figure 29.26, then again open Raspberry Pi Software Configuration Tool from Tera Term. Go to *Display Options > Resolution* and change the resolution of Raspberry Pi to anything other than *Default*. As the PC from which we'll remotely control the Raspberry Pi has 1920×1080 resolution, we'll choose *DMT Mode 82 1920×1080 60Hz 16:9* as shown in Figure 29.40. Click Enter and reboot the Raspberry Pi for the change to take place. Now, you can access the Raspberry Pi from the VNC viewer without any error. If you are opening the Raspberry Pi GUI for the first time, then you might need to configure some initial settings whose procedure is the same as explained in the previous method.

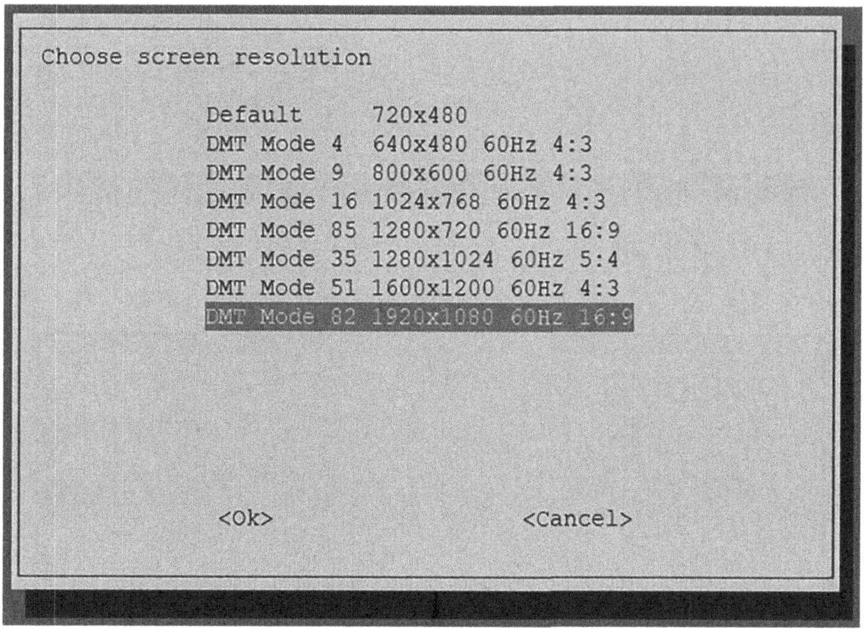

FIGURE 29.40   Changing the screen resolution

29.5 HARDWARE REQUIRED

TABLE 29.5   Hardware required for the project

HARDWARE	QUANTITY
Raspberry Pi 4B	1
Raspberry Pi OS flashed Micro SD card	1
Resistor (220 Ω)	1
LED	1
5 V DC Adapter	1
USB A to USB C cable	1
Jumper wires	As required

29.6 CONNECTIONS

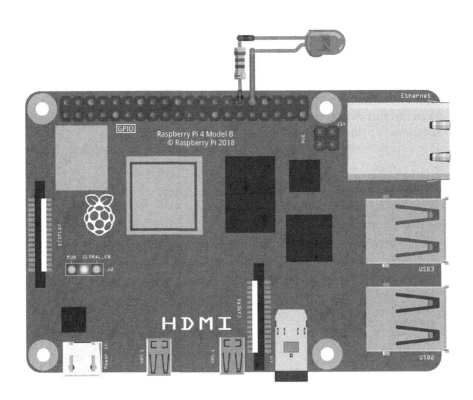

FIGURE 29.41   Connection layout

TABLE 29.6   Connection details

RASPBERRY PI 4B	RESISTOR (220 Ω)	LED
GPIO 12		+ve pin (long pin)
	Pin 1	−ve pin (short pin)
GND	Pin 2	

The connection diagram of the components tabulated in Table 29.5 is shown in Figure 29.41, which will help you get a visualization of the connections. Table 29.6 presents the pin-wise connection details.

In the case of Arduino Uno and NodeMCU, you know that even if a resistor is not connected in series with the LED, it doesn't cause any damage. But what happens in the case of Raspberry Pi? Explore it on your own. Raspberry Pi doesn't have any onboard analog pins and cannot be used for controlling analog sensors without the use of external hardware. For example, LDR which is an analog sensor can be interfaced with Raspberry Pi using a capacitor.

29.7 INTRODUCTION TO PYTHON PROGRAMMING LANGUAGE

If you haven't updated the Raspberry Pi previously, then you can update it from the terminal using the command sudo apt-get update. After updating the Raspberry Pi, install Python 3 and IDLE from the terminal using the command sudo apt-get install python3 idle3. Similar to Arduino IDE, IDLE is an editor for python. To open IDLE from the terminal, idle command can be used. The libraries for python can also be installed from the terminal using pip.

Python is one of the easiest programming languages to code. It has a simple syntax that makes it easy to read, understand and remember. If you know any programming language, then you can easily learn or at least understand the python programming language. To get started, the following are some of the basic syntax and their usage in python.

is used for a single comment and a pair of """ is used for multi-line comment.

```
#This is a single line comment
""" This is the first line of a multi-line comment.
This is the second line of multi-line comment. """
```

To declare a variable in python, the datatype need not be mentioned.

In C:

```
int varI = 3;
float varF = 5.0;
char varC = 'c';
```

In python:

```
varI = 3
varF = 5.0
varC = 'c'
```

Some of the allowed python variable declarations:

```
varAllowed = 3
var_Allowed = 5.0
varAllowed03 = 'c'
```

Some of the NOT allowed python variable declarations:

```
1varNotAllowed = 3
var-notAllowed = 5.0
var notAllowed = 'c'
```

To know the datatype of a variable type() can be used. int(), float() and str() is used for in-place conversion of a variable to integer, float and string data type, respectively.

For printing data, print() is used.

```
print("Data to be printed")
num = 2
print("This is " + str(num) + " print statement.")
```

By default, `print()` statement will add a line break. To print the following data on the same line, `end=""` needs to be used.

```
print("Data will be printed ", end="")
print("on the same line")
```

For iterating from 1 to 9,

```
for i in range(1,10):
    print("Iter value: " + str(i))

i = 1
while(i < 10):
    print("Iter value: " + str(i))
    i += 1
```

It has to be remembered that, indentation is very important in python. To detect a block of code, python uses indentation whereas C language uses curly brackets,{}.

29.8 PYTHON CODE FOR LED BLINK

For controlling the GPIO pins of Arduino Uno or NodeMCU using the Arduino IDE, it is not necessary to explicitly import any extra libraries. They are imported in the background when we select the appropriate board before uploading the code. But, for controlling the GPIO pins of Raspberry Pi using Python, we need to use `RPi.GPIO` library. It can be installed using the command `pip install RPi.GPIO` from the terminal. `time` library is used for delay operations in python. It need not be installed separately as it is one of the pre-installed libraries in python. There are two modes to refer the pins of Raspberry Pi in the python code:

- `GPIO.BCM` is used to refer according to the Physical pin numbers
- `GPIO.BOARD` is used to refer according to the GPIO pin numbers

If the mode is set as `GPIO.BCM`, then the `LEDPin` should be defined as 12 (GPIO number). If the mode is set as `GPIO.BOARD`, then the `LEDPin` should be defined as 32 (Physical pin number). Similar to the NodeMCU, the GPIO pin numbers and the physical pin numbers of Raspberry Pi are different. `GPIO.setup()` is used for setting the direction of operation of a pin. It is similar to `pinMode()` in Arduino IDE. As we need to control the LED, we'll set the direction as `GPIO.OUT`.

```
# Import required libraries
import RPi.GPIO as GPIO
import time

# Used for numbering the pins according to GPIO pin number
GPIO.setmode(GPIO.BCM)
LEDPin = 12
# Used for numbering the pins according to the physical pin number
#GPIO.setmode(GPIO.BOARD)
#LEDPin = 32

# Used for setting the direction of operation of a pin.
GPIO.setup(LEDPin, GPIO.OUT)
```

try-except block in python tries to execute the code in `try` block. If it fails to execute due to an Interrupt or Error, then the code in `except` block is executed. The code written inside `finally` block is used for executing after the `try-except` block. In the `try` block, we'll continuously blink an LED.

`while(1)` is used for running code inside it infinitely. It is similar to `void loop()` in Arduino IDE. `GPIO.output()` is used for digitally writing voltages on the pins. It is similar to `digitaWrite()` in Arduino IDE. `GPIO.output()` takes two arguments i.e., pin number (LEDPin) and the digital voltage to be written. `True` is treated as 3.3V and `False` is treated as 0V. These are similar to `HIGH` and `LOW` in Arduino IDE, respectively. `time.sleep()` is used for pausing the program. It is similar to `delay()` in Arduino IDE but `time.sleep()` considers the time in seconds not in milliseconds.

If a `KeyboardInterrupt` is detected, then `"Blink code is stopped"` is printed on the console. `KeyboardInterrupt` can be generated by pressing Ctrl+C on the keyboard. After executing `try` and `except` blocks, `finally` block is executed. Until now, the LEDPin (GPIO pin 12) resource is allocated for blinking the LED. To release its allocation and have a clean exit, `GPIO.cleanup()` is used.

```
try:
    while(1):
        # Used for writing digital voltage on a pin.
        GPIO.output(LEDPin, True)
        # Used for pausing the program.
        time.sleep(1)
        GPIO.output(LEDPin, False)
        time.sleep(1)

except KeyboardInterrupt:
    print("Blink code is stopped.")

finally:
    # This helps in switching off all the active channels
    # and ensures a clean exit
    GPIO.cleanup()
```

29.9 PYTHON CODE FOR CONTROLLING BRIGHTNESS OF LED

Importing required libraries and declaring the pins is same as the previous code.

```
# Import required libraries
import RPi.GPIO as GPIO
import time

# Used for numbering the pins according to GPIO pin number
GPIO.setmode(GPIO.BCM)
LEDPin = 12
# Used for numbering the pins according to the physical pin number
#GPIO.setmode(GPIO.BOARD)
#LEDPin = 32

# Used for setting the direction of operation of a pin.
GPIO.setup(LEDPin, GPIO.OUT)
```

There are four hardware PWM pins or two hardware PWM channels on Raspberry Pi.

- PWM0 channel – GPIO 12 (Pin 32) and GPIO 18 (Pin 12)
- PWM1 channel – GPIO 13 (Pin 33) and GPIO 19 (Pin 35)

The hardware PWM pins on Raspberry Pi are shared with audio subsystem. So, you can use either hardware PWM output or audio output. But all the pins of Raspberry Pi can be configured as PWM pins using software. So, if you want to output audio from the Raspberry Pi, then you can use software PWM pins.

GPIO.PWM() is used for setting the frequency of a PWM pin. It accepts two arguments: PWM pin number and frequency. PWM pin is LEDPin (GPIO 12), and frequency is 500 Hz. Depending on the dutyCycle value, the brightness of LED can be increased or decreased. Initially, it is set to 0, and then it is gradually increased by fadeAmount for every step.

while(1) is used for running the code inside it infinitely. For every step, the dutyCycle value is changed by fadeAmount. ChangeDutyCycle() is similar to analogWrite() in Arduino IDE. It is used for changing the duty cycle of the PWM pin. If the dutyCycle value goes below 0 or goes above 100, then the sign of fadeAmount is reversed. After every step, the program is paused for 100 milliseconds.

When a KeyboardInterrupt occurs "Brightness Control code is stopped." is printed on the console. After executing try and except blocks, finally block is executed to release the LEDpin allocation and for a clean exit.

```
# Set the frequency of PWM pin.
# Frequency -> 500
LED = GPIO.PWM(LEDPin, 500)
dutyCycle = 0
fadeAmount = 5
LED.start(dutyCycle) # Initially the duty cycle is 0

try:
    # Brightness of LED increases and then decreases
    while(1):
        dutyCycle = dutyCycle + fadeAmount
        LED.ChangeDutyCycle(dutyCycle)

        # Change the direction of fadeAmount
        if (dutyCycle <= 0 or dutyCycle >= 100):
            fadeAmount = -fadeAmount

        time.sleep(0.1) # Pauses for 100ms

except KeyboardInterrupt:
    print("Brightness Control code is stopped.")

finally:
    # This helps in switching off all the active channels
    # and ensures a clean exit
    LED.stop()
    GPIO.cleanup()
```

You can find all the resources of this project at this Github link (https://github.com/anudeep-20/30IoTProjects/tree/main/Project%2029).

In this project, you have learnt about Raspberry Pi and two different ways of setting it up. You have also learnt about PWM in Raspberry Pi and controlled an LED using Raspberry Pi and Python. In the next project, you'll build a door system that operates on biometric authentication.

Let's build the final project!

Biometric Authenticated Door Using Raspberry Pi

30

Biometric authentication makes use of human features that make an individual different from others to authenticate. It relies on the physical characteristics of a person like fingerprint, facial patterns, or retinal patterns for verification. Since this data is unique to individual users, this form of authentication is generally more secure than the other traditional forms of multi-factor authentication. Usage of a single characteristic or multiple characteristics depends on the infrastructure as well as the level of security needed. This project will help you build an automated door system based on facial recognition, a category of biometric security.

30.1 MOTIVE OF THE PROJECT

In this project, you'll build an automated door system that operates based on facial recognition. In place of a physical door, we'll use a servo motor to mimic the opening and closing of the door.

30.2 Pi CAMERA

Similar to a webcam for a PC, a pi camera is an add on hardware for the Raspberry Pi to perform camera-related operations. It can be used for capturing photos, video streaming, image processing and many more. Previously, only the pi camera V1 was available, but now you can get many types of pi cameras like high-quality lens pi camera, lens adjustable pi camera, etc. Recently, pi camera V2 is also released. This project can be completed using a 5MP pi camera V1 as shown in Figure 30.1. But you can also use a higher resolution pi camera depending on the use case. Pi cameras are connected to the Raspberry Pi using an FFC cable as shown in Figure 30.2. Alternatively, a USB webcam can also be used for camera-related operations of the Raspberry Pi.

DOI: 10.1201/9781003147169-30

FIGURE 30.1   5MP pi camera V1

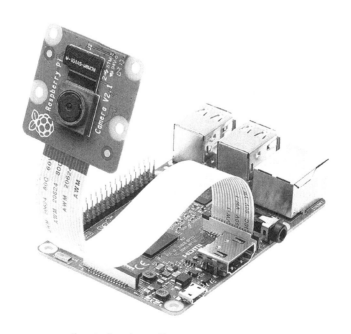

FIGURE 30.2   Pi camera connection to Raspberry Pi

30.3 ALGORITHM

For any recognition project, first, we need to detect and then recognize. Even for building a face recognizer, we need to first detect a human face and then recognize it. Face detection and face recognition are both different. Face detection has the objective of finding faces (location and size) in a given image. After a human face is detected, face recognition is responsible for finding characteristics that best describe a particular person.

For detecting a human face, we will use a frontal face haar cascade. Haar cascade is a machine learning–based classifier that is trained with a large number of positive and negative images. In case of a frontal face haar cascade, it is trained with a large number of images containing human faces and an equally large number of images that don't contain humans. Haar cascades are not the most accurate algorithms, but they are certainly fast and easy to use.

For recognizing a person's face, we will use LBPH (Local Binary Pattern Histogram) algorithm. It is one of the oldest and most popular algorithms. It uses four parameters: radius, neighbours, Grid X and Grid Y. The radius parameter is used to build a circular LBP around the central pixel. The number of pixels to be taken from the circular LBP is defined by the neighbours parameter. The computational cost increases as the number of neighbours increases. Consider a sample image of resolution 9×9 as shown in Figure 30.3. The darkened pixels shown in Figure 30.4 are the selected pixels for different values of radius and neighbours.

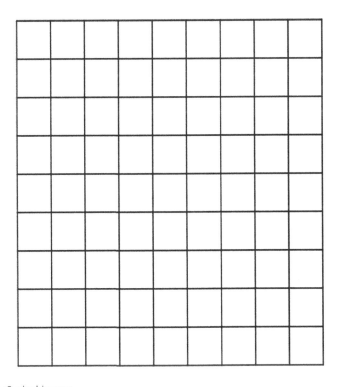

FIGURE 30.3 A 9×9 pixel image

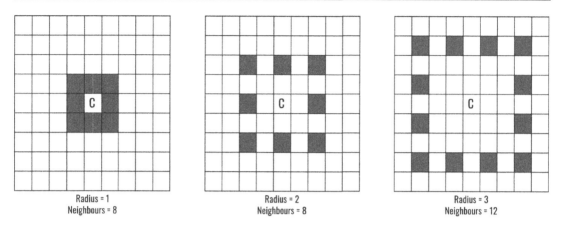

FIGURE 30.4   Selected pixels for different values of radius and neighbours parameters

An actual grayscale image consists of not just one 9×9 pixel box but many such pixel boxes. So, we'll use a sliding window with a radius of 1 and neighbours as 8 to traverse across the image. Each pixel in a grayscale image varies from 0 to 255. 0 being pure black and 255 being pure white. Consider anyone sliding window (3 ×3) region in Figure 30.5. It consists of 9 pixels and converts each of them to their respective intensity value. Depending on the central pixel's intensity value, convert all the other pixels into a binary value. If a pixel's intensity value is less than the central pixel's intensity value, then set it as 0 or else set it as 1. Convert the binary matrix into a binary value in clockwise direction with the rightmost pixel as LSB. So, the resultant binary value will be 10010101 which is equivalent to the decimal value 149. Some may also use other methods to convert the binary matrix into binary value like moving in anticlockwise direction, concatenating the columns, etc. In any case, there will be no difference in the end result. The resultant decimal value is set to the central pixel and re-converted to grayscale. The same procedure is repeated on the whole image resulting in an LBP image as shown in Figure 30.6.

FIGURE 30.5   One sliding window region in a grayscale image

FIGURE 30.6   LBP image

For extracting histograms from the LBP image, the other two parameters Grid X and Grid Y are used. Grid X is used to divide the image in the horizontal direction, whereas Grid Y is used to divide the image in the vertical direction. For a Grid X value of 8 and a Grid Y value of 8, the LBP image will be divided into 256 cells as shown in Figure 30.7. From each cell, a histogram is extracted and all the individual histograms are concatenated together to form a large histogram. This contains all the features and minute details of the image.

One image is not sufficient to recognize a person. So, we'll use several images of the same person from different angles for accurate recognition. All these images are converted into histograms and concatenated together to form a large histogram. If there are multiple persons, then each person is given a unique ID and the same procedure is repeated for each person. All the final histograms of all the persons are stored in a *yml* file along with the person's ID. The number of images used for training the algorithm should not be too less or too high. They have to be optimal. In our case, we'll use 150 images of each person.

For an input image, we'll perform the same steps and convert it into a histogram. The input image's histogram is compared to each person's histogram. The ID corresponding to the closet matched histogram is returned. The comparison between two histograms can be done in many ways like absolute value, euclidean distance, etc. It will also return a confidence score which tells the similarity between the person in the input image and the person corresponding to the returned ID. A lower confidence score indicates higher similarity and vice versa.

FIGURE 30.7   Dividing the LBP image into cells

30.4 INSTALLATION OF OpenCV IN RASPBERRY Pi

OpenCV is the acronym for Open Computer Vision. It is an open-source library that provides computer vision, image processing and machine learning functionalities to C, C++, Python and Java programming languages. We'll use this library to collect multiple images of a person, train them using the LBPH algorithm and use the trained model to recognize the person.

OpenCV installation is one of the most difficult, time-consuming and tiring processes. It has a lot of dependencies that need to be installed. It also consumes a lot of space. So, if you are using a micro SD card less than or equal to 16GB, then it's recommended to claim all the SD card space by expanding the file system. Enter sudo raspi-config in the terminal to open Raspberry Pi software configuration tool. Go to *Advanced* Options, select *Expand Filesystem* as shown in Figure 30.8 and reboot the Raspberry Pi to ensure all the space in the micro SD card is available.

Before installing the OpenCV dependencies, ensure the Raspberry Pi is up to date using the command sudo apt-get update && sudo apt-get upgrade. Next, install CMake using the following command:

```
sudo apt-get install build-essential cmake pkg-config.
```

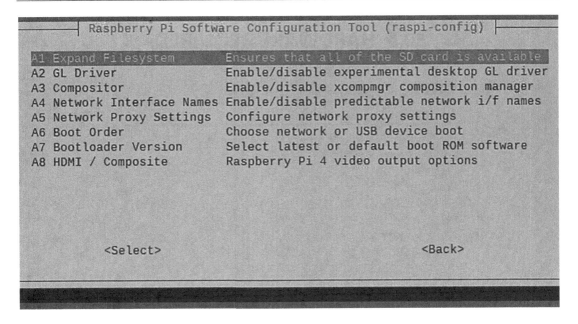

FIGURE 30.8   Expand filesystem

The following three commands are used to install image and video I/O packages that help us load different image and video file formats:

```
sudo apt-get install libjpeg-dev libtiff5-dev libjasper-dev libpng-dev
sudo apt-get install libavcodec-dev libavformat-dev libswscale-dev libv4l-dev
sudo apt-get install libxvidcore-dev libx264-dev
```

For displaying the images on our screen, OpenCV uses `highgui` submodule. For its prerequisites and compilation, run the following commands:

```
sudo apt-get install libfontconfig1-dev libcairo2-dev
sudo apt-get install libgdk-pixbuf2.0-dev libpango1.0-dev
sudo apt-get install libgtk2.0-dev libgtk-3-dev
```

The following commands are used to install some optimization functions and dependencies:

```
sudo apt-get install libatlas-base-dev gfortran
sudo apt-get install libhdf5-dev libhdf5-serial-dev libhdf5-103
sudo apt-get install libqtgui4 libqtwebkit4 libqt4-test python3-pyqt5
```

Finally, update NumPy and install OpenCV using the following commands:

```
sudo pip install -U numpy
sudo pip install opencv-contrib-python
```

During the installation, you might get an *Unable to fetch* or *Failed to fetch* error. It might be due to disturbances in the network connection. Try again. It will work.

30.5 HARDWARE REQUIRED

TABLE 30.1   Hardware required for the project

HARDWARE	QUANTITY
Raspberry Pi 4B	1
Raspberry Pi OS flashed Micro SD card	1
Pi camera	1
FFC cable	1
Servo motor	1
5 V DC Adapter	1
USB A to USB C cable	1
Jumper wires	As required

30.6 CONNECTIONS

The connection diagram of the components is tabulated in Table 30.1, and the connection layout in Figure 30.9 will help you get a visualization of the connections. Table 30.2 presents the pin-wise connection details. Also, connect the pi camera to the camera slot on Raspberry Pi using an FFC cable.

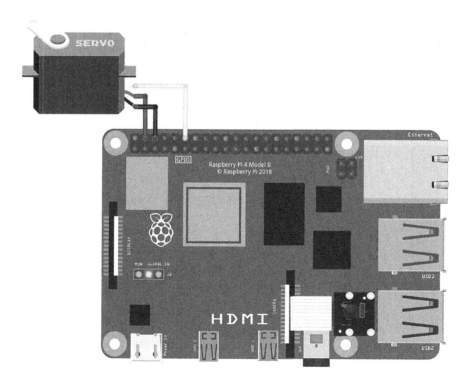

FIGURE 30.9   Connection layout

TABLE 30.2   Connection details

RASPBERRY PI 4B	PI CAMERA	SERVO
5 V		VCC (Red)
GPIO 18		OUT (Yellow)
GND		GND (Black)

30.7 DATASET CREATION

Before starting the implementation, create a folder architecture in the Raspberry Pi as shown in Figure 30.10. To use the pi camera, you need to enable *Camera* from *Applications Menu > Preferences > Raspberry Pi Configuration > Interfaces* tab as shown in Figure 30.11 and reboot the Raspberry Pi.

Face_Recognition_LBPH
| —— —— dataset
⠀⠀⠀⠀⠀| —— (*Stores people facial images*)
| —— —— model
⠀⠀⠀⠀⠀| —— (*Stores the trained yml file*)
| —— —— dataset_creator.py
| —— —— face_recognizer.py
| —— —— haarcascade_frontalface.xml
| —— —— train_model.py

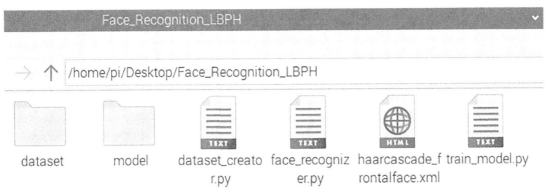

FIGURE 30.10   Folder architecture

The first step of the recognition problem is dataset collection. If you have enough number of facial images (around 150) of the person to be detected, then you can directly place them in the dataset folder. But they must be cropped such that only the person to be detected should be there in the image, and they must be named as *ID.X.jpg* where *ID* is the unique ID of the person and *X* is the image number. If you don't have any dataset or a sufficient number of images, then you can use the following code to generate. If you are using the following code to generate dataset, then expose the face of the person to be detected with good lighting while the camera captures images. It helps to obtain images with good clarity. Also, rotate the face in all possible directions so that the model will be trained more efficiently and gives accurate results. Remember to collect image dataset of at least two persons or else the face recognizer model will recognize every person as the same.

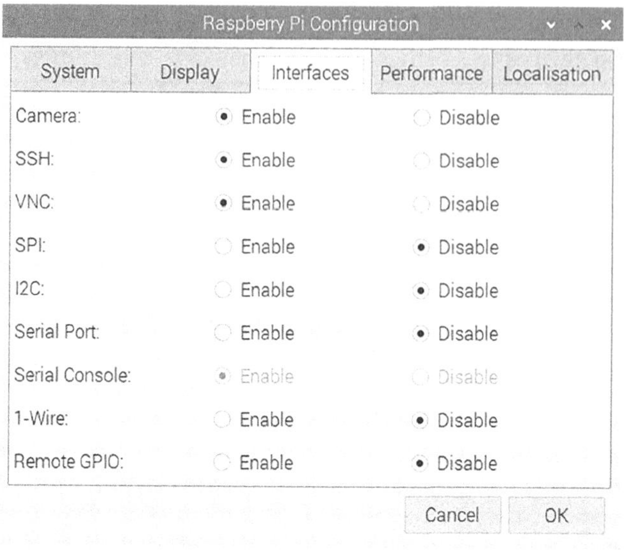

FIGURE 30.11   Enable camera

cv2 is used for importing OpenCV. Numpy is the acronym for numerical python. It is the commonly used library for all the numerical operations in python. It is a general convention to import numpy as np. So, all the functions in NumPy library can be accessed by using the word np. cv2. CascadeClassifier() is used to load haar cascades for object detection. Every person whose dataset is being collected must be assigned a unique ID. cv2.VideoCapture() can be used to capture video from video files, camera or image sequences. For capturing video from a camera, the id of the video capturing device must be given as the argument. To capture video from the default camera 0 must be given as the argument. If you have multiple cameras, then the value might differ. numSamples stores the number of image samples taken for each person.

```
# Import required libraries
import cv2
import numpy as np
import time

# Create a frontal detection classifier using haar cascade
faceCascade = cv2.CascadeClassifier('haarcascade_frontalface.xml')
ID = input("Please enter a unique ID: ")
# Starts recording video from the camera
camera = cv2.VideoCapture(0)
sampleIter = 0
numSamples = 150 # Number of sample faces for training
```

camera.read() gets an individual image frame from the video feed. ret stores whether an image frame is captured or not. cameraFeed stores the image frame. In OpenCV, a colour image is known as BGR (Blue, Green and Red) image. cv2.COLOR _ BGR2GRAY is used to convert a BGR image to grayscale. grayCameraFeed is used to store the grayscale converted image frame. faceCascade. detectMultiScale() is used to detect human faces in the grayscale image with the help of frontal face haar cascade. It returns the coordinates, height and width of the detected face in the image. In a cartesian plane, the origin will be at the bottom-left corner, whereas in OpenCV, by default, the origin will be at the top-left corner. `cv2.rectangle()` is used to draw a rectangle on the image. It accepts multiple arguments. The first one is the image on which rectangle must be drawn. The next two arguments are the diagonal coordinates of the rectangle which must be given as a tuple. The fourth argument is the colour of the rectangle, and the last argument is the line thickness of the rectangle.

`cv2.imwrite()` is used to save the image. It accepts two arguments; the name of the image and the image to be saved. In python, colon (:) is used as an array slicing operator. For example, `str = "I LOVE LEARNING"` then `str[2:9]` stores `"LOVE LE"`. `cv2.waitKey()` is used to wait for a keypress event infinitely or for some milliseconds. If no argument is passed, then it'll wait infinitely until a key is pressed. If any positive number is given as the argument, then it waits for at least that number of milliseconds for the user to press a key. `cv2.imshow()` displays an image frame using the OpenCV GUI. It accepts two arguments, the window name and the image to be displayed. If the number of sample images collected is equal to `numSamples` or if q is pressed by the user, then the video capture from the camera will be stopped and all the windows created by OpenCV will be closed.

```python
while (True):
    # Returns image frames from the video
    ret, cameraFeed = camera.read()
    # Converts BGR frames to grayscale frames
    grayCameraFeed = cv2.cvtColor(cameraFeed, cv2.COLOR_BGR2GRAY)
    # Detect faces in the image frame
    faces = faceCascade.detectMultiScale(grayCameraFeed, 1.3, 5)

    for (x, y, w, h) in faces:
        # Draw a rectangle around the face
        cv2.rectangle(cameraFeed, (x, y), (x+w, y+h), (255, 0, 0), 2)
        # Store the face in the dataset folder
        cv2.imwrite("dataset/" + str(ID) + "." + str(sampleIter)+".jpg",
grayCameraFeed[y:y+h, x:x+w])
        sampleIter += 1
        cv2.waitKey(25) # Waits for 25ms

    # Shows the frame
    cv2.imshow("Face Detector", cameraFeed)
    # If required number of samples are collected break from the loop
    if (sampleIter == numSamples):
        print("Dataset with "+ str(numSamples) +" faces is successfully
created.")
        break

    # If q is pressed then break from the loop
    if cv2.waitKey(1) & 0xFF == ord('q'):
        break
# Stops recording video from the camera
camera.release()
cv2.destroyAllWindows() # Closes all the windows opened by OpenCV
```

After collecting the required number of image samples, open the *dataset* folder and glance through all the saved images. You might find some images that don't have any face like the one shown in Figure 30.12. Though the number of such images is less (usually <5%), they play a key role in reducing the overall accuracy. So, manually remove those images. This is an important step in dataset collection known as data cleaning or pre-processing. To minimize the number of non-facial images saved, expose the face of the person to be detected with good lighting while collecting the data.

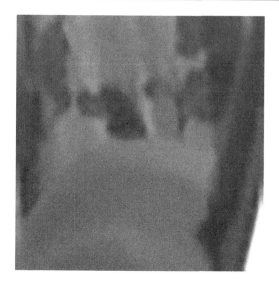

FIGURE 30.12   Non-facial redundant image

30.8 TRAINING THE MODEL

The next step is to train the LBPH face recognizer model with the collected dataset. This step requires an extra library known as Pillow. It can be installed from the terminal using the command `python3 -m pip install --upgrade Pillow`. It is imported as `PIL` in python. You can either import the whole library or just a few of its functions. As we only require the functions in the `Image` class of Pillow library, we'll import it alone. `os` is another inbuilt library in python which provides functions for interacting with the operating system.

cv2.face.LBPHFaceRecognizer _ create() creates an untrained LBPH face recognizer model. It accepts four arguments: `radius`, `neighbors`, `grid _ x` and `grid _ y`. These parameters are clearly explained in the Algorithm section. By default, their values are 1, 8, 8 and 8, respectively. `datasetPath` stores the path of the folder path in which the image dataset is stored. `imgPaths` stores the path of all the images in the dataset.

```
# Import required libraries
import cv2
from PIL import Image
import numpy as np
import os

# Create an empty LBPH face recognizer model
model = cv2.face.LBPHFaceRecognizer_create()
# Folder path in which dataset is stored
datasetPath = 'dataset'

# Stores complete path of the images
imgPaths = [os.path.join(datasetPath,imgFile) for imgFile in
os.listdir(datasetPath)]
faces = [] # Stores the facial images in the form of a numpy array
IDs = [] # Stores the respective IDs of the faces
```

`Image.open()` is used to open an image. It accepts the path of the image as an argument. `convert('L')` converts a colour image to grayscale using the ITU-R 601-2 luma transform. `np.array()` is used to convert an array to a numpy array. It accepts two arguments, the array to be converted to numpy array and the datatype of the resultant array. All the converted image numpy arrays are appended to the `faceArr`. IDs stores the unique IDs corresponding to the images. Using the `faces` and `IDs` numpy arrays, the model is trained and saved in the `model` folder.

```
# Iters through all the images
for imgPath in imgPaths:
    # Converts any BGR image to grayscale image
    faceImg = Image.open(imgPath).convert('L')
    # Converts facial image into numpy array
    faceArr = np.array(faceImg, 'uint8')
    faces.append(faceArr)
    # Obtains the ID of the Image from the Image file name
    ID = int(os.path.split(imgPath)[-1].split('.')[0])
    IDs.append(ID)
    # Shows each image
    cv2.imshow("Model Trainer", faceArr)
    cv2.waitKey(10) # Waits for 10ms

# Trains the model using the facial images array
model.train(faces, np.array(IDs))
# Saves the trained model in model folder
model.save('model/trainedModel.yml')
print("Trained model is saved in model folder")
cv2.destroyAllWindows() # Closes all the windows opened by OpenCV
```

30.9 FACIAL RECOGNITION

The last step is to use the trained LBPH face recognizer model to recognize the faces. Again create an empty LBPH face recognizer model and import the previously trained *yml* file. `font` stores the font of the text written on the image. Servo motors also use PWM to rotate. So, the OUT pin of servo motor is connected to the GPIO 18 of Raspberry Pi, a hardware PWM pin. As we are controlling the servo motor, its direction is set as `GPIO.OUT`. The `servoPin` is initialized as a PWM pin with a frequency of 50Hz, and its initial duty cycle is set to 0. `timeDoorOpen` stores the number of seconds the door must be open, i.e, the servo should stop rotating after opening the door.

```
# Import required libraries
import cv2
import RPi.GPIO as GPIO
import numpy as np
import time

# Create an empty LBPH face recognizer model
faceRecognizer = cv2.face.LBPHFaceRecognizer_create()
# Imports the trained model
faceRecognizer.read('model/trainedModel.yml')
# Create a frontal detection classifier using haar cascade
faceCascade = cv2.CascadeClassifier("haarcascade_frontalface.xml")
```

```
font = cv2.FONT_HERSHEY_SIMPLEX # Font for writing on the Image

# Used for numbering the pins according to GPIO pin number
GPIO.setmode(GPIO.BCM)
servoPin = 18
# Used for numbering the pins according to the physical pin number
#GPIO.setmode(GPIO.BOARD)
#servoPin = 12

# Used for setting the direction of operation of a pin.
GPIO.setup(servoPin, GPIO.OUT)

# Set the frequency of PWM pin.
# Frequency -> 50
servo = GPIO.PWM(servoPin,50)
servo.start(0) # Initially the duty cycle is set to 0
# Time for which door must be open
timeDoorOpen = 20
```

openDoor() is a custom function used for opening and closing the door. For opening the door, the servo motor is rotated from 0° to 180°, and for closing the door, it is rotated back from 180° to 0°. For keeping the servo motor still, the duty cycle must be set to 0. For rotating the servo motor from 0° to 180°, the duty cycle must be changed from 2 to 12. Similarly, for rotating the servo motor from 180° to 0°, the duty cycle must be changed from 12 to 2.

```
def doorControl():
    dutyCycle = 2

    # Rotate the servo from 0° to 180°
    print("Opening the Door")
    while (dutyCycle <= 12):
        servo.ChangeDutyCycle(dutyCycle)
        time.sleep(1)
        dutyCycle += 1

    servo.ChangeDutyCycle(0) # Keeps the servo still
    time.sleep(timeDoorOpen)

    # Rotate the servo from 180° to 0°
    print("Closing the Door")
    while (dutyCycle >= 2):
        servo.ChangeDutyCycle(dutyCycle)
        time.sleep(1)
        dutyCycle -= 1

    servo.ChangeDutyCycle(0) # Keeps the servo still
```

Each frame from the camera feed is converted to grayscale and human face in it are detected. All the detected faces are recognized using faceRecognizer.predict(). If any face is recognized, then its corresponding ID and confidence score are returned. In our case, we'll only allow the entry for ID 1 person and restrict the entry for ID 2 person. So, whenever the person corresponding to ID 1 is detected, the doorControl() function is called. You can also increase the number of persons allowed entry according to your use case. ID = -1 at the end clears the previously detected ID. So, avoid having negative numbers as the unique ID for the persons to be detected.

You might observe something different in the code implementation compared to the dataset creation code implementation. Previously, the cv2.VideoCapture(0) is given outside the loop and not inside it. But here it is given inside the loop. It is because camera.read()has a buffer memory that can store image frames even when the program is executing something else. If the person with ID 1 is detected, then the doorControl() function is being called multiple times. To avoid it and read the frames only from real-time camera feed, the `camera.release()` is used at the end to drop any buffer image frames.

```python
while (True):
    # Detect a face in the camera feed
    camera = cv2.VideoCapture(0)
    ret, cameraFeed = camera.read()
    grayCameraFeed = cv2.cvtColor(cameraFeed, cv2.COLOR_BGR2GRAY)
    faces = faceCascade.detectMultiScale(grayCameraFeed, 1.2, 5)

    for (x, y, w, h) in faces:
        # Draw rectangle around the detected face
        cv2.rectangle(cameraFeed, (x, y), (x + w, y + h), (0, 260, 0), 7)
        # Recognize the face
        ID, conf = faceRecognizer.predict(grayCameraFeed[y:y+h, x:x+w])

        if(ID == 1):
            # Write text on the image frame
            # Replace ID_1 with the name of 1st Person
            cv2.putText(cameraFeed, "ID_1", (x, y-40), font, 1, (0, 255,
255), 2)

            doorControl() # Opens the door

        if(ID == 2):
            # Write text on the image frame
            # Replace ID_2 with the name of 2nd Person
            cv2.putText(cameraFeed, "ID_2", (x, y-40), font, 1, (0, 255,
255), 2)

            print ("Entry Restricted!!")

    # Shows the camera feed
    cv2.imshow("Face Recognizer", cameraFeed)
    camera.release() # Stops recording video from the camera
    ID = -1

    # If q is pressed then break from the loop
    if cv2.waitKey(1) & 0xFF == ord('q'):
        break
# Closes all the windows opened by OpenCV
cv2.destroyAllWindows()
# This helps in switching off all the active channels and
# ensures a clean exit
servo.stop()
GPIO.cleanup()
```

You can find all the resources of this project at this Github link (https://github.com/anudeep-20/30IoTProjects/tree/main/Project%2030).

In this project, you have learnt about different pi cameras and various steps involved in object recognition. You have also learnt about the LBPH algorithm, OpenCV and installed it on your Raspberry Pi. Finally, you have built a human face recognizer that can be used for authentication.

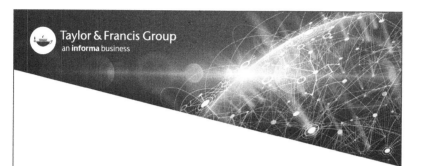

Printed in the United States
by Baker & Taylor Publisher Services